SpringerWienNewYork

Powertrain

Edited by Helmut List

Scientific Board
K. Kollmann, H. P. Lenz, R. Pischinger
R. D. Reitz, T. Suzuki

Hermann Hiereth
Peter Prenninger

Charging the Internal Combustion Engine

Powertrain

SpringerWienNewYork

Dipl.-Ing. Dr. Hermann Hiereth
Esslingen, Federal Republic of Germany

Dipl.-Ing. Dr. Peter Prenninger
AVL List GmbH, Graz, Austria

Translated from the German by Klaus W. Drexl.
Originally published as *Aufladung der Verbrennungskraftmaschine*
© 2003 Springer-Verlag, Wien

This work is subject to copyright.
All rights are reserved, whether the whole or part of the material is concerned, specifically those of, translation, reprinting, re-use of illustrations, broadcasting, reproduction by photocopying machines or similar means, and storage in data banks. Product liability. The publisher can give no guarantee for all the information contained in this book. This also refers to that on drug dosage and application thereof. In each individual case the respective user must check the accuracy of the information given by consulting other pharmaceutical literature.
The use of registered names, trademarks, etc., in this publication does not imply, even in the absence of a specific statement, that such names are exempt from the relevant protective laws and regulations and therefore free for general use.
© 2007 Springer-Verlag, Wien
Printed in Austria
SpringerWienNewYork is a part of Springer Science + Business Media
springeronline.com

Typesetting: Thomson Press (India) Ltd., Chennai, India
Printing: Druckerei Theiss GmbH, 9431 St. Stefan im Lavanttal, Austria
Printed on acid-free and chlorine-free bleached paper
SPIN 11686729

With 370 Figures

Library of Congress Control Number 2007927101

ISSN 1613-6349
ISBN 978-3-211-33033-3 SpringerWienNewYork

Preface

Supercharging the reciprocating piston internal combustion engine is as old as the engine itself. Early on, it was used to improve the high-altitude performance of aircraft engines and later to increase the short-term peak performance in sporty or very expensive automobiles. It took nearly 30 years until it reached economic importance in the form of the efficiency-improving exhaust gas turbocharging of slow- and medium-speed diesel engines. It took 30 more years until it entered high-volume automotive engine production, in the form of both mechanically driven displacement compressors and modern exhaust gas turbocharging systems.

Since, in spite of promising alternative developments for mobile applications, the internal combustion engine will remain dominant for the foreseeable future, its further development is essential. Today many demands are placed on automobile engines: on the one hand, consumers insist on extreme efficiency, and on the other hand laws establish strict standards for, e.g., noise and exhaust gas emissions. It would be extremely difficult for an internal combustion engine to meet these demands without the advantages afforded by supercharging. The purpose of this book is to facilitate a better understanding of the characteristics of superchargers in respect to their physical operating principles, as well as their interaction with piston engines. This applies both to the displacement compressor and to exhaust gas turbocharging systems, which often are very complex.

It is not intended to cover the layout, calculation, and design of supercharging equipment as such – this special area is reserved for the pertinent technical literature – but to cover those questions which are important for an efficient interaction between engine and supercharging system, as well as the description of the tools necessary to obtain an optimal engine–supercharger combination.

Special emphasis is put on an understandable depiction of the interrelationships in as simple a form as possible, as well as on the description and exemplified in-depth discussion of modern supercharging system development processes. As far as possible, the principal interactions are described, and mathematical functions are limited to the necessary minimum, without at the same time disregarding how indispensable simulation and layout programs today are for a fast, cost-efficient, and largely application-optimized engine–supercharger adaptation.

This book is written for students as well as engineers in research and development, whom we presume to be significantly more knowledgeable about the basics of the internal combustion engine than about supercharging systems.

When compiling the bibliography, we – due to the extensive number of relevant publications – have emphasized those texts which influence or support the descriptions and statements within the book.

We have to thank a large number of persons and companies that have enabled this book via their encouragement and who provided us with illustrations.

Our special thanks go to the editor of the series "Der Fahrzeugantrieb/Powertrain", Prof. Helmut List, who encouraged us to tackle this book and who actively supported the editing and the preparation of the illustrations. We thank the companies ABB, DaimlerChrysler, Garrett-Honeywell, 3K-Warner, and Waertsilae-New Sulzer Diesel for permitting us to use extensive material with results and illustrations and the *Motortechnische Zeitschrift* for their permission to republish numerous illustrations.

We thank Univ.-Prof. Dr. R. Pischinger and Dipl.-Ing. G. Withalm for their useful suggestions and systematic basic research. For special hints and additions in regard to fluid mechanics we thank Dipl.-Ing. S. Sumser, Dipl.-Ing. H. Finger and Dr.-Ing. F. Wirbeleit. Also, for their extensive simulation and test results we thank the highly committed colleagues from the AVL departments Thermodynamics as well as Diesel and Gasoline Engine Research. We thank Dipl.-Ing. N. Hochegger for the excellent preparation of the illustrations.

Without the kind assistance of all companies and individuals mentioned above this book would not have been possible. We thank Springer Wien New York for the professional execution and production of this book.

H. Hiereth, P. Prenninger

Contents

Symbols, indices and abbreviations XII

1 Introduction and short history of supercharging 1

2 Basic principles and objectives of supercharging 5
2.1 Interrelationship between cylinder charge and cylinder work as well as between charge mass flow and engine power output 5
2.1.1 Interrelationship between cylinder charge and cylinder work 5
2.1.2 Interrelationship between charge mass flow and engine power output 6
2.2 Influence of charge air cooling 8
2.3 Definitions and survey of supercharging methods 9
2.4 Supercharging by means of gasdynamic effects 9
2.4.1 Intake manifold resonance charging 9
2.4.2 Helmholtz resonance charging 11
2.5 Supercharging with supercharging units 13
2.5.1 Charger pressure–volume flow map 13
2.5.2 Displacement compressor 14
2.5.3 Turbo compressor 15
2.6 Interaction between supercharger and internal combustion engine 17
2.6.1 Pressure–volume flow map of the piston engine 17
2.6.2 Interaction of two- and four-stroke engines with various superchargers 20

3 Thermodynamics of supercharging 23
3.1 Calculation of charger and turbine performance 23
3.2 Energy balance of the supercharged engines' work process 24
3.2.1 Engine high-pressure process 24
3.2.2 Gas exchange cycle low-pressure processes 24
3.2.3 Utilization of exhaust gas energy 25
3.3 Efficiency increase by supercharging 26
3.3.1 Characteristic values for the description of the gas exchange and engine efficiencies 26
3.3.2 Influencing the engine's total efficiency value via supercharging 30
3.4 Influence of supercharging on exhaust gas emissions 31
3.4.1 Gasoline engine 33
3.4.2 Diesel engine 33
3.4.3 Methods for exhaust gas aftertreatment 34
3.5 Thermal and mechanical stress on the supercharged internal combustion engine 34

3.5.1		Thermal stress 34
3.5.2		Mechanical stress 35
3.6		Modeling and computer-aided simulation of supercharged engines 36
3.6.1		Introduction to numeric process simulation 36
3.6.2		Cycle simulation of the supercharged engine 37
3.6.3		Numeric 3-D simulation of flow processes 48
3.6.4		Numeric simulation of the supercharged engine in connection with the user system 49
4		Mechanical supercharging 51
4.1		Application areas for mechanical supercharging 51
4.2		Energy balance for mechanical supercharging 52
4.3		Control possibilities for the delivery flow of mechanical superchargers 53
4.3.1		Four-stroke engines 53
4.3.2		Two-stroke engines 55
4.4		Designs and systematics of mechanically powered compressors 55
4.4.1		Displacement compressors 55
4.4.2		Turbo compressors 59
5		Exhaust gas turbocharging 60
5.1		Objectives and applications for exhaust gas turbocharging 60
5.2		Basic fluid mechanics of turbocharger components 60
5.2.1		Energy transfer in turbo machines 60
5.2.2		Compressors 61
5.2.3		Turbines 65
5.3		Energy balance of the charging system 74
5.4		Matching of the turbocharger 75
5.4.1		Possibilities for the use of exhaust energy and the resulting exhaust system design 75
5.4.2		Turbine design and control 82
5.4.3		Compressor design and control 89
5.5		Layout and optimization of the gas manifolds and the turbocharger components by means of cycle and CFD simulations 92
5.5.1		Layout criteria 92
5.5.2		Examples of numeric simulation of engines with exhaust gas turbocharging 97
5.5.3		Verification of the simulation 101
6		Special processes with use of exhaust gas turbocharging 105
6.1		Two-stage turbocharging 105
6.2		Controlled two-stage turbocharging 106
6.3		Register charging 108
6.3.1		Single-stage register charging 108
6.3.2		Two-stage register charging 110
6.4		Turbo cooling and the Miller process 113
6.4.1		Turbo cooling 113
6.4.2		The Miller process 114

6.5	Turbocompound process 116
6.5.1	Mechanical energy recovery 117
6.5.2	Electric energy recovery 119
6.6	Combined charging and special charging processes 121
6.6.1	Differential compound charging 121
6.6.2	Mechanical auxiliary supercharging 122
6.6.3	Supported exhaust gas turbocharging 124
6.6.4	Comprex pressure-wave charging process 125
6.6.5	Hyperbar charging process 128
6.6.6	Design of combined supercharging processes via thermodynamic cycle simulations 129
7	Performance characteristics of supercharged engines 133
7.1	Load response and acceleration behavior 133
7.2	Torque behavior and torque curve 134
7.3	High-altitude behavior of supercharged engines 135
7.4	Stationary and slow-speed engines 137
7.4.1	Generator operation 138
7.4.2	Operation in propeller mode 139
7.4.3	Acceleration supports 140
7.4.4	Special problems of turbocharging two-stroke engines 141
7.5	Transient operation of a four-stroke ship engine with register charging 143
8	Operating behavior of supercharged engines in automotive applications 144
8.1	Requirements for use in passenger vehicles 144
8.2	Requirements for use in trucks 145
8.3	Other automotive applications 146
8.4	Transient response of the exhaust gas turbocharged engine 146
8.4.1	Passenger car application 147
8.4.2	Truck application 148
8.5	Exhaust gas turbocharger layout for automotive application 151
8.5.1	Steady-state layout 151
8.5.2	Transient layout 154
8.5.3	Numerical simulation of the operating behavior of the engine in interaction with the total vehicle system 158
8.6	Special problems of supercharged gasoline and natural gas engines 159
8.6.1	Knocking combustion 159
8.6.2	Problems of quantity control 161
9	Charger control intervention and control philosophies for fixed-geometry and VTG chargers 162
9.1	Basic problems of exhaust gas turbocharger control 162
9.2	Fixed-geometry exhaust gas turbochargers 163
9.2.1	Control interaction possibilities for stationary operating conditions 163
9.2.2	Transient control strategies 166
9.2.3	Part-load and emission control parameters and control strategies 170
9.3	Exhaust gas turbocharger with variable turbine geometry 173

9.3.1	General control possibilities and strategies for chargers	173
9.3.2	Control strategies for improved steady-state operation	173
9.3.3	Control strategies for improved transient operation	175
9.3.4	Special control strategies for increased engine braking performance	177
9.3.5	Special problems of supercharged gasoline and natural gas engines	179
9.3.6	Schematic layout of electronic waste gate and VTG control systems	179
9.3.7	Evaluation of VTG control strategies via numerical simulation models	181
10	Instrumentation for recording the operating data of supercharged engines on the engine test bench	184
10.1	Measurement layout	185
10.2	Engine torque	185
10.3	Engine speed	186
10.4	Turbocharger speed	187
10.5	Engine air mass flow	188
10.6	Fuel mass flow	189
10.7	Engine blowby	189
10.8	Pressure and temperature data	189
10.9	Emission data	191
11	Mechanics of superchargers	194
11.1	Displacement compressors	194
11.1.1	Housing and rotors: sealing and cooling	194
11.1.2	Bearing and lubrication	195
11.2	Exhaust gas turbochargers	195
11.2.1	Small chargers	195
11.2.1.1	Housing: design, cooling and sealing	195
11.2.1.2	Rotor assembly: load and material selection	198
11.2.1.3	Bearing, lubrication, and shaft dynamics	199
11.2.1.4	Production	200
11.2.2	Large chargers	202
11.2.2.1	Design, housing, cooling, sealing	202
11.2.2.2	Rotor assembly	205
11.2.2.3	Production	207
12	Charge air coolers and charge air cooling systems	208
12.1	Basics and characteristics	208
12.2	Design variants of charge air coolers	209
12.2.1	Water-cooled charge air coolers	211
12.2.2	Air-to-air charge air coolers	212
12.2.3	Full-aluminum charge air coolers	212
12.3	Charge air cooling systems	213
13	Outlook and further developments in supercharging	215
13.1	Supercharging technologies: trends and perspectives	215
13.2	Development trends for individual supercharging systems	215
13.2.1	Mechanical chargers	215

13.2.2	Exhaust gas turbochargers 216	
13.2.3	Supercharging systems and combinations 217	
13.3	Summary 221	

14	Examples of supercharged production engines 222	
14.1	Supercharged gasoline engines 222	
14.2	Passenger car diesel engines 233	
14.3	Truck diesel engines 242	
14.4	Aircraft engines 245	
14.5	High-performance high-speed engines (locomotive and ship engines) 245	
14.6	Medium-speed engines (gas and heavy-oil operation) 248	
14.7	Slow-speed engines (stationary and ship engines) 251	

Appendix 255
References 259
Subject index 265

Symbols, indices and abbreviations

Symbols

a	speed of sound [m/s]; Vibe parameter; charge coefficient	$\dot{m}_F$	fuel mass flow [kg/s], [kg/h]
		$\dot{m}_{red}$	reduced mass flow [kg$\sqrt{K}$/s bar]
A	(cross sectional) area [m^2]	mep	mean effective pressure [bar]
A_{min}	minimum air requirement	mp	mean pressure [bar]
A_{St}	stoichiometric air requirement (also other units) [kg/kg]	n	number; (engine) speed [s^{-1}, min^{-1}]
		n_C	compressor speed [s^{-1}, min^{-1}]
B	bore [m]	n_{cyl}	number of cylinders [−]
bmep	brake mean effective pressure [bar]	n_E	engine speed [s^{-1}, min^{-1}]
bsfc	brake specific fuel consumption [kg/kW h]	p	pressure, partial pressure [Pa, bar]
c	specific heat capacity, $c = dq_{rev}/dT$ [J/kg K]; absolute speed in turbo machinery [m/s]	P	power output [W], [kW], [PS, hp]
		p_0	standard pressure, $p_0 = 1{,}013$ bar
c_m	medium piston speed [m/s]	p_{con}	control pressure
c_v, c_p	specific heat capacity at v = const. or p = const. [J/kg K]	P_{eff}	specific power [kW]
		p_{ign}	ignition pressure
d_{cyl}	cylinder diameter [m]	Q, q	heat [J]
d_v	valve diameter [m]	Q_{diss}	removed heat quantity
d_{vi}	inner valve diameter [m]	Q_{ext}	external heat [J]
D	(characteristic) diameter [m]	Q_F	supplied fuel heat [J]
D_C	compressor impeller diameter [m]	$Q_{F.u}$	fuel energy not utilized
D_T	turbine rotor diameter [m]	$dQ_F/d\varphi$	rate of heat release [J/°CA]
E	enthalpy [J]	Q_{fr}	frictional heat [J]
e_{ext}	specific external energy [J/kg]	Q_{low}	net calorific value (lower heating value) [kJ/kg]
F	force [N]	Q_{rev}	reversible heat [J]
fmep	friction mean effective pressure [bar]	$\dot{Q}$	heat flow [W]; heat transfer rate
h	specific enthalpy [J/kg]	r	crank radius [m]; reaction rate of a compressor stage or of an axial turbine stage [−]
I	polar moment of inertia [kg m^2]; electric current [A]		
		R	specific gas constant [J/kg K]; distance radius [cm]
imep	indicated mean effective pressure [bar]		
k	coefficient of heat transfer [W/m^2 K]	S	entropy [J/K]; turbine blade speed ratio [−]; stroke [m]
L_v	valve lift [m]		
m	mass [kg]; shape coefficient (of the Vibe rate of heat release) [−]; compressor slip factor [−]	S_P	piston stroke [m]
		sfc	specific fuel consumption (usually in g/kW h) [kg/J]
m_A	air mass [kg]		
m_F	fuel mass [kg]	t	time [s]; temperature [°C]
m_{fA}	fresh air mass remaining in cylinder [kg]	T	temperature [K]; torque [Nm]; turbine trim [%]
m_{in}	total aspirated fresh charge mass [kg]	u	specific internal energy [J/kg]; circumferential speed of the rotor [m/s]
m_{out}	total outflowing gas mass [kg]		
m_{RG}	residual gas mass [kg]	U	voltage [V]; internal energy [J]
m_S	scavenging mass [kg]	v	specific volume [m^3/kg]; (particle) speed [m/s]; velocity [mph, km/h]
$\dot{m}$	mass flow [kg/s]		
$\dot{m}_A$	air mass flow [kg/s], [kg/h]	V	volume [m^3]

V_c	compressed volume [m³]	η_{inc}	efficiency of real combustion process [−]
V_{cyl}	displacement of one cylinder [m³]	η_m	mechanical efficiency [−]
V_{tot}	engine displacement [m³]	η_ρ	efficiency of density recovery [−]
V_φ	cylinder volume at crank angle φ [m³]	$\eta_{s-i,C}$	internal isentropic compressor efficiency [−]
$\dot{V}$	volume flow	$\eta_{s-i,T}$	internal isentropic turbine efficiency [−]
$\dot{V}_s$	scavenge part of total volume flow	η_{TC}	turbocharger efficiency [−]
w	specific work [J/kg]; relative medium velocity in the rotor [m/s]	η_{th}	thermodynamic efficiency (of the ideal process with combined combustion) [−]
W	work [J]	$\eta_{th\omega}$	thermodynamic efficiency of the ideal process with constant-volume combustion [−]
W_{eff}	effective work [J]		
W_{fr}	friction work [J]	κ	adiabatic exponent [−]
W_i	indicated work [J]	λ	thermal conductivity, thermal conductivity coefficient [W/m K]; air-to-fuel ratio
W_t	technical work [J]		
W_{th}	theoretical comparison cycle work	λ_a	air delivery ratio [−]
α	heat transfer number [W/m² K]; heat transfer coefficient [W/m² K]	λ_f	wall friction coefficient
		λ_{fr}	pipe friction coefficient [−]
Γ	scavenging efficiency [−]	λ_S	scavenging ratio [−]
δ	wall thickness [m]	λ_{vol}	volumetric efficiency [−]
δ_0	start of combustion (SOC) [−]	μ	flow coefficient, overflow coefficient [−]
$\Delta\delta_d$	combustion duration	μ_σ	port flow coefficient [−]
Δ	difference between two values	ξ	loss coefficient [−]
ϵ	compression ratio [−]	Π	pressure ratio [−]
η_C	efficiency of Carnot process [−]	ρ	density [kg/m³]
η_{CAC}	charge air cooler efficiency [−]	ρ_1, ρ_2	density pre-compressor or pre-inlet port [kg/m³]
η_{com}	combustion efficiency	φ	crank angle [deg]
η_{cyc}	cycle efficiency factor [−]	φ_{RG}	amount of residual gas
η_{eff}	effective efficiency [−]	ψ	mass flow function [−]
η_F	fuel combustion rate [−]	ω	angular speed [s⁻¹]
η_i	indicated efficiency [−]		

Further indices and abbreviations

0	reference or standard state; start	CFD	computational fluid dynamics
1	condition 1, condition in area 1, upstream of compressor	CG	combustion gas
		ChA	charge air
2	condition 2, condition in area 2, downstream of compressor	circ	circumference
		CS	compression start
2′	upstream of engine (downstream of charge air cooler)	CT	constant throttle
		CVT	continuously variable transmission
3	upstream of turbine	cyl	cylinder
4	downstream of turbine	d	duration
		DI	direct injection
A	air	diss	dissipated (heat); extracted (heat)
abs	absolute	dyn	dynamic
AF	air filter	E	engine
add	added (heat)	E.c.	exhaust closes
amb	ambient	ECU	electronic control unit
b	burned (region)	eff	effective
BDC	bottom dead center	EGC	exhaust gas cooler
C	compression; compressor; coolant	EGR	exhaust gas recirculation
CA	crank angle [°]	EGT	exhaust gas throttle
CAC	charge air cooler, intercooler	E.o.	exhaust opens
CAT	catalyst	EP	exhaust manifold, port; plenum

Ex	(cylinder-) outlet, exhaust gas	OP	opacity
f	fresh	opt	optimum
F	fuel	out	outside, outer; (plenum-) outlet, exhaust
fA	fresh air	p	with $p = $ const.
FD	start of fuel delivery	P	pump, piston
FE	finite elements	Pl	plenum
FL	full load	PL	partial load
fr	friction	PT	power turbine
GDI	gasoline direct injection	PWC	pressure wave charger
geo	geometric, geometry	red	reduced
GEX	gas exchange cycle (low-pressure cycle)	rel	relative
h	height	RG	residual gas
HP	high-pressure phase	Rot	axial compressor rotor
i	internal, indicated; index $(i \ldots n)$	RON	research octane number
I.c.	inlet closes	s	isentropic, with $s = $ const.; scavenge
IDC	ignition dead center	scg	scavenging
IDI	indirect diesel injection	stat	static
idle	idle	T	turbine
Imp	impeller	TC	exhaust gas turbocharger
Int	(cylinder-; turbine-) inlet, intake, inflowing	TDC	top dead center
		th	theoretical, thermodynamic
I.o.	inlet opens	Th	throttle
IP	intake port or manifold	tot	total
IS	injection start	u	unburned (region)
leak	leakage, blowby	V	valve; volume
med	medium	Volute	turbine volute
max	maximum	VTG	variable turbine geometry
meas	measurement	W	wall (heat); water
min	minimum	WC	working cycle
mix	mixture	WG	waste gate
neck	turbine neck area	X	control rack travel

1 Introduction and short history of supercharging

Very likely, the future of the internal combustion engine can be described within the energy-sociopolitic environment as follows: For the foreseeable future, crude oil will still be the main energy source for internal combustion engines in automotive and other mobile applications; natural gas and, to a limited extent synthetic fuels (methanol and similar fuels), as well as, in the very long run, hydrogen, will additionally gain in importance. Internal combustion engines for these fuels are reciprocating or rotational piston combustion engines and gas and steam turbines. These engines are employed, under consideration of the particular requirements and according to their development status, in aircraft, locomotives, ships, stationary powerplants, and in road vehicles.

In aircraft design, the demand has always been for highest power density, i.e., smallest volume and highest power-to-weight ratio. The reciprocating piston internal combustion engine was the first power source to fulfill these requirements. With this, it actually enabled the engine-powered airplane and dominated this application until the end of the forties. Nowadays, with the exception of applications in small airplanes, it is superseded by the gas turbine, which as propeller turbine or as pure jet engine makes far higher power densities possible.

The classic power unit for train propulsion was the piston steam engine, which, in 2-, 3-, and 4-cylinder designs, lasted the longest for this use. Today, the steam locomotive is superseded by the electric or by the diesel locomotive, where diesel traction is more efficient for long hauls and stretches on which trains run infrequently. Diesel engines of high power density with hydraulic or electric power transfer today dominate diesel locomotive design. Repeatedly, the gas turbine was tested for this application – also as a short-time booster power unit – but could not prevail due to fuel economy and durability reasons.

In ship building, after the classic piston steam engine, first the steam turbine and then the gas turbine seemed to best accommodate the highly increasing power demands. In fast ships, also warships, where fuel consumption and fuel quality are not as decisive as power density and performance, the gas turbine even today occupies a niche market. But the highly supercharged, high-speed diesel engine, mostly in multiple engine configuration, is capturing this market to an increasing degree. In merchant shipping, due to its good fuel economy and the possibility to use even the cheapest heavy oils, the medium-speed and the slow-speed diesel heavy-oil engine have penetrated the market widely.

In large power plants with an output of 100 MW or more, the steam turbine still dominates. The extent to which smaller, decentralized electric power generating or heat and power cogeneration plants with internal combustion engines can take hold, remains to be seen. To cover peak power demands, the gas turbine has gained increased importance for this application.

For passenger cars as well as for trucks nowadays practically only the high-speed internal combustion engine is used, for reasons of its power density, durability, and cost, but especially for its ease of control and its flexibility in transient operation. Additionally, in the last decade extensive development work has led to reduced exhaust emissions with simultaneously improved efficiency. For truck engines, exhaust gas turbocharging in combination with charge air cooling has contributed decisively to attain both goals. From the heaviest truck down to transporters with about 4-tons payload, today practically only the exhaust gas turbocharged, charge air cooled, direct-injection diesel engine is used. In passenger cars as well, this engine configuration is gaining increased importance due to its extraordinary efficiency. In regard to supercharging, the passenger car gasoline engine remains problematical, due to its high exhaust gas temperature as well as to the requirement that an acceptable driving performance must be attained. This even more since also very narrow cost targets have to be met. But also here new approaches to technical solutions can be observed, so that it can be presumed that in 10 to 20 years supercharged combustion engines will totally dominate the market.

The history of supercharging the internal combustion engine reaches back to Gottlieb Daimler and Rudolf Diesel themselves.

Supercharging the high-speed gasoline engine is as old as it itself. Already Gottlieb Daimler had supercharged his first engines, as his patent DRP 34926 obtained in 1885 shows (Fig. 1.1). In this case, the piston's bottom was used, which in the four-stroke engine works as a mixture pump with double work-cycle frequency and therefore delivers a greater mixture volume than the work cylinder could aspirate.

Transferring the charge from the crankcase cavity into the work cylinder was performed by a valve in the piston bottom. The reason for Daimler's bold design was his desire for a possible speed and charge increase of the engines, despite the fact that at that time only very small intake and exhaust valves were feasible. The problems, especially with the piston bottom valve, however soon forced Daimler to abandon this intrinsically correct idea in favor of larger valves as well as the application of multiple-valve cylinder heads, which were designed by his co-worker Maybach.

Supercharging found its first series application in aircraft engines, especially to increase high-altitude performance. In the years from 1920 to 1940, turbo compressors were continuously improved, in aerodynamics as well as in the circumferential speed of the impellers.

Supercharging of gasoline engines experienced its first absolute peak in regard to power and high-altitude performance increases in aircraft engines during World War II. Brake mean effective pressure values of up to 23 bar were reached with mechanically powered turbo compressors. The last U.S. gasoline aircraft engines were the first series production compound engines, such as the 18-cylinder dual-radial compound engine from Curtiss Wright with a takeoff power of 2420 kW (see Fig. 6.22).

From about 1920, automotive supercharged engines for racing, but also for the short-term power increase of sport and luxury vehicles, were equipped with mechanically powered and engageable displacement compressors. In most cases they were one- or two-stage Roots blowers. Figure 1.2 shows such a passenger car engine with 40/60 hp from 2.6 liter displacement, built in 1921 by Daimler.

Exhaust gas turbocharged gasoline engines were first introduced into the U.S. market around 1960, e.g., the Chevrolet Corvair [76]. For the supercharging of gasoline engines, the big breakthrough towards large-scale series production, with the exception of use in airplanes, only happened very recently, with, e.g., the 2.3 liter compressor engine from DaimlerChrysler in its SLK and C class, or the exhaust gas turbocharged engines from Audi, Opel, and Saab.

Introduction

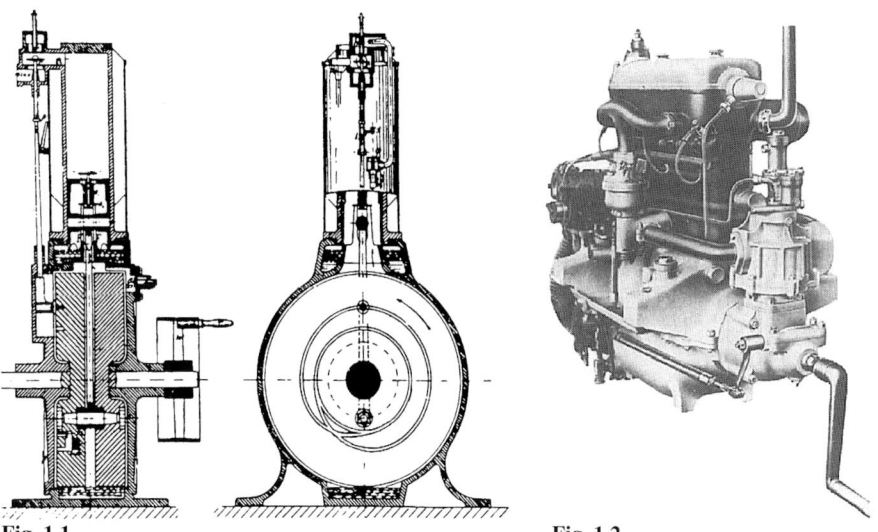

Fig. 1.1. Patent DRP 34926 from 1885 for the high-speed gasoline engine, by Gottlieb Daimler

Fig. 1.2. 40/60 hp passenger car compressor engine with Roots blower from 1921, by Daimler

Rudolf Diesel also got involved with supercharging very early, as his patent DRP 95680 demonstrates (Fig. 1.3). In his cross-head engine he used the piston bottom as a two-stroke charge pump. This patent also describes a process for cooling the air in a downstream plenum.

With his layout, Diesel achieved a power increase of 30%. However, since he was primarily concerned about the efficiency of his engine and it dramatically deteriorated – due to a totally incorrect size of the intake valve and the downstream plenum, he stopped these tests. This type of

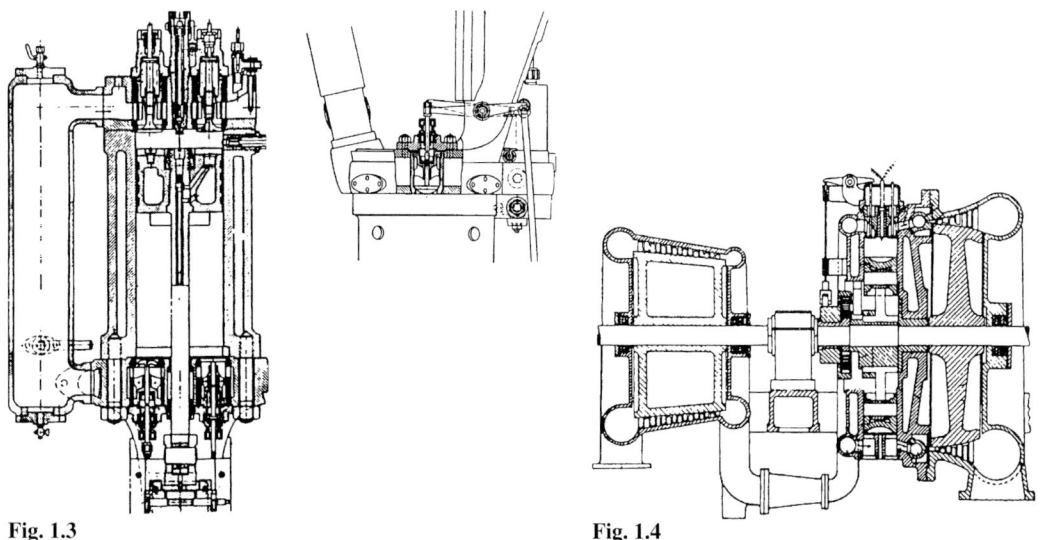

Fig. 1.3. Patent DRP 95680 by Rudolf Diesel for a diesel engine with supercharging by the lower side of the piston

Fig. 1.4. Buechi's patent drawing DRP 204630 for a turbocompound diesel engine

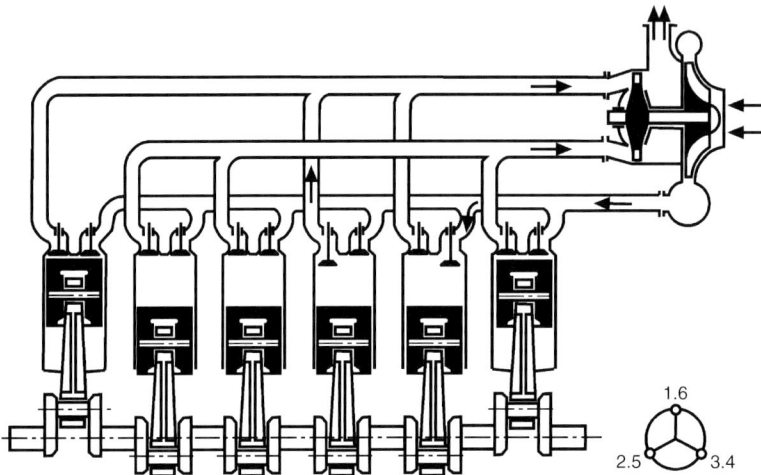

Fig. 1.5. Buechi's patent from 1925 for pressure-wave or pulse turbocharging via flow division

supercharging was, with correct dimensioning of the components, very successfully used 30 years later in marine diesel engines (e.g., by Werkspoor).

The development of exhaust gas turbocharging is closely connected with the name and patents of the Swiss engineer Alfred Buechi. As early as 1905, in patent DRP 204630 (Fig. 1.4) he described a turbocompound diesel engine – although not meaningful in the proposed form. But it still took until 1925 for the first exhaust gas turbocharged diesel engines to be introduced into the market, in the form of engines for two passenger ships and one stationary diesel engine from MAN and the Maschinenfabrik Winterthur. In both cases, the exhaust gas turbochargers were still located beside the engine. All chargers were designed by Buechi.

In the MAN marine engines, the mean effective pressure was increased by 40% to 11 bar, and important insights were gained:

Exhaust gas turbocharged engines are very overload capable.
The turbocharger group controls itself during operation.

In order to overcome the problem of a negative pressure gradient between charge pressure and exhaust gas backpressure, i.e., a negative scavenging gradient, which happened with these early exhaust gas turbochargers due to their low overall efficiency, in 1925 Buechi applied for another patent for a pressure-wave or pulse-charging layout. This was to be achieved by separating the exhaust manifolds and combining the cylinders with ignition intervals of more than 240° crank angle, as well as narrow exhaust manifold areas (Fig. 1.5).

The first tests at the Schweizer Lokomotiv- and Maschinenfabrik Winterthur on a 4- and a 6-cylinder engine with BBC charger were very promising. A power increase of 100% could be achieved with good thermodynamic results, and a third insight was gained:

Exhaust manifolds not only must have a small area but also must be as short as possible.

With that, flow and heat losses are minimized. Consequently, today exhaust gas turbochargers are mounted directly on the engine as a part of the exhaust manifold. Since then, the system described has been called Buechi-charging and is the basis for the exhaust gas turbocharging of all automotive engines.

2 Basic principles and objectives of supercharging

The objective of supercharging is to increase the charge density of the working medium (air or air-fuel mixture), by any means and with the help of a suitable system, before it enters the work cylinder, i.e., to precompress the charge. In doing so, the temperature of the working medium should not be markedly raised, since this would adversely influence the temperature profile of the high-pressure work cycle.

The density increase of the working medium increases the power density and can also be used to improve the combustion process with the aim to achieve lower exhaust gas and/or noise emissions. The interrelationships between mean effective pressure or power output and density of the cylinder air or mixture charge will be discussed below.

2.1 Interrelationship between cylinder charge and cylinder work as well as between charge mass flow and engine power output

In all internal combustion engines, work and power are generated through the transformation of the chemical energy stored in the fuel via combustion or oxidation and subsequent conversion of the heat energy into mechanical energy. The oxygen necessary for the combustion is extracted from the air introduced into the working chamber. Therefore, the power output of any internal combustion engine in which the processed air is used as combustion partner for the fuel, depends on the air quantity present in the cylinder.

2.1.1 Interrelationship between cylinder charge and cylinder work

The air-aspirating reciprocating piston engine is a volume pump and the maximum amount of air volume that can be introduced into the cylinder is

$$V_A = V_{cyl} \text{ and } m_A = V_{cyl}\rho_{A,cyl}. \tag{2.1}$$

The cylinder air charge, multiplied by the density of the air, results in the cylinder air mass, which determines the fuel mass that can be combusted in it and with which work can be gained via the increases in pressure and temperature taking place during combustion.

On the one hand, the indicated work W_i in the cylinder is the product of force times displacement as well as of the piston-area times stroke times pressure,

$$W_i = \frac{d_{cyl}^2 \pi}{4} S_P \text{imep}. \tag{2.2}$$

On the other hand, work is the product of added heat quantity times process efficiency,

$$W_i = Q_{\text{add,cyl}} \eta_i, \tag{2.3}$$

where $Q_{\text{add,cyl}}$ is the added heat quantity per cylinder charge, and η_i is the process efficiency, itself the quotient of mechanical work and added heat energy.

The heat quantity that can be added to the cylinder depends on the amount of fuel that is introduced into it, and that again depends on the amount of oxygen present in the cylinder. The amount of oxygen stands in a fixed relation to the air mass in the cylinder – and not to the cylinder volume. If we simplify and neither consider the incomplete charge of the cylinder, the volumetric efficiency, nor excess air that may be necessary for combustion, this heat quantity will be

$$Q_{\text{add,cyl}} = m_F Q_{\text{low}} = \frac{V_{\text{cyl}} \rho_{\text{A,cyl}} Q_{\text{low}}}{A_{\min}}, \tag{2.4}$$

where m_F is the added fuel quantity, $A_{\min}$ the minimum air requirement, Q_{low} the net calorific value of the fuel, and $\rho_{\text{A,cyl}}$ is the air density in the cylinder.

Keeping Q_{low} and $A_{\min}$ constant, it is directly derived that

$$Q_{\text{add,cyl}} \sim \rho_{\text{A,cyl}}. \tag{2.5}$$

The air mass $m_{\text{A,cyl}}$ in the cylinder is directly proportional to the air density $\rho_{\text{A,cyl}}$, so that also the heat quantity that can be added is directly proportional to this air mass in the cylinder and consequently must approximate the charge density of the engine. With this, the cylinder work in a given engine is directly dependent on the density of the air in the working cylinder at the end of the intake stroke and gas exchange.

Combining the equations above results in

$$V_{\text{cyl}} \text{imep} = \frac{V_{\text{cyl}} \rho_{\text{A,cyl}} Q_{\text{low}} \eta_i}{A_{\min}}, \tag{2.6}$$

with the consequence

$$\text{imep} \sim \rho_{\text{A,cyl}}. \tag{2.7}$$

Therefore, with the internal efficiency considered constant (i.e., unchanged combustion process and unchanged losses in the high-pressure process), the medium indicated pressure of a work cylinder is proportional to the charge density in the cylinder at the beginning of the compression stroke.

2.1.2 Interrelationship between charge mass flow and engine power output

After the cylinder work has been determined, the engine power output can easily be related to the air mass flow. It must be proportional to the swept volume of the whole engine (according to the total number of its work cylinders) as well as, depending on the working process, the number of power cycles in a given time.

$$P_i = V_{\text{tot}} \text{imep} n_{\text{WC}}, \tag{2.8}$$

where V_{tot} is the displacement of the engine, imep the indicated mean effective pressure, and n_{WC} the number of working cycles. The latter still has to be defined in detail. Only in a two-stroke engine, where every revolution represents a working cycle, is it identical to the measured speed. If we introduce an index i between the number of engine revolutions n and the number of working cycles n_{WC}, for a two-stroke engine, $i = n/n_{WC} = 1$. In the four-stroke engine, on the other hand, combustion takes place during every second revolution only, and therefore in a four-stroke engine, $i = n/n_{WC} = 2$.

With this, the indicated engine power output can be determined as follows:

$$P_i = V_{tot} \text{imep} \frac{n}{i}, \quad \text{where } V_{tot} = n_{cyl} V_{cyl}. \tag{2.9}$$

Including the proportionality of imep and $\rho_{A,cyl}$, we find:

$$P_i \sim V_{tot} \rho_{A,cyl} \frac{n}{i} \quad \text{or} \quad P_i \sim \dot{m}_{A,cyl}. \tag{2.10}$$

We now have tied the engine power output to the air mass flow through the engine.

If an internal combustion engine is supposed to generate power output for more than a single work cycle, the exhaust gas has to be removed from the cylinder and after each such work cycle be replaced with fresh air in the case of a diesel engine or fresh mixture in the case of a gasoline engine.

In the ideal engine, which we have looked at up to now, this happens without losses and completely. For the real engine, the gas exchange process has to be described in more detail. It is important since it influences the engine characteristics considerably. The following requirements apply for the layout of the gas exchange:

- the exhaust gas present in the cylinder at the end of the working stroke has to be removed as completely as possible,
- the fresh air or fresh charge quantity required must be exactly prepared to the requirements of the engine, e.g., regarding cooling or exhaust gas quality,
- the aspirated fresh charge must fill the cylinder as completely as possible.

In practice, this means that the total fresh charge mass flowing into the cylinder, m_{In}, and the fresh charge mass remaining in the cylinder, m_{fA}, usually are not identical. They differ by that fraction of charge mass which, during the simultaneous opening of the inlet and exhaust devices (the so-called overlap period), without participating in the combustion, directly flows into the exhaust, i.e., the scavenging mass m_S.

$$m_S = m_{in} - m_{fA}. \tag{2.11}$$

In a naturally aspirating four-stroke engine, due to the small valve areas during the overlap period, the scavenging mass is insignificant. In most cases it is also not very significant in supercharged engines with a larger valve overlap. In some engine types (medium-speed supercharged natural gas engines as well as large slow-speed two-stroke engines), the scavenging air portion is systematically used to cool the combustion chamber. For that it is necessary to generate a positive pressure gradient throughout the engine (high turbocharger efficiencies), which results in larger scavenging air quantities during valve overlap. Especially since supercharging is common today, in two-stroke engines with their very large overlap areas of the gas exchange control devices, a careful layout and design is very important for the optimization of the scavenging process. Altogether, in a two-stroke engine an attempt must be made to achieve a good gas exchange with small scavenging air masses, so that the exhaust gas mass remaining in the cylinder, m_{RG}, stays as small as possible. Exhaust gas mass and the fresh mixture mass remaining in the cylinder per cycle, m_{fA}, thus constitute the cylinder charge mass m_{cyl} which is in the cylinder at the beginning of compression,

$$m_{cyl} = m_{fA} + m_{RG}. \tag{2.12}$$

The exhaust gas mass m_{Ex} exiting into the exhaust per work cycle also contains the scavenging mass directly scavenged into the exhaust manifold during the overlap period, and for the mixture-aspirating gasoline engine, it is identical to the inflowing fresh charge mass. In the air-aspirating diesel engine it is larger than the aspirated air mass by the amount of injected fuel mass m_F per work cycle,

$$m_{Ex} = m_{In} + m_F. \tag{2.13}$$

2.2 Influence of charge air cooling

Independent of its design, in any compressor the compression of the intake air results in a temperature increase, which primarily depends on the desired pressure ratio, i.e., the supercharging factor, and the compressor efficiency:

$$T_2 = T_1 \left\{ 1 + \frac{1}{\eta_{s\text{-}i,C}} \left[\left(\frac{p_2}{p_1} \right)^{(\kappa-1)/\kappa} - 1 \right] \right\}. \tag{2.14}$$

Here, T_1 and T_2 represent the temperatures upstream and downstream of the compressor in kelvin, $\eta_{s\text{-}i,C}$ the isentropic compressor efficiency, and p_1 and p_2 the pressures upstream and downstream of the compressor.

At constant charge pressure, this temperature increase diminishes the inflowing fresh charge corresponding to the density change caused by it, and downstream causes higher process temperatures with all its associated disadvantages.

As an example for the efficiency of charge air cooling, let us consider an ideal engine with the following characteristics:

intake pressure $p_1 = 1$ bar charge pressure ratio $p_2/p_1 = 2.5$
intake temperature $T_1 = 293$ K (20 °C) compressor efficiency $\eta_{s\text{-}i,C} = 0.70$

This results in a final charge temperature of $T_2 = 418$ K (145 °C).

In the following comparison, the combustion air ratio is kept constant, i.e., the fuel mass and with it the power output are determined according to the charge mass.

With above data, the aspirated engine has the air density
$\rho_1 = \rho_2 = 1.19$ kg/m^3 (=100%).
The supercharged engine without charge air cooling has the charge density
$\rho_2 = 2.09$ kg/m^3 (=175%).
The charge air cooled engine with a cool-down to 40 °C enables a density increase to
$\rho_2 = 2.78$ kg/m^3 (=234%).

In this example, we see the enormous effect of charge air cooling, since at a constant pressure ratio a density increase of 2.78/2.09, i.e., an increase of 33% is obtained, combined with a process start temperature which is about 190 °C lower.

Charge air cooling therefore has the following advantages:
– a further **power increase** of supercharged engines at constant pressure ratio due to the increased charge density;

- a lower charge temperature at process start with lower process temperatures, resulting in **lower thermal stress for the components**;
- lower NO_x emissions due to the lower process temperatures;
- a decisive improvement in the knocking tendency of supercharged gasoline engines; only with charge air cooling, gasoline engines can achieve acceptable fuel consumption.

2.3 Definitions and survey of supercharging methods

Here we will define possible types of pre-compression processes and the characteristic properties of chargers or compressors.

Supercharging by means of gasdynamic effects
- The exploitation of the pressure waves in the intake and exhaust systems via pulse or variable intake systems and tuned exhaust manifold lengths
- Supercharging via Helmholtz resonator intake manifold layouts
- Pressure-wave charging via direct pressure exchange between exhaust gas and charge air (Comprex, register-resonance charger)

Supercharging with mechanically driven chargers
- Displacement or rotary piston charger without internal compression (e.g., Roots blower)
- Displacement or screw-type charger with internal compression (Lysholm, Wankel, spiral charger)
- Turbo compressors (radial compressor, axial compressor)

Supercharging systems with exhaust gas energy recovery
- Coupling a turbo compressor with a turbine – both located on the same shaft –, called an exhaust gas turbocharger
- Coupling of a displacement compressor with an expander located on the same shaft (Wankel)

Supercharging via combination of the components mentioned above
- Turbocompound system, consisting of an exhaust gas turbocharger with downstream energy recovery turbine
- Combined systems of resonance charger and exhaust gas turbocharger
- Combination of a mechanical charger with an exhaust gas turbocharger

2.4 Supercharging by means of gasdynamic effects

We begin our detailed examination of the various supercharging possibilities with the widely used **pressure wave charging** via **pulse** or **variable intake systems**. In addition, with the help of tuned exhaust pipe lengths during the overlap period a lower pressure can be achieved in the exhaust system than in the cylinder; this results in an improved scavenge process of the residual gas. Lastly, increases in volumetric efficiency are possible via so-called resonance charging with Helmholtz resonator and resonance manifold combinations (Cser supercharging).

2.4.1 Intake manifold resonance charging

This type of precompression uses the dynamics of the pressure waves in the intake and exhaust manifolds of high-speed engines. It is therefore a dynamic pressure increase in the intake system without the use of a compressor.

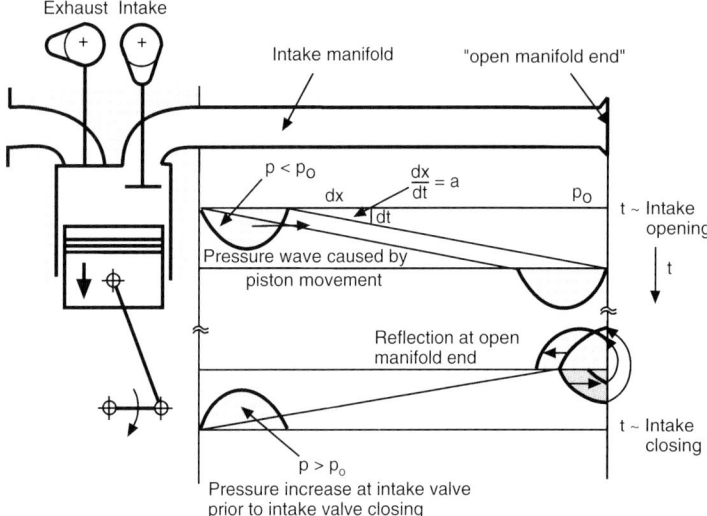

Fig. 2.1. Excitation and propagation characteristics of air pressure waves in an intake manifold, and pulse charge effect obtainable with them

The periodical opening of the intake and exhaust valves of a reciprocating piston engine excites the corresponding gas columns in the intake and exhaust manifolds, which results, depending on the phase position, frequency and engine speed, in manifold pressures at the valves which are significantly different from the ambient pressure. With every opening of the intake or exhaust valve a lower or higher pressure wave enters into the corresponding manifold system and is reflected at its end (manifold or muffler) as a high or low pressure wave (Fig. 2.1) respectively.

If the lengths of the intake and exhaust manifolds are tuned correctly, shortly before "intake closes" a higher pressure wave arrives at the intake valve, which increases the pressure in the combustion chamber. Correspondingly, shortly after "intake opens" and before "exhaust closes", in the so-called valve overlap phase, a lower pressure wave reaches the exhaust valve and thus creates a positive scavenging gradient relative to the intake manifold, with corresponding improvement of the combustion chamber scavenging process or an improved expulsion of remaining exhaust gases. Physically the aspirating work of the piston is transformed into compression work. Both effects

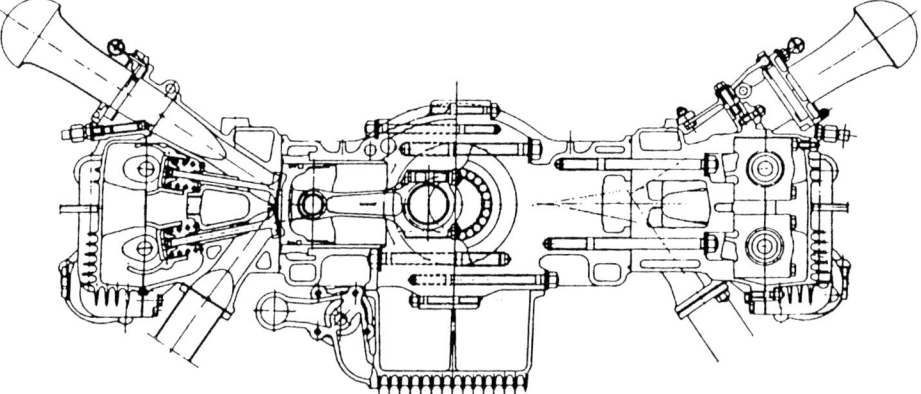

Fig. 2.2. Sports engine (Ferrari) with pulse intake manifolds

2.4 Supercharging by means of gasdynamic effects

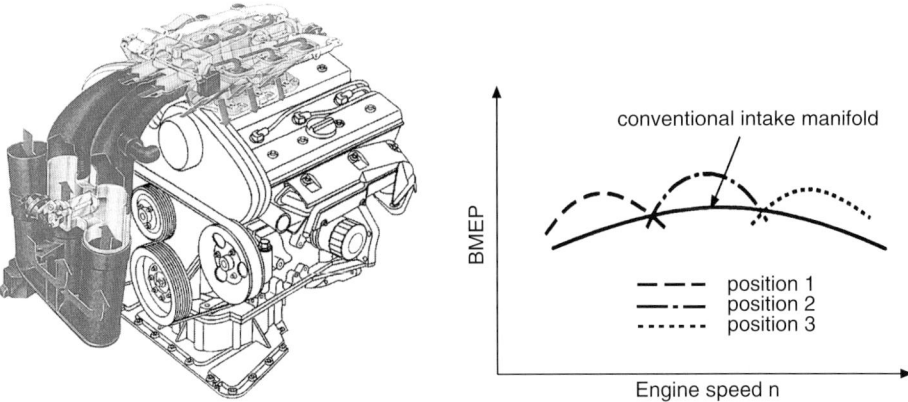

Fig. 2.3. Three-stage variable intake system (Opel) with achievable torque increases

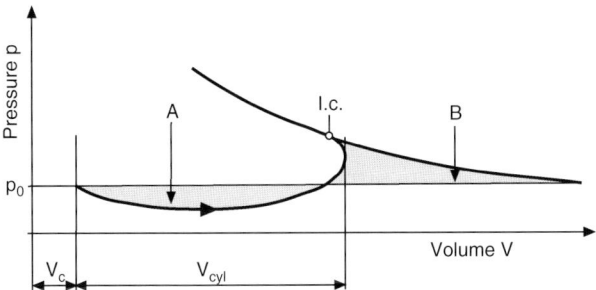

Fig. 2.4. pV diagram of the gas exchange work for resonance charging

combined are preferred in sports or racing engines, since in those the necessary wave propagation time, due to the very high engine speed, is shortened and with that also the necessary manifold length. Figure 2.2 shows a sports engine (Ferrari) with pulse intake manifolds. If the exhaust system is also included in this pulse tuning – as is usual in today's racing engines – air delivery ratios of maximum 1.25–1.3 and a significant charging effect are achieved.

On the intake side, today so-called variable intake systems are frequently used for series production engines, which operate with variable reflection lengths, as shown in Fig. 2.3 with the three-stage Opel intake manifold as an example. This layout increases the volumetric efficiency in the lower speed range and improves the torque curve in the medium speed range. Additionally, a rise in volumetric efficiency is gained in the area of rated horsepower.

In any case, with all these systems the gas cycle work is increased, because – due to the generation of the aspirating wave in the intake system – the pressure in the cylinder is decreasing further than with regular intake manifold layouts. Figure 2.4 shows this effect in the pV diagram. With the possibility of a continuous adjustment of the intake manifold length (e.g., in Formula 1), an increase in volumetric efficiency can be achieved in the entire full load speed range.

2.4.2 Helmholtz resonance charging

To obtain Helmholtz resonance charging, a plenum-manifold system (Helmholtz resonator) is connected to the intake side to several cylinders, with a layout in which the aspiration cycle periods of these cylinders correspond to the eigenfrequency of the plenum-manifold system. With this

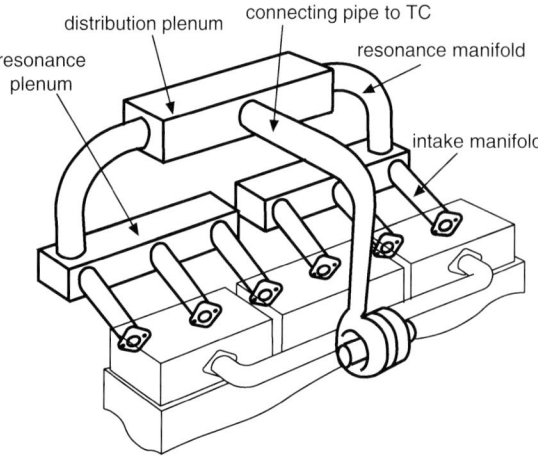

Fig. 2.5. Helmholtz resonance charging using discrete resonance plenums (Saurer)

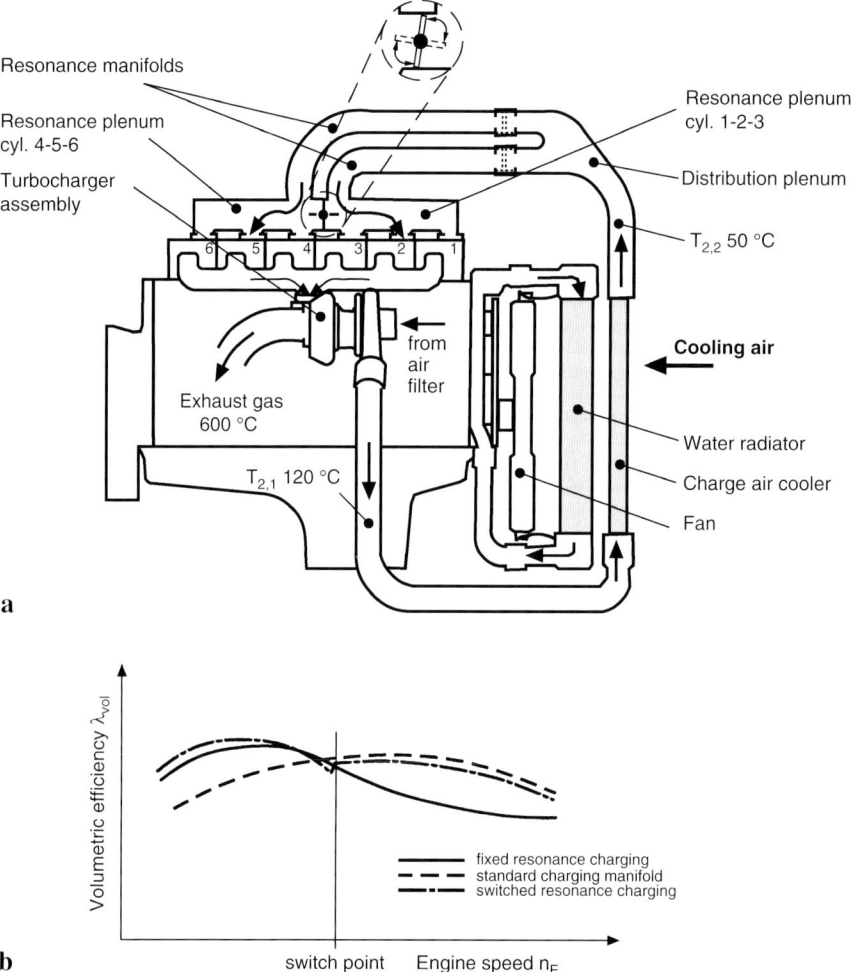

Fig. 2.6 a, b. Switched resonance charging with volumetric efficiency curves for standard and switched versions

arrangement, supercharging is obtained at the resonance speed or in a limited speed range. The disadvantage of this layout is that, if it is not designed variable (Fig. 2.5), the intended volumetric efficiency increase in the lower speed range is reached only with a loss in the upper speed range. This disadvantage can be mostly avoided if the layout is made switchable via a simple blocking valve in the charge air manifold (Fig. 2.6a).

Figure 2.6b shows in principle the volumetric efficiency curves of a standard charging manifold system compared with a fixed and a switched resonance charging system.

The layout and optimization of the gasdynamic systems described here is usually done on the basis of numeric cycle simulations, which allow the evaluation of the system variants so that the most promising can be selected and optimized. Before such systems are optimized on an engine test bench – especially in combination with a suitable control algorithm – it is advantageous to first evaluate complex three-dimensional assemblies in the course of their detailed design in view of gasdynamic behavior with the aid of 3-D CFD (computational fluid dynamics) simulations. The 3-D simulation area may be evaluated independently of the complete engine, where the boundary conditions for the simulation can be provided by the above mentioned cycle calculations. On the other hand, if it is necessary to take the retroactive effects of the 3-D simulation area on the operation characteristics of the complete engine into consideration [40] (e.g., distribution of exhaust gas recirculation in an air plenum), various commercial software systems offer the possibility of a direct integration of the CFD simulation area into the thermodynamic engine simulation model (AVL-BOOST/FIRE, WAVE/STAR-CD, GT-POWER/VECTIS).

2.5 Supercharging with supercharging units

The various compressor principles were already mentioned in Sect. 2.3. However, it is important to note that compressors basically can be divided into the following two categories, depending on the mechanisms employed to compress the gas:

- **displacement type** superchargers, e.g., reciprocating piston, rotary piston and rotating piston chargers
- **flow type** superchargers, e.g., turbo machineries such as radial and axial compressors

The displacement type compressors or chargers can additionally be distinguished by their working principle, i.e., whether they use internal compression (e.g., reciprocating piston compressor) or simple gas delivery without internal compression (e.g., Roots blower). As is shown in Fig. 2.7, the use of internal compression can reduce the specific work needed for gas compression, which significantly improves compressor efficiency, especially at higher pressure ratios. Today, applications with relatively low compression ratios up to 1:1.7 are widespread. However, up to this pressure ratio the advantage gained by internal compression is relatively small. On the other hand, charger designs without internal compression (e.g., Roots blower) can be manufactured more easily and therefore are more cost efficient, which is the reason why this design is often preferred.

2.5.1 Charger pressure–volume flow map

The behavior of the supercharger designs discussed here can be best explained in a pressure–volume flow map (Fig. 2.8), in which are plotted

- on the **x-coordinate** the **pumped volume flow** and thus the mass flow,
- on the **y-coordinate** the **pressure ratio** of the particular compressor.

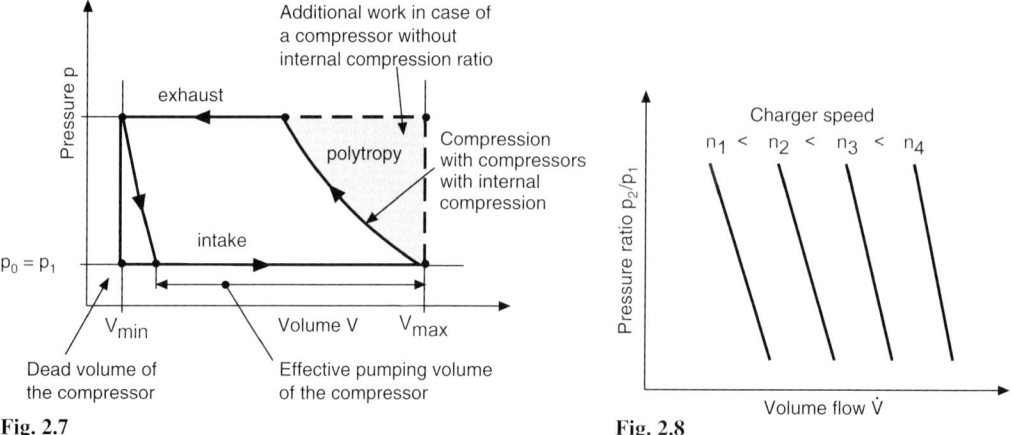

Fig. 2.7

Fig. 2.8

Fig. 2.7. Specific gas compression work of a displacement compressor with and without internal compression

Fig. 2.8. Principle pressure–volume flow map of a displacement (piston) charger at given charger speeds

Customarily this map is augmented by

- curves with constant **charger speed**,
- curves of constant isentropic or **total efficiencies**.

Although the various layouts and design principles strongly affect the performance map, it allows us to show and compare the characteristics of displacement and turbo compressors very well.

2.5.2 Displacement compressor

The simplest example of this design is the reciprocating piston charger, which, however, nowadays is only used for slow-speed two-stroke engines in parallel or series layout with the exhaust gas turbocharger. But it is very well suited to deduct the characteristics in the performance map.

With the help of the pV diagram for this charger type we will discuss the effects on its efficiency and influences of the real process management on the compressor work. Figure 2.9 shows the pV diagram of a reciprocating piston compressor. Here we clearly see the influence of the dead space and the value of the desired boost pressure on the real intake volumes and therefore the delivery

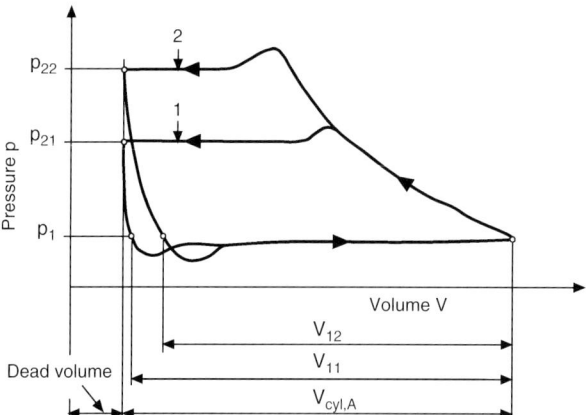

Fig. 2.9. pV diagram of a reciprocating piston compressor with varying boost pressures

2.5 Supercharging with supercharging units

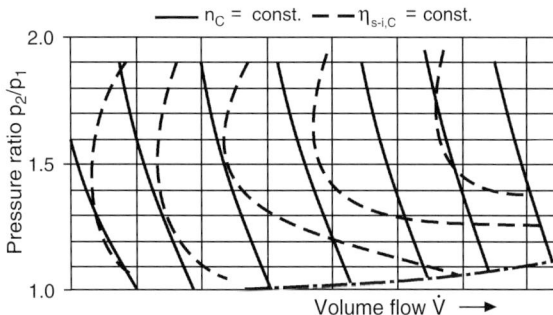

Fig. 2.10. Pressure–volume flow map of a displacement compressor with delivery curves and efficiencies

quantity. The charging efficiency and with it the delivery quantity thus decrease with increasing pressure.

An example of a real pressure–volume flow map of a displacement compressor is shown in Fig. 2.10. For all displacement compressors the volume flow decreases, with increasing boost pressure p_2, and therefore the volume flow curves in the map at constant speeds are slightly tilted to the left. The curves showing the efficiency η_{s-i} or η_{tot} strongly depend on the charger type. The map characteristic shown above is very similar to that of the rotating piston chargers which are commonly used today because of their small installation space and cost advantages, as well as to that of rotary (Wankel) chargers, Roots blowers, and the Lysholm screw-type compressor. However, since the Roots blower cannot offer internal compression, it should primarily be used for applications with low boost pressure.

It must be pointed out that all **displacement compressors**, in contrast to flow compressors, more or less **deliver discontinuously**. Depending on the degree of internal compression, they therefore cause pressure waves in the charge air manifolds, which results in uneven cylinder volumetric efficiencies or can lead to noise problems in the engines.

The characteristics of displacement compressors can be summarized as follows.

There is **no unstable area** in the pressure–volume flow diagram, i.e., the total delivery range indicated by the charger dimensions (V_{cyl} and n) can be utilized.

The **achievable pressure ratio** is **independent of the supercharger speed**. But it is decisively dependent on the design conditions, such as dead volume, leakage, installed size, and design type. Nowadays, p_2/p_1 reaches actual values of 1.8–2.

Relatively steep characteristics are obtained for constant charger speeds, i.e., with increasing boost pressure they are slightly tilted to the left.

This behavior influences and naturally affects the control strategies of such charging systems, because with boost pressure changes, only small increases or decreases in delivery quantities are achieved. This can be easily controlled, e.g., via a simple flow bypass.

The delivery quantities achievable are approximately proportional to the charger displacement.

At constant pressure ratio, the delivery quantity is approximately proportional to the charger speed.

2.5.3 Turbo compressor

For applications with reciprocating piston engines, the most important turbo compressor is the radial compressor, which derives its name from the radial exit direction of the delivery medium out of the compressor impeller. The intake of the delivery medium occurs axially.

Since the radial compressor will be discussed in detail in Chap. 5 in connection with exhaust gas turbocharging and as part of the exhaust gas turbocharger, at this point its function will only be addressed as the basis for its map characteristics.

All flow compressors are based on the physical principle of the **transformation of kinetic energy**, which is supplied to the medium in the impeller, into a **pressure rise via flow deceleration**, partially in the impeller, partially in a diffuser. The complete process between compressor inlet and outlet can be clearly described using the first thermodynamic theorem for open systems:

$$w_C = \frac{v_2^2}{2} - \frac{v_1^2}{2} + h_2 - h_1, \qquad (2.15)$$

where w_C is the added specific compressor work, v_i are the medium absolute flow speeds at the intake (1) and outlet (2), and h_i are the corresponding enthalpies. The latter describe the gas condition, which enables, directly from Eq. (2.15), the calculation of the pressure and temperature at the compressor outlet or the compressor work.

The danger of flow stalling exists in the flow compressor, as in the diffuser. Therefore, in a single compressor stage, only a limited pressure ratio can be achieved. Since the **radial compressor** enables the **highest per-stage pressure ratios**, it is the preferred choice for a compressor in exhaust gas turbochargers. In this layout, the chargers can be of very compact design. Their disadvantage in comparison to axial compressors is lower efficiency.

From all these facts it is clear that flow compressors show totally different map characteristics compared with displacement compressors.

Additionally, all **turbo compressors deliver continuously**, except for the speed fluctuation at the compressor impeller exit caused by the finite blade thicknesses. Although they thus generally feature a better acoustic quality, radial compressors are also sometimes equipped with silencer systems to eliminate these high-frequency noise excitations.

The map characteristics of turbo compressors can, then, be predicted as follows (Fig. 2.11).

There is an **unstable area** in the delivery map, which is located in the left sector of low flow rates and which widens at higher pressure ratios. The pressure ratio obtainable also depends on the delivery quantity. The borderline between stable and unstable delivery is called the surge limit.

The **achievable pressure ratio** will be **about proportional to the speed squared** and will thus be limited by the maximum possible charger speed and by the maximum circumferential speed, which itself is determined by the mechanical rigidity of the impeller.

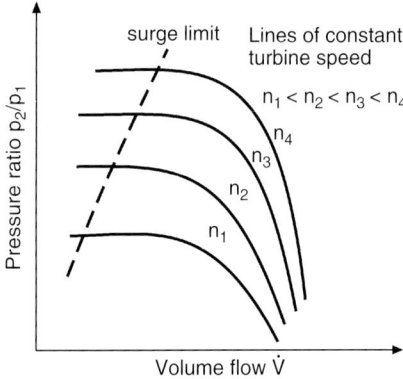

Fig. 2.11. Principle pressure–volume flow map of a turbo compressor at given charger speeds, with surge limit

The characteristic curves of constant charger speed reach the same pressure ratio in a wide range, and thus they run horizontally despite different delivery quantity. The achievable pressure ratio will decrease only with further increasing flow rates, due to incorrect flow into the impeller and, if installed, diffuser blades. The speed curves drop in an increasingly steep decline to a maximum flow rate value without pressure increase. This maximum value, also called choke limit, is attained when the speed of sound is reached at the compressor intake.

It is important to note that in a turbo compressor, contrary to a displacement compressor, a **pressure increase must always be associated with a speed increase**, and the **maximum pressure ratio is always reached at maximum speed of the compressor**.

With this, the essential characteristics of displacement compressors and flow compressors are defined, so that now the interaction with a reciprocating piston combustion engine can be examined.

2.6 Interaction between supercharger and internal combustion engine

In order to be able to evaluate the interaction between the charger and the reciprocating piston engine, it is necessary to develop the engine map similar to the charger map, i.e., how its air flow depends on engine speed and charge pressure.

2.6.1 Pressure–volume flow map of the piston engine

In the pressure–volume flow map of the engine (Fig. 2.12), the x-coordinate also represents the volume flow or the mass flow rate through the engine, and on the y-coordinate the pressure ratio between cylinder and outside pressure at the start of compression is plotted.

Therefore, it is also of practical use to reference this engine map, that is, its pressure–volume flow diagram, to the **state at charger intake**. Since in this scale the pressure–volume flow map of the charger (or the supercharging system) and that of the engine (to be supercharged) are identical, the interaction between charger and engine can be shown and evaluated in it.

Two-stroke engine

The two-stroke engine has a relatively simple map, since both inlet and exhaust are open simultaneously for extended periods of its gas exchange, i.e., around the bottom dead center. This causes a flow-through or scavenge process which can be described rather easily. The inlet

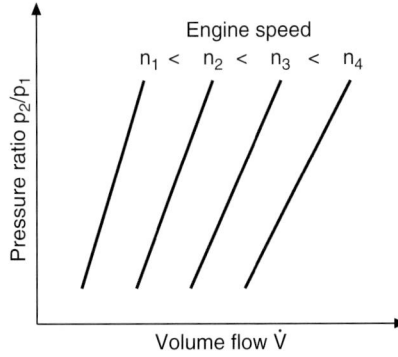

Fig. 2.12. Principle pressure–volume flow map of a reciprocating piston engine for given engine speeds

and exhaust port areas are substituted with a so-called equivalent area, which can be calculated as follows:

$$A_{\text{red}} = \frac{A_{\text{In}} A_{\text{Ex}}}{\sqrt{A_{\text{In}}^2 + A_{\text{Ex}}^2}}, \quad (2.16)$$

where A_{In} describes the intake port area, A_{Ex} the exhaust port area, and A_{red} is the equivalent port area. Further, a common flow coefficient μ_{red} is defined in such a way that it results in the same flow resistance as the series-connected inlet and exhaust areas. When the equivalent port area $\int A_{\text{red}} \, d\varphi$ is integrated over the engine cycle, which is 360° crank angle in the case of the two-stroke engine, the mass flow function describes the volume flow map:

$$\dot{V}_1 = \psi_{23} \frac{\rho_2}{\rho_1} \sqrt{2RT_2} \mu_{\text{red}} \frac{\int A_{\text{red}} \, d\varphi}{360} \quad (2.17)$$

with the flow rate function

$$\psi_{23} = \sqrt{\frac{\kappa}{\kappa - 1} \left[\left(\frac{p_3}{p_2}\right)^{2/\kappa} - \left(\frac{p_3}{p_2}\right)^{(\kappa+1)/\kappa} \right]}, \quad (2.18)$$

where μ_{red} is the flow coefficient associated with the equivalent area A_{red}, p_2 the charge or scavenge pressure, and p_3 the exhaust backpressure at the engine flange.

As can be seen from Eqs. (2.17) and (2.18), the scavenged air or mixture mass depends only on the backpressure at the exhaust port p_3 and the supercharger efficiency η_{TC}, at given geometric relations of the gas exchange ports and at a certain boost pressure (which influences the charge density via T_2).

Additionally, if the influence of the speed-dependent pulsation in the inlet and exhaust manifolds on the pressure upstream and downstream of the equivalent area A_{red} is neglected, there is no difference if, within a cycle's time period, the ports are opened seldom slowly or often rapidly. This results in an approximately speed-independent air or mixture mass flow and therefore, at a given backpressure, one singular engine operating curve only. Figure 2.13 schematically shows the volume flow through a two-stroke engine, depending on the boost pressure ratio p_2/p_1 and the backpressure p_3 as parameters. For a specified power output, a specific air or mixture volume flow $\dot{V}_1$ is needed. However, if the pressure p_{Ex} in the exhaust manifold changes, differing boost pressures or boost pressure ratios must compensate for this to maintain the necessary pressure gradient between inlet and exhaust, i.e., to assure $\dot{V}_1$ under all conditions.

The bold line shown in Fig. 2.13 schematically represents the operating curve of a two-stroke engine with exhaust gas turbocharging. With this type of supercharging, the exhaust backpressure increases with increasing boost pressure, which is the reason for the steeper slope of the curve compared to the case with constant backpressures obtained with mechanical supercharging.

Four-stroke engine

During the gas exchange process, the four-stroke engine works as a displacement compressor. Therefore, its volume flow is also calculated based on speed, swept volume, volumetric efficiency, and density ratio. However, its swallowing characteristics show a behavior contrary to that of a turbine: The volume flow increases with increasing boost pressure, since aspiration takes place at the precompression pressure p_2. This is why in this map the swallowing-capacity functions for constant engine speed are tilted to the right. For the four-stroke engine, the volume flow is

3 Thermodynamics of supercharging

3.1 Calculation of charger and turbine performance

Basic knowledge of thermodynamic processes in combustion engines is assumed for full understanding of the following chapter. Only interrelations important for supercharging itself will be discussed.

In general, a change in state during the (pre)compression of combustion air, i.e., a polytropic compression, leads to an increase in the temperature of the charge due to
- the isentropic temperature increase during compression, and
- the losses associated with the compressor efficiency, which finally will result in a polytropic change of state for the actual compression process.

For technical compressors, this temperature increase is used to calculate efficiency.

$$T_{2s} = T_1 \left(\frac{p_2}{p_1}\right)^{(\kappa-1)/\kappa}, \tag{3.1}$$

$$\Delta T = \frac{T_{2s} - T_1}{\eta_{s\text{-i},C}} \tag{3.2}$$

or

$$\eta_{s\text{-i},C} = \frac{h_{2s} - h_1}{h_{2\text{eff}} - h_1}, \tag{3.3}$$

and under the simplifying assumption of an ideal gas with constant specific heat, the following applies:

$$\eta_{s\text{-i},C} = \frac{T_{2s} - T_1}{T_{2\text{eff}} - T_1}. \tag{3.4}$$

The isentropic specific compression work can be calculated by applying the fundamental laws of thermodynamics as

$$w_{s\text{-i},C} = \frac{\kappa}{\kappa - 1} R T_1 \left[\left(\frac{p_2}{p_1}\right)^{(\kappa-1)/\kappa} - 1\right]. \tag{3.5}$$

Then, the real compressor power output can be determined as

$$P_C = \frac{\dot{m}_C w_{s\text{-i},C}}{\eta_{s\text{-i},C} \eta_{m,C}}, \tag{3.6}$$

where $\eta_{m,C}$ is the mechanical efficiency of the compressor (bearing, transmission, sealing).

To describe the **pressure ratio** p_2/p_1, i.e., the ratio between start and end pressure of the compression, the symbol Π is frequently used:

$$\Pi = p_2/p_1. \tag{3.7}$$

3.2 Energy balance of the supercharged engines' work process

3.2.1 Engine high-pressure process

Now we will examine the actual thermodynamic process, the so-called high-pressure process of the engine, in which the mechanical cylinder work is generated. The constant-volume cycle serves as thermodynamically ideal reference cycle. Then heat is supplied instantaneously and completely at top dead center of the piston movement. This cycle yields the maximum attainable efficiency of a combustion engine at a given compression ratio.

$$\eta_{th\omega} = 1 - 1/\varepsilon^{\kappa-1} \qquad (3.8)$$

or

$$\eta_{th\omega} = 1 - \left(\frac{p_1}{p_2}\right)^{(\kappa-1)/\kappa}. \qquad (3.9)$$

It can be seen that in this case the thermal cycle efficiency depends only on the compression ratio, and not on the supplied heat quantity and therefore the engine load. For the analysis of the real engine nowadays so-called thermodynamic cycle simulations are commonly used (see Sect. 3.6).

3.2.2 Gas exchange cycle low-pressure processes

These processes, or cycle parts, describe the charge exchange as well as the exhaust gas energy utilization for charge precompression and thus the technical processes of related supercharging. With the principle layout in mind, looking at the pV- and the TS-diagram (Fig. 3.1) of a mechanically supercharged ideal engine, three significant facts can be identified.

As a consequence of the cycle, at the end of the expansion (working) stroke (4) the pressure in the cylinder of a supercharged four-stroke engine is higher than the ambient pressure p_1 (5-6). However, this higher pressure cannot be transformed into work directly in the cylinder, due to the fact that the end of expansion is given by its geometric limitation. Therefore, an attempt must be made to exploit this pressure outside of the work cylinder.

Since the boost pressure is higher than ambient pressure, the gas exchange itself positively contributes to the engine work.

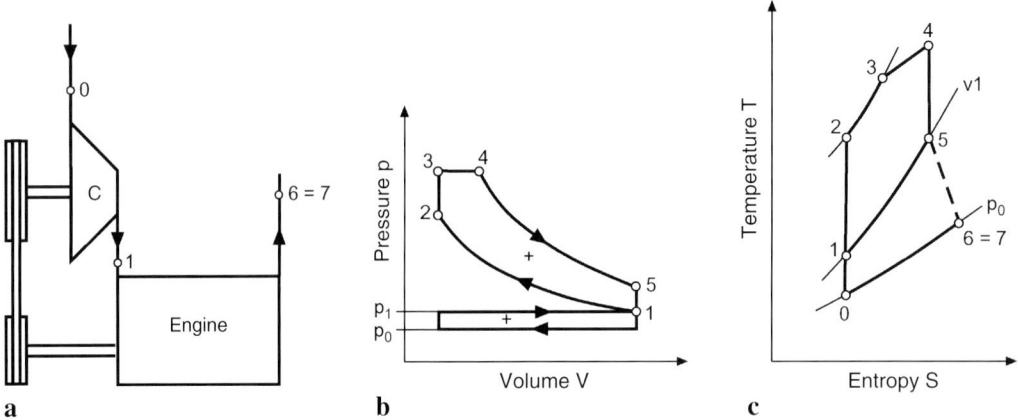

Fig. 3.1. Principle layout (**a**), pV (**b**) and TS diagram (**c**) of a mechanically supercharged ideal engine

3.2 Energy balance of the supercharged engines' work process

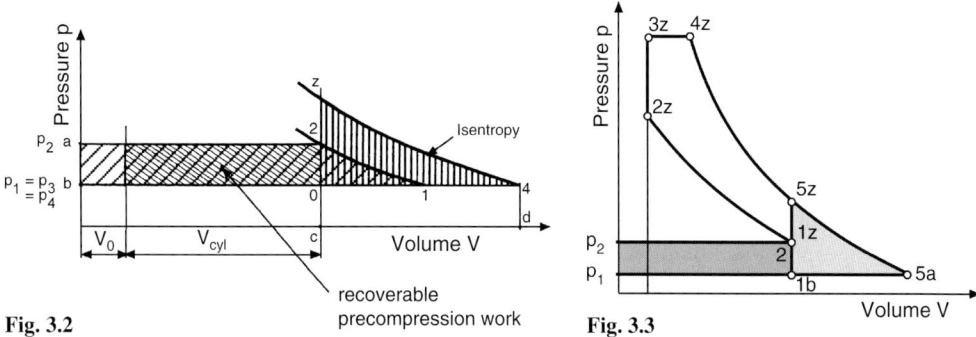

Fig. 3.2. Recovery of a part of the precompression work as crankshaft work

Fig. 3.3. pV diagram of a supercharged engine illustrating the reclaimable exhaust gas energy (area 5z-5a-1b)

Without efficiency losses, this work would approximately correspond to the compression work (charge exchange loop 1-5-6-7).

In return, however, the compressor work must be provided by the engine itself. The specific compression work which has to be employed is calculated for an isentropic ideal case according to Eq. (3.5), while – also idealized – the gas exchange work gained, w_{GEX}, is calculated with Eq. (3.10):

$$w_{GEX} = (p_2 - p_1)V_{cyl} \tag{3.10}$$

Accordingly, in the case of mechanical supercharging not the total charger work Δw will be lost, but only the difference

$$w_{s\text{-}i,C} - w_{GEX} = \Delta w. \tag{3.11}$$

This process can be understood as positive work output of the working piston during the intake stroke, during which the boost pressure p_2 (which is higher than the ambient pressure) acts on the piston. Thus a part of the precompression work can be recovered as crankshaft work, as Fig. 3.2 shows schematically.

3.2.3 Utilization of exhaust gas energy

Due to the geometrically given piston movement in a reciprocating piston combustion engine on the one hand, and on the other due to the thermodynamic cycle of the combustion process, the pressure at the end of the expansion stroke (5z) is significantly higher than the pressure at compression start of the high-pressure cycle (1z), as was described in Sect. 3.2.1 and shown in Fig. 3.3.

The energy available in the exhaust gas at the end of expansion in the high-pressure cycle (5z, 5a, 1b) therefore cannot be utilized in the working cylinder of the combustion engine itself but rather in a suitable downstream process.

Such a downstream process favored today is the recovery of the remaining exhaust gas energy via a so-called exhaust gas turbine. In it, a flow turbine uses the exhaust gas expansion energy to power a flow compressor located on the same shaft, which itself precompresses the combustion air before intake into the work cylinder.

There are several possibilities for the use of the remaining exhaust gas energy. The energy transport from the cylinder to the turbine is important, i.e., the design of the exhaust manifold.

With a careful layout of the exhaust system, the utilization of the exhaust gas energy can be maximized.

The corresponding optimization of such systems, i.e., the complex flow conditions around the exhaust valve, including the area of the exhaust port of a two stroke engine, demand comprehensive tests and/or simulations. Only today's availability of three-dimensional (3-D) mathematical simulation models with sufficient precision makes it possible to study these topics with adequate accuracy by means of numeric methods.

The aim is the optimum layout of the valve arrangement in combination with an exhaust manifold designed under gasdynamic aspects, so that maximum pressure recovery can be obtained, while at the same time the pressure gradient upstream of the turbine is minimized.

The complex issue of exhaust gas energy utilization via exhaust gas turbocharging is a very central and substantial item in the field of supercharging. Therefore, the simulation-related themes are covered intensively in Sect. 3.6, and those of the thermodynamic as well as flow design in Chap. 5.

3.3 Efficiency increase by supercharging

3.3.1 Characteristic values for the description of the gas exchange and engine efficiencies

Chain of engine efficiencies

In order to clarify those relations, which ultimately will lead to the actual, so-called effective efficiency of a combustion engine, in the following the efficiency definitions of internal combustion engines are described.

The brake or effective efficiency η_{eff},

$$\eta_{\text{eff}} = W_{\text{eff}}/Q_F, \qquad (3.12)$$

covers the sum of all losses in an internal combustion engine and can therefore be defined as the ratio between the brake effective work delivered and the mechanical work equivalent of the added fuel. In order to be able to evaluate and, if needed, minimize the losses individually, this total efficiency is generally subdivided into the following subefficiencies.

The fuel combustion rate η_F

$$\eta_F = \frac{Q_F - Q_{F,u}}{Q_F}, \qquad (3.13)$$

is defined as the ratio of burned fuel energy to added fuel energy, Q_F. It is especially useful for gasoline engines, which are operated at rich air-to-fuel ratios. The fuel energy not utilized is called $Q_{F,u}$.

The indicated efficiency η_i,

$$\eta_i = W_i/Q_F, \qquad (3.14)$$

is the ratio between the indicated work (based on the cylinder pressure curve) and the heat equivalent of the added fuel.

The process efficiency η_{th},

$$\eta_{\text{th}} = \frac{Q_{\text{add}} - Q_{\text{diss}}}{Q_{\text{add}}}, \qquad (3.15)$$

reflects to what extent the added heat could be converted in a theoretical reference cycle, e.g., in a constant-volume cycle or a mixed constant-volume–constant-pressure cycle (Seiliger cycle). Here Q_{add} describes the added heat and Q_{diss} the removed heat quantity. Thus, the theoretical efficiency characterizes the maximum of mechanical work which would be extractable from a given heat quantity, $Q_F \eta_{\text{th}} = W_{\text{th}}$.

The cycle efficiency factor η_{cyc},

$$\eta_{\text{cyc}} = W_i / W_{\text{th}}, \qquad (3.16)$$

contains all internal losses of the high-pressure as well as the low-pressure or gas exchange cycles, e.g., the influence of the real instead of the ideal gas characteristics, the residual gas, wall heat, and work gas losses as well as the gas exchange losses. Due to the latter, it is nowadays mostly further subdivided into a cycle efficiency factor for the high-pressure part of the cycle and one for the gas exchange cycle, i.e., the low-pressure part, with $\eta_{\text{cyc,HP}}$ as the term for the high-pressure cycle and $\eta_{\text{cyc,GEX}}$ as the term for the gas exchange. As a benchmark for comparison, again the work W_{th} attainable in the theoretical comparison cycle is used. The cycle efficiency factor describes to what extent the efficiency of the real process approaches the value of the theoretical reference cycle.

The mechanical efficiency η_m,

$$\eta_m = \frac{\text{bmep}}{\text{imep}} = \frac{\text{bmep}}{\text{bmep} + \text{fmep}}, \qquad (3.17)$$

is defined as the ratio of effective to indicated power or work and thus is also defined as the ratio of brake to indicated mean effective pressure. Finally, the following **chain of efficiencies** is obtained:

$$\eta_{\text{eff}} = \eta_F \eta_{\text{th}} \eta_{\text{cyc}} \eta_m. \qquad (3.18)$$

Gas exchange characteristics

The charge or gas exchange cycle significantly affects the operating behavior of the engine. In a four-stroke engine, this process primarily takes place during the exhaust and intake strokes, in a two-stroke engine close to the piston bottom dead center, while the ports are opened. In order to describe the quality and the characteristics of this process, ratios are defined which enable a comparison of the gas exchange cycles of various engines. These ratios, which characterize the volumetric filling of the cylinder with fresh gas, can be measured only in part directly or indirectly, often with great difficulty, and in part they can be calculated only.

The **air delivery ratio** λ_a represents an important factor, since it compares the total effective volume flow through the engine with the theoretical flow, which is calculated from the displacement and the number of combustion cycles per unit of time.

$$\lambda_a = \frac{\dot{V}}{V_{\text{tot}} n_{\text{WC}}} = \frac{m_{\text{in}}}{m_{\text{th}}}, \qquad (3.19)$$

where $n_{\text{WC}} = n$ for two-stroke engines and $n_{\text{WC}} = n/2$ for four-stroke engines.

This volume flow $\dot{V}$ can now be measured directly at the intake into the engine air supply system, e.g., with calibrated gas meters (normally in combination with large compensating plenums; see Chap. 10). Since the state of gas at this engine intake is practically identical to the ambient conditions, while on the other hand, especially in supercharged engines, pressures and temperatures are significantly different in the intake plenum, from which the engine aspirates the fresh charge, we differentiate between an ambient-related and an intake manifold-related air delivery ratio. In the former case, the volume flow at ambient conditions is measured directly, in the latter case the mean pressure and temperature in the plenum are used for the calculation of this ratio. The conversion from the ambient-related to the manifold-related value can be done in the following way:

$$\dot{V}_{IP} = \dot{V}_{amb} \frac{\rho_{amb}}{\rho_{IP}}. \qquad (3.20)$$

It is important to choose the measuring point in the intake manifold or the air plenum (IP) carefully so that representative conditions are measured (no local heat increases, no areas with flow separation, etc.). With the intake manifold-referenced air delivery ratio determined in the described manner, it is possible to compare measured results from various engines and also to compare simulated values with test bench data. In regard to their gas cycle quality, it is even possible to compare supercharged engines – with and without charge air cooling – to naturally aspirating engines.

It must be considered, however, that the differing temperature level of the fresh gas, both of engines with and without charge air cooling, may lead to differing heat flows in the intake manifold and port. Thus, in a highly supercharged engine without charge air cooling, the gas temperature can be significantly higher in some cases than the manifold wall temperature, so that the charge is cooled down between intake plenum and intake valve, which influences the volume flow at the valve significantly. In engines with charge air cooling, the fresh gas temperature will possibly be close to the water temperature and thus the intake port temperature, while in naturally aspirating engines the charge may be significantly heated up in the intake manifold and intake port, especially at low speeds.

Finally, in gasoline engines, the type of mixture formation – carburetor, single- or multipoint injection, and cylinder direct injection – and the layout of the mixture formation components have to be considered. Since in the real engine, fresh gas losses may occur during the gas exchange (in the four-stroke engine and especially in the two-stroke engine, the inlet and outlet control devices are in part open simultaneously), the air delivery ratio alone cannot adequately describe the quality of the gas exchange. For that the **volumetric efficiency** λ_{vol} can be used, which compares the fresh gas mass captured in the cylinder – again related to ambient or intake manifold conditions – with the cylinder displacement,

$$\lambda_{vol} = m_{fA}/m_{th}. \qquad (3.21)$$

This value characterizes the remaining fresh gas mass after the gas exchange cycle and thus is, among other things, a decisive factor for the attainable power. Especially for gasoline engines with external mixture formation, this value is additionally influenced by the added fuel vapor or inert gas (due to exhaust gas recirculated), so that a so-called **mixture-related volumetric efficiency** has to be distinguished from the **air volumetric efficiency**. The relationship between the two values is determined by the mass fraction of the fuel and the corresponding density of this medium at intake manifold conditions. Under the assumption of identical density for combustion gas or vapor and

3.3 Efficiency increase by supercharging

fresh air, the mixture volumetric efficiency can be approximately calculated by modifying the air delivery ratio according to the fuel mass fraction corresponding with the fuel-to-air ratio:

$$\lambda_{\text{vol,mix}} = \frac{m_{\text{A,cyl}} + m_{\text{F,cyl}}}{m_{\text{th}}}. \quad (3.22)$$

But all these volumetric efficiency values can be determined experimentally, directly or indirectly, only with great difficulty (e.g., concentration measurements using tracer gases). On the other hand, cycle and CFD (computational fluid dynamics) simulations can provide very detailed information about these values. With such simulations it is also possible to optimize the gas exchange cycle in regard to those very relevant figures. Further ratios that are also very relevant for the gas exchange as well as the operational behavior of the engine, are the following:

scavenging ratio

$$\lambda_S = \frac{m_{\text{fA}}}{m_{\text{fA}} + m_{\text{RG}}}, \quad (3.23)$$

amount of residual gas

$$\varphi_{\text{RG}} = \frac{m_{\text{RG}}}{m_{\text{fA}} + m_{\text{RG}}}, \quad (3.24)$$

scavenging efficiency of the engine

$$\Gamma = \frac{m_{\text{fA}}}{m_{\text{fA}} + m_S}. \quad (3.25)$$

The scavenging ratio (not to be mixed up with the scavenging air delivery ratio used for the description of the scavenging cycle of two-stroke engines [126]) specifies the ratio of the fresh gas mass trapped in the cylinder to the total cylinder charge mass. The amount of residual gas specifies the ratio of the gas remaining in the cylinder after the gas exchange process to the total cylinder charge mass. And the scavenging efficiency specifies that part of the total aspirated fresh gas mass which is captured in the cylinder after the gas exchange. Thus, the latter term represents a very characteristic value for the two-stroke scavenging process. Here, high scavenging efficiencies have to be aimed for to optimally utilize the fresh gas provided by the scavenging pump or blower.

It should be mentioned that the amount of residual gas is decisively dependent on the design and firing order of the engine. In both a two-stroke and a four-stroke engine, the amount of residual gas is strongly influenced by the blow down pressure pulses of the cylinders following in the firing order. The amount of residual gas can be significantly reduced, e.g., by means of an optimized exhaust manifold layout (connection of cylinders with sufficient angular firing distance, pulse converter, resonance exhaust manifold). Especially in gasoline engines, in view of knocking stability, the achievable engine brake mean effective pressure can be increased by such measures. On the other hand, in modern gasoline and diesel engines, increased amounts of residual gas are desirable in order to achieve a dethrottling effect at partial load (gasoline engines), as well as to influence the combustion temperature and fuel combustion rate with regard to the NO_x formation in the engine. For this as well, the amount of residual gas may be used as a suitable characteristic figure.

Finally, the relationship between volumetric efficiency, air delivery ratio, and scavenging efficiency Γ is as follows:

$$\Gamma = \lambda_{\text{vol}}/\lambda_a. \quad (3.26)$$

3.3.2 Influencing the engine's total efficiency value via supercharging

On the basis of these efficiency relationships we can now answer the question why, for a particular power output, a supercharged engine has a better effective efficiency than a naturally aspirated engine. A decisive factor is that for many reasons – e.g., the hydrodynamics of bearing and piston lubrication – the friction mean effective pressure increases with increasing speed, but only to a small extent with increasing load. Already on the basis of the equation for the mechanical efficiency (3.17), its dependence on the engine load is very obvious. This will be demonstrated with the following simple example.

We assume two engines of identical horsepower at a given speed, one of which is a naturally aspirated engine which reaches the required horsepower at a brake mean effective pressure of bmep = 10 bar. The other is a correspondingly smaller supercharged engine which reaches the same horsepower at a brake mean effective pressure of 20 bar. For the naturally aspirated engine, let the friction mean effective pressure be fmep = 2 bar. For the supercharged engine, due to the larger dimensions of its bearings etc. corresponding to the increased cylinder pressures associated with supercharged operation, let the friction mean effective pressure be 2.2 bar.

The result of this is:
- naturally aspirated engine: $\eta_m = 10/(10+2)[\text{bar}] = 83\%$
- supercharged engine: $\eta_m = 20/(20+2.2)[\text{bar}] = 90\%$

As a result of the higher specific load, the calculated mechanical efficiencies show a significantly better value for the supercharged engine, as is also shown in Fig. 3.4. Therefore, a very important relationship can be established between engine load and the effective efficiency:

The higher the load – read: the brake mean effective pressure – required for an engine to reach a given horsepower, the better its effective efficiency. Figure 3.5 shows this interrelationship for two medium-speed diesel engines of equal horsepower, with and without supercharging and at two speeds.

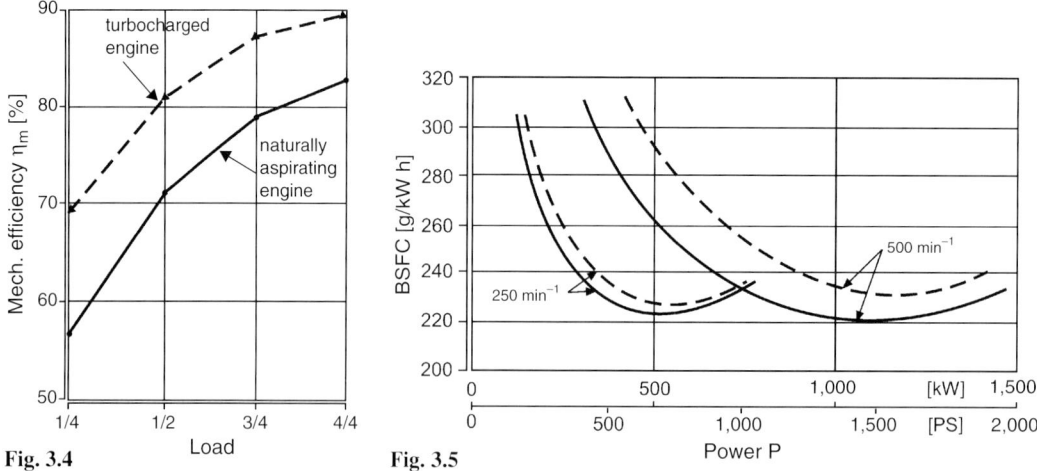

Fig. 3.4.

Fig. 3.5.

Fig. 3.4. Advantage in mechanical efficiency of the supercharged engine in comparison to the naturally aspirated engine

Fig. 3.5. Fuel consumption values of two medium-speed diesel engines of equal horsepower with (solid curves) and without (dash curves) supercharging, showing significant advantages for the supercharged engine [159]

3.4 Influence of supercharging on exhaust gas emissions 31

In comparison to this, the other efficiency factors are barely influenced by supercharging, since, due to the change of density of the intake air, the flow and thermodynamic conditions are influenced only to a minor extent.

3.4 Influence of supercharging on exhaust gas emissions

It must be considered that, especially for a diesel engine, the combustion cycle and, thus, the achievable efficiency of the engine are more and more influenced by the exhaust gas emission limits regulated by law. It is therefore necessary to briefly discuss the various test procedures which are used for different vehicle categories in various countries to quantify their pollutant emission level.

For passenger cars and light-duty trucks (LDV, light-duty vehicle), transient tests with the complete vehicle, derived from actual driving patterns, are used today, like the so-called FTP

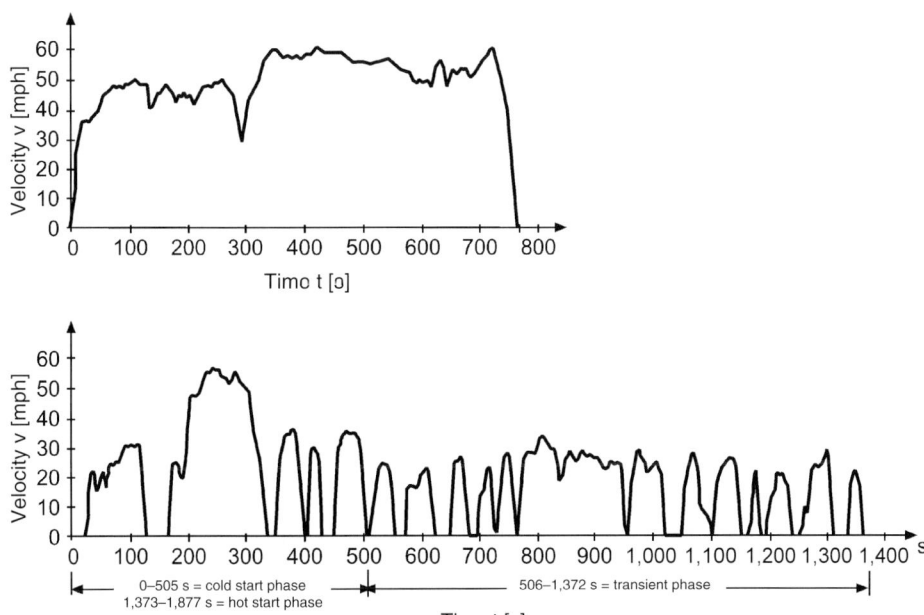

Fig. 3.6. FTP Cycle from the U.S. exhaust emission regulations for passenger cars and light-duty trucks

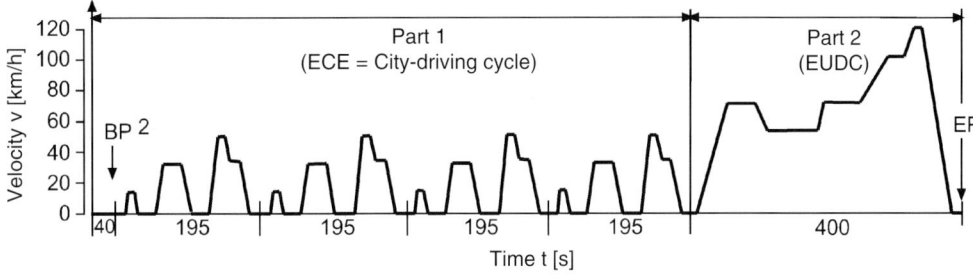

Fig. 3.7. European NEDC for passenger cars

Cycle (Federal Test Procedure; Fig. 3.6) or the European NEDC (New European Driving Cycle; Fig. 3.7).

Due to the wide variety of designs, pure engine test cycles are used for medium and heavy trucks, some stationary, like ECE R 49 (Fig. 3.8) and the new Euro-3-test (Fig. 3.9), some transient, like the Fige-3-transient test – an enhancement to the Euro-3-test – for engines with particulate filter or for gas engines (Fig. 3.10). Under these test conditions, the following statements generally valid can be formulated for the various combustion processes.

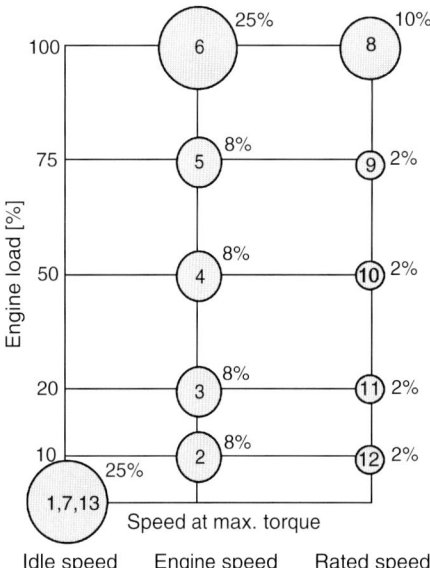

Fig. 3.8. ECE-R-49 stationary test cycle for trucks until 1999 (Euro 0 to Euro II)

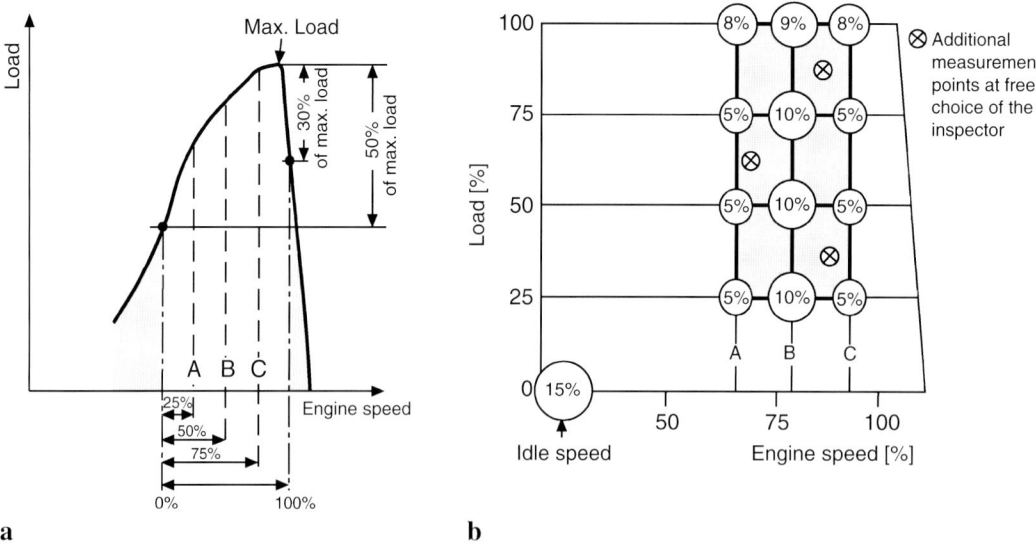

Fig. 3.9. Euro-3 truck test cycle: **a** test speeds, **b** load points with weighting

3.4 Influence of supercharging on exhaust gas emissions

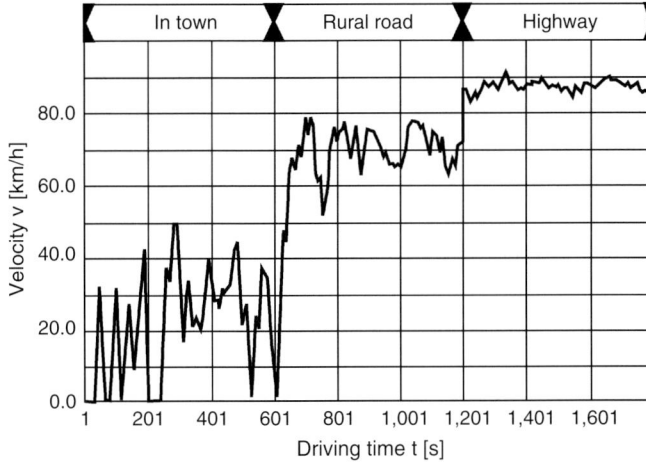

Fig. 3.10. Fige-3 transient test cycle for engines with particulate filter or for gaseous-fuel engines, or generally as of Euro IV

3.4.1 Gasoline engine

For the gasoline engine, the problem of exhaust gas aftertreatment has been solved to a major extend by the introduction of the λ-controlled three-way catalyst (TWC). Further emission reductions, down to SULEV (super ultra low emission vehicle) specifications, can be achieved mainly by improving the cold start phase, in which today about 80–85% of the total cycle emissions are generated, by means of an improved catalyst light-off, and by reduced raw emissions during the cold start.

In a gas engine, at least for trucks, lean operation can be a fuel-efficient alternative. However, λ values of at least 1.6–1.8 must be drivable reliably and with low residual methane emissions, i.e., with good combustion quality. The gasoline direct-injection engine (GDI), which was introduced to series production at the end of the nineties, essentially shows the same exhaust gas problem areas as the direct-injection diesel engine.

3.4.2 Diesel engine

The classic diesel combustion process – like the GDI process just mentioned – always operates with (sometimes substantial) excess air. This eliminates the possibility of using three-way catalysts as described above. Critical emissions are particulate matter (PM), NO_x as well as CO and HC emissions.

In heterogeneous combustion, soot must and will always result to some extent as a combustion end product, so that substantial generation of particulate matter cannot be avoided. The soot emission, and with it a part of the particulate matter emission, depend on the combustion air ratio. With a suitable layout of the supercharging system, a supercharged engine can be operated with high excess air ratios in all load ranges – even at full load – so that the preconditions for low particulate operation are better with a supercharged engine.

With excess oxygen, the flame temperatures are also always high, inevitably leading to high nitrogen oxide formation. Since the NO_x generation depends to the power of 4 on the temperature prevailing at the point of its formation, primarily local temperature peaks in the combustion chamber must be avoided to prevent NO_x emissions. This can best be done by operating the engine with high excess air ratios or by diluting the charge with inert gas. In the diesel engine, this can best be realized through the recirculation of cooled, oxygen-depleted exhaust gas.

Furthermore, since supercharged engines are operated with relatively high compression end pressures and temperatures, they can be operated with significantly later injection start and longer injection duration than naturally aspirated engines of the same power. This also contributes to the avoidance of locally high combustion chamber temperatures, without significantly increasing fuel consumption.

In a diesel engine, CO and HC emissions are uncritically low.

The test procedures and emission standards for passenger cars, trucks, and stationary engines valid in Europe, the United States and Japan are summarized in the appendix – Fig. A.1 and Tables A.2 to A.5. For additional information, due to the extensive nature of the regulations as well as test procedures and measurement instructions, it is referred to special literature and Codes of Regulations.

3.4.3 Methods for exhaust gas aftertreatment

Regarding the methods for exhaust gas aftertreatment as well, we must refer the reader to the broad spectrum of special literature, unless technical aspects specially related to supercharging demand otherwise. This is the case when water injection, particulate filters as well as oxidation or NO_x storage catalysts are applied.

With water injection, not only the temperature of the exhaust gases is lowered due to the vaporization of the water in the combustion chamber but also the volume flow through the turbine is increased. This results in a significant increase of the enthalpy of the turbine intake gases, which itself can be used for a further increase in boost pressure or for a turbo-compound operation.

If particulate filters are located in the high-pressure exhaust stream, upstream of the turbine, they represent a considerable heat sink with undesirable consequences for load changes of the engine. The same is valid for the application of oxidation or NO_x storage catalysts if, for whatever reasons, they are also located upstream of the turbine.

Locating all these aftertreatment systems downstream of the exhaust gas energy recovery device, like an exhaust gas turbocharger or a compound turbine, at the most slightly increases the exhaust gas backpressure and thereby reduces the reclaimable exhaust gas expansion pressure ratio. Other disadvantages, especially during transient operation of such engines, also have to be taken into account (e.g., extended warm-up periods).

3.5 Thermal and mechanical stress on the supercharged internal combustion engine

3.5.1 Thermal stress

With increasing fuel quantity, i.e., energy, added to the cylinder, naturally the amount of heat to be dissipated increases as well. The heat flows through the engine increase correspondingly. Additionally, as is shown in Fig. 3.11, at higher degrees of supercharging and without charge air cooling, the temperature of the charge air increases significantly, which results in further increased engine thermal loads. Therefore, simultaneous to the strength calculations for new engine layouts with the finite-elements (FE) method, numerical CFD simulation tools must be used for the analysis of the coolant and heat flows.

3.5 Thermal and mechanical stress on supercharged engine

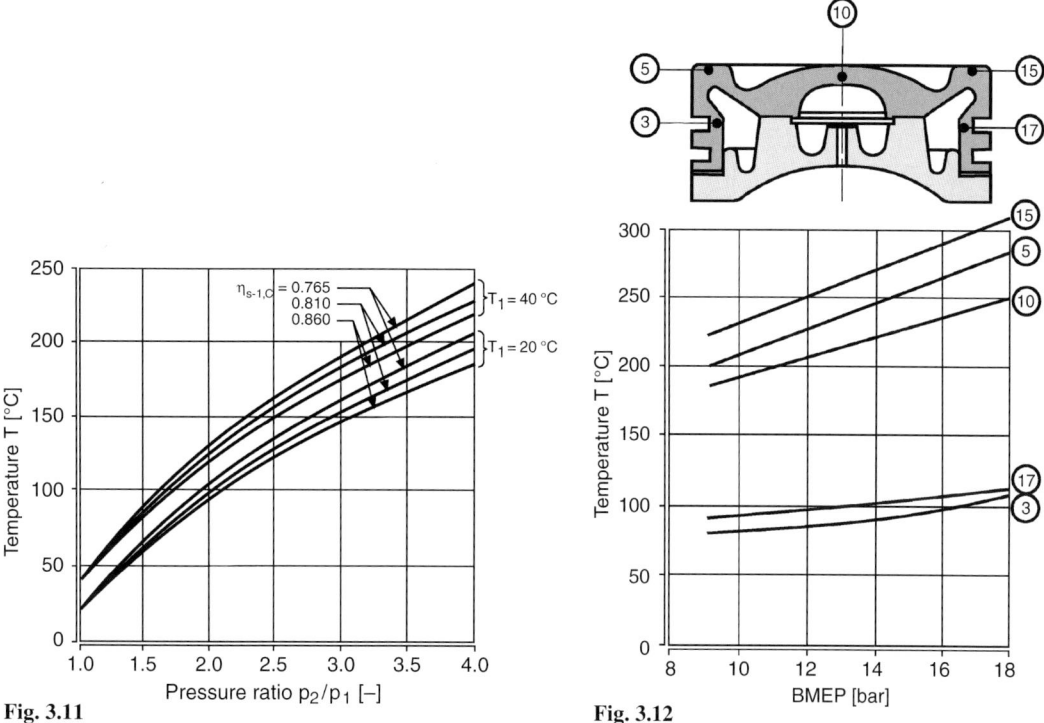

Fig. 3.11. Temperature of the charge air depending on the pressure ratio, for varying intake temperatures and compressor efficiencies, without charge air cooling

Fig. 3.12. Maximum temperatures for an assembled force-cooled piston for a medium-speed diesel engine

Only after consideration and analysis of all interactions by means of simulations, an optimum overall concept can be achieved regarding weight and load capacity combined with sufficient cooling at the smallest coolant circulation quantity possible.

The most important engine parts, besides the complete powertrain structure, are those loaded with high heat flow density, i.e., the cylinder head, the piston, and the cylinder liner. Figure 3.12 shows the maximum operating temperatures of an assembled and force-cooled piston for a medium-speed diesel engine.

3.5.2 Mechanical stress

With increasing boost pressure, compression end pressure and peak firing pressure are also increased, as shown in Fig. 3.13 in a pV and a TS diagram for a naturally aspirated and an exhaust gas turbocharged engine. The increasing pressures require the strengthening of certain parts or to approach their limit of strength, e.g., connecting rod, piston, cylinder head and bearings. The optimization of the entire powertrain of supercharged engines with regard to its strength becomes more and more important and mandatory as the brake mean effective pressure increases. Today, new engine designs are no longer feasible without the help of modern numerical simulations.

The strength-related optimization does not mean that supercharged engines have to be significantly heavier than naturally aspirated engines with comparable displacement.

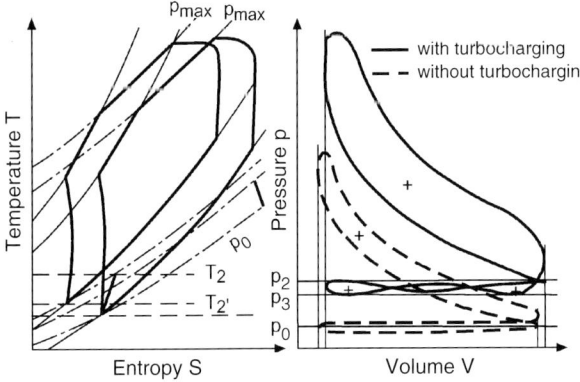

Fig. 3.13. *pV* and *TS* diagrams for a naturally aspirated and an exhaust gas turbocharged engine with significantly higher peak pressures for the turbocharged engine [74]

Very often the design of, e.g., housing wall thicknesses was in the past and still is decided not by strength, but by casting considerations.

During the layout of the powertrain, especially the dimensions of the bearings, the situation has to be evaluated in a very differentiated manner depending on the speed and mean effective pressure values. Inertial forces and gas forces may partially compensate for each other regarding their effect on bearing loads. Thus, a passenger car gasoline engine designed for high speeds could occasionally be supercharged without changing the bearing dimensions. In any case, high mean effective pressure values at low engine speeds are extremely critical for the bearings and therefore must be investigated in detail.

3.6 Modeling and computer-aided simulation of supercharged engines

3.6.1 Introduction to numeric process simulation

Given the advanced state of research with regard to detailed understanding of the most important physical processes in an internal combustion engine, it is possible to describe these relations by means of mathematical models. In doing so, these models may be based on fundamental basic equations (e.g., laws of conservation for mass, momentum and energy) as well as on empirical formulations for the description of complex processes (e.g., wall heat losses, heat release functions).

As a matter of principle, when simulating real processes, all physical relations mentioned above must be described correctly. Starting from the microscopic processes of thermal fluid dynamics – which encompasses gas dynamics, chemical reaction kinetics as well as, in the broader sense, the mechanics and heat conductance of continua, via the mesoscopic processes on the subsystem level, e.g., the thermodynamics of the entire internal combustion engine – all parts must be considered in a macroscopic simulation of the total system.

Basically, all complete systems can be reduced to the level of their microscopic subprocesses. The enormous time and effort necessary for modeling, as well as the extreme requirements regarding computing capacity, justify such an approach only in exceptional cases (such as aerospace), where, e.g., the verification of subsystem simulations is not possible by experiments. But for many technical applications, including the area of engine supercharging, it is desirable to define subsystems and to develop and use numeric simulation tools in accordance with the actual task.

3.6 Modeling and computer-aided simulation of supercharged engines

The three simulation levels mentioned above can be classified here as follows:

1. 3-D flow simulation in the gas conducting components including compressors and turbines (with consideration of thermal boundary conditions)
2. engine process simulation including the supercharging components
3. simulation of the total system behavior in actual use (e.g., modeling of the complete powertrain of a vehicle)

In this list, the engine process simulation constitutes the most important tool for layout of the engine and its supercharging components. For this reason, the corresponding physical and mathematical relationships will be described in more detail.

Regarding levels 1 and 3, only the most important methods will be mentioned and suitable software packages will be referred to.

3.6.2 Cycle simulation of the supercharged engine

In a thermodynamic cycle simulation, the essential processes of the complete engine system are described by means of mathematical equations in such a way that the physical states, such as pressures, temperatures and mass flows, can be calculated in all volumes and at all instants. Processes on a microscopic scale are described by analogous models and/or measurement data maps. In this way, on the one hand, highest simulation accuracies (e.g., calculation of engine mass flows with accuracy within 1–2% of the actual values) and, on the other hand, acceptable computing times (within minutes) can be achieved.

However, due to the simplifying approach, some subtasks cannot be analyzed with these methods in sufficient detail, e.g., the flow optimization of the inlet flow to a catalyst located a short distance downstream of the turbine outlet. For such tasks, the mentioned tools of level 1 – in this case, e.g., the 3-D flow simulation – must be used.

But let us concentrate in this section on the engine process simulation. The individual parts of the supercharged engine can be categorized into modules. The physical processes in the various modules are then mapped via mathematical models. With the use of master control programs, the individual modules can be finally interconnected. From the previous sections it is obvious that the following components are especially relevant for supercharged engines:

- cylinder for the description of the gas exchange and the high-pressure cycle
- pipe elements
- plenum elements
- turbocharger, consisting of flow compressor and turbine
- displacement compressor
- charge air cooler

Cylinder

For the mathematical description of the physical processes in the cylinder we can distinguish between the gas exchange and the high-pressure phase. Only during the gas exchange, mass flows occur between the cylinder and the connected pipes and manifolds.

Accordingly, for the **high-pressure cycle** the first law of thermodynamics for closed systems can be used assuming a simplified 1-zone model as follows:

$$\frac{d(m_{cyl} \cdot u)}{d\alpha} = -p_{cyl} \cdot \frac{dV}{d\alpha} + \frac{dQ_F}{d\alpha} - \sum \frac{dQ_W}{d\alpha} - h_{leak} \cdot \frac{dm_{leak}}{d\alpha} \qquad (3.27)$$

where $d(m_{cyl} \cdot u)/d\alpha$ describes the gradient of the internal energy in the cylinder, $(-p_{cyl} \cdot dV/d\alpha)$ the piston work, $dQ_F/d\alpha$ the heat release rate due to the added fuel energy, $\sum dQ_W/d\alpha$ the wall heat flow, and $h_{leak} \cdot dm_{leak}/d\alpha$ the enthalpy loss caused by the blowby gas flow (with the fuel energy Q_F, wall heat losses Q_W, and the blowby mass flow $dm_{leak}/d\alpha$).

It is apparent that the condition in the cylinder – characterized by the internal energy – is changed by the piston work, the heat energy released during combustion, the wall heat losses and by the enthalpy drain caused by the blowby mass flow. Equation (3.27) is to be used in general for engines both with internal and with external mixture formation. However, those terms expressing the change of gas composition during the high-pressure cycle must be described using the following specific characteristics for the differing mixture formation methods.

Internal mixture formation:
– instantaneous combustion of the added fuel corresponding to the heat release gradient
– instantaneous complete mixture of the combustion products with the remaining cylinder charge (1-zone model)
– continuous reduction of the air/fuel ratio during combustion

External mixture formation:
– homogeneous fuel–air mixture in the entire cylinder
– constant air-to-fuel ratio of the unburned charge during combustion
– same pressure and same temperature for both the burned and the unburned charge fraction

With these assumptions, Eq. (3.27) can be reformulated as follows:
– internal mixture formation:

$$\frac{dT_{cyl}}{d\alpha} = \frac{1}{m_{cyl}(\partial u/\partial T + (\partial u/\partial p)p_{cyl}/T_{cyl})} \left[\frac{dQ_F}{d\alpha}\left(1 - \frac{u_{cyl} + (\partial u/\partial p)p_{cyl}}{Q_{low}}\right) - \frac{dQ_W}{d\alpha} \right.$$
$$\left. - \frac{dm_{leak}}{d\alpha}\left(h_{leak} - u_{cyl} - p_{cyl}\frac{\partial u}{\partial p}\right) - m_{cyl}\frac{\partial u}{\partial \lambda}\frac{d\lambda}{d\alpha} - p_{cyl}\frac{dV_{cyl}}{d\alpha}\left(1 - \frac{\partial u}{\partial p}\frac{m_{cyl}}{V_{cyl}}\right) \right], \quad (3.28)$$

– external mixture formation:

$$\frac{dT_{cyl}}{d\alpha} = \frac{1}{(m_{cyl}\partial u/\partial T + (m_V p_{cyl}/T_{cyl})\partial u_V/\partial p)}$$
$$\times \left[\frac{dQ_F}{d\alpha}\left\{1 + \frac{1}{Q_{low}}\left[u_F + \lambda A_{st}u_A - (1 + \lambda A_{st})\right]\left(u_V + p_{cyl}\frac{\partial u_V}{\partial p}\right)\right\} \right.$$
$$\left. - \frac{dQ_W}{d\alpha} - p_{cyl}\frac{dV_{cyl}}{d\alpha}\left(1 - \frac{m_V}{V_{cyl}}\frac{\partial u_V}{\partial p}\right) - \frac{dm_{leak}}{d\alpha}\left(h_{leak} - u_{cyl} - p_{cyl}\frac{m_V}{m_{cyl}}\frac{\partial u_V}{\partial p}\right) \right]. \quad (3.29)$$

For both equations, the mathematical models for the heat energy released during combustion and for the wall heat losses still must be defined. As examples out of a multitude of models, here the 1-zone Vibe function and the wall heat loss model developed by Woschni will be discussed:

$$\frac{dx}{d\alpha} = \frac{a}{\Delta\delta_d}(m+1)y^m \exp(-ay^{m+1}), \quad (3.30)$$

$$dx = dQ/Q, \quad (3.31)$$

$$y = \frac{\alpha - \delta_0}{\Delta\delta_d}. \quad (3.32)$$

where Q is the total added fuel energy, δ_0 the start of combustion, $\Delta\delta_d$ the combustion duration, m the shape coefficient, and a the Vibe parameter ($a = 6.9$ for complete combustion).

The modified Woschni wall heat loss model [151] for the high-pressure phase reads as

$$Q_{\text{Wi}} = A_i \alpha_W (T_{\text{cyl}} - T_{\text{Wi}}), \qquad (3.33)$$

where, if

$$\frac{C_1 V_{\text{cyl}} T_{\text{cyl},1}}{C_2 p_{\text{cyl},1} V_{\text{cyl},1}}(p_{\text{cyl}} - p_{\text{cyl},o}) \geq 2c_m \left(\frac{V_c}{V_{\text{cyl}}}\right)^2 \text{imep}^{-0.2},$$

then

$$\alpha_W = 130 d^{-0.2} p_{\text{cyl}}^{0.8} T_{\text{cyl}}^{-0.53} \left[C_1 c_m + C_2 \frac{V_{\text{cyl}} T_{\text{cyl},1}}{p_{\text{cyl},1} V_{\text{cyl},1}}(p_{\text{cyl}} - p_{\text{cyl},o}) \right]^{0.8}, \qquad (3.34)$$

otherwise

$$\alpha_W = 130 d^{-0.2} p_{\text{cyl}}^{0.8} T_{\text{cyl}}^{-0.53} \left[C_1 c_m \left\{ 1 + 2 \left(\frac{V_c}{V_{\text{cyl}}}\right)^2 \text{imep}^{-0.2} \right\} \right]^{0.8}. \qquad (3.35)$$

In these equations, Q_{Wi} describes the wall heat flows (cylinder, piston, liner), A_i the corresponding surfaces or T_{Wi} the wall temperatures. The constants C_1 and C_2 can be calculated as follows:
$C_1 = 2.28 + 0.308 \cdot c_u/c_m$
$C_2 = 0.00324$ for direct-injection (DI) engines
$C_2 = 0.00622$ for indirect-injection (IDI) engines
Further well-known models for modeling the heat release are the approaches by Woschni and Anisits [153], Hiroyasu [70], and Spicher [76].

Regarding wall heat losses, the literature includes modeling approaches by Hohenberg [73], Annand [7], and Bargende [16].

Combined with the gas equation

$$p_{\text{cyl}} = \frac{1}{V} m_{\text{cyl}} R_o T_{\text{cyl}}, \qquad (3.36)$$

where R_o is the gas constant, the complete system of equations can be solved for any crank angle iteration by methods such as the Runge–Kutta method; this way, the conditions in the cylinder during the high-pressure cycle can be determined.

During the **gas exchange** the incoming and outgoing mass or enthalpy flows have to be considered in the equation of the first law of thermodynamics:

$$\frac{d(m_{\text{cyl}} \cdot u)}{d\alpha} = -p_{\text{cyl}} \cdot \frac{dV}{d\alpha} - \sum \frac{dQ_W}{d\alpha} + \sum \frac{dm_{\text{in}}}{d\alpha} \cdot h_{\text{in}} - \sum \frac{dm_{\text{out}}}{d\alpha} \cdot h_{\text{out}}. \qquad (3.37)$$

Again under consideration of the fact that for external and internal mixture formation the gas properties of the incoming mass flows are different, Eq. (3.37) can be reformulated into the two subsequently combined equations:

- internal mixture formation:

$$\frac{dT_{cyl}}{d\alpha} = \frac{1}{m_{cyl}(\partial u/\partial T + (\partial u/\partial p)p_{cyl}/T_{cyl})}\left\{ -\sum \frac{dQ_W}{d\alpha} - p_{cyl}\left(1 - \frac{\partial u}{\partial p}\frac{m}{V}\right)\frac{dV}{d\alpha} \right.$$

$$\left. -\left(u_{cyl} + \frac{\partial u}{\partial p}p - h_{in}\right)\frac{dm_{in}}{d\alpha} + \left(u_{cyl} + \frac{\partial u}{\partial p}p - h_{out}\right)\frac{dm_{out}}{d\alpha} - m_{cyl}\frac{\partial u}{\partial \lambda}\frac{d\lambda}{d\alpha} \right\}; \quad (3.38)$$

- external mixture formation:

$$\frac{dT_{cyl}}{d\alpha} = \frac{1}{m_{cyl}(\partial u_{cyl}/\partial T + (m_V p_{cyl}/T_{cyl})\partial u_V/\partial p)}\left\{ -\sum \frac{dQ_W}{d\alpha} - p_{cyl}\left(1 - \frac{m_V}{V}\frac{\partial u_V}{\partial p}\right)\frac{dV}{d\alpha} \right.$$

$$\left. -\left[\left(u_V + p_{cyl}\frac{\partial u_V}{\partial p}\right)\frac{dm_V}{d\alpha} + u_A\frac{dm_A}{d\alpha} + u_F\frac{dm_F}{d\alpha}\right] + h_{in}\frac{dm_{in}}{d\alpha} - h_{out}\frac{dm_{out}}{d\alpha} \right\}. \quad (3.39)$$

As a representative example for the wall heat losses during the gas exchange, the corresponding model equation according to Woschni will again be cited:

$$\alpha_W = 130 d^{-0.2} p_{cyl}^{0.8} T_{cyl}^{-0.53}(C_3 c_m)^{0.8}, \quad (3.40)$$

$$C_3 = 6.18 + 0.417 c_u/c_m.$$

For the determination of the mass flows through the intake and exhaust ports, a model with an equivalent throttle concentrated in the valve gap is used. Thus, all flow losses occurring in the port and valve gap are represented by this equivalent throttle, which, depending on the valve position, opens varying effective flow areas and therefore mass flows at given pressure gradients between cylinder and the connected port.

$$\frac{dm}{dt} = A_{eff} p_{0,1} \sqrt{\frac{2}{R_o T_{0,1}}} \psi, \quad (3.41)$$

where A_{eff} is the effective open flow area, $p_{0,1}$ is the static pressure upstream of the throttle, and $T_{0,1}$ the static temperature upstream of the throttle.

For subsonic flow, the mass flow function (see also (2.18)) now reads

$$\psi = \sqrt{\frac{\kappa}{\kappa - 1}\left[\left(\frac{p_2}{p_{0,1}}\right)^{2/\kappa} - \left(\frac{p_2}{p_{0,1}}\right)^{(\kappa+1)/\kappa}\right]}, \quad (3.42)$$

and for sonic flow (e.g., during the exhaust blow down phase), it reads

$$\psi = \psi_{max} = \left(\frac{2}{\kappa + 1}\right)^{1/(\kappa-1)} \sqrt{\frac{\kappa}{\kappa + 1}}. \quad (3.43)$$

The effective flow area mentioned above is the result of complex three-dimensional flow processes especially in the valve area. Thus, it must either be calculated using 3-D flow simulations or it has

3.6 Modeling and computer-aided simulation of supercharged engines

to be determined from flow bench tests. Moreover, it can be related to a specific reference area – usually the inner valve seat area (Fig. 3.14) – so that the efficiency of a port can be described by a flow coefficient.

$$A_{\text{eff}} = \mu_\sigma \frac{d_{\text{Vi}}^2 \pi}{4}, \qquad (3.44)$$

where μ_σ is the dimensionless port flow coefficient and d_{Vi} the inner valve seat diameter (reference diameter). This approach moreover allows a comparison of the port efficiencies of different engines.

Besides their quantity, the gas composition of the exhausting mass flows is also a relevant parameter for the gas exchange. In both four-stroke engines and, especially, in two-stroke engines, phases occur during the gas exchange, during which both the inlet and exhaust ports or valves are open. The scavenging process occurring during that period can be characterized using one of the following models (Fig. 3.15).

– Perfect displacement scavenging: The entering fresh gas displaces the exhaust gas present in the cylinder without mixing and without scavenging losses of fresh gas through the exhaust ports.
– Perfect mixing scavenging: The entering fresh gas mixes immediately and completely with the gas present in the cylinder, so that the leaving gas mixture contains both fresh gas and exhaust gas. This scavenging model is preferentially used for the relatively short overlap period in four-stroke engines; for two-stroke engines with relatively low scavenging quality it can also be used with success if more precise scavenging models are not available.
– Shortcut scavenging: The entering fresh gas immediately leaves the cylinder via the exhaust ports, without influencing the composition of the cylinder mass.
– Real engine scavenging: Especially for two-stroke engines, it is desirable to chose and develop engine layout and port shapes whose scavenging behavior approaches the perfect

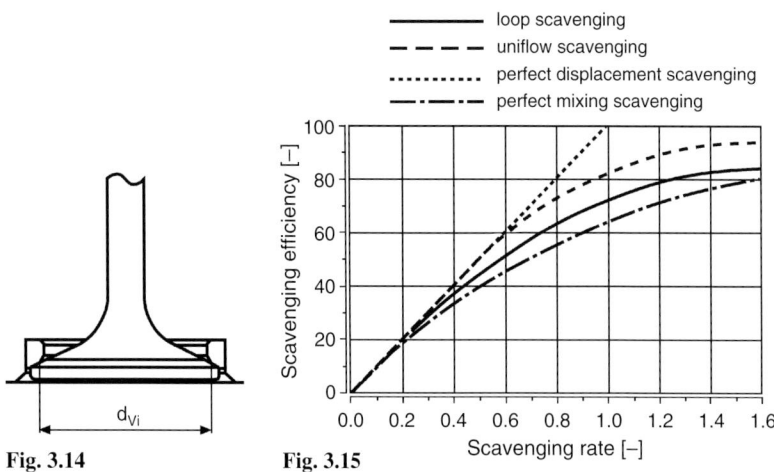

Fig. 3.14   Fig. 3.15

Fig. 3.14. Definition of the inner valve seat diameter

Fig. 3.15. Comparison of possible scavenging processes

scavenging process as much as possible. Uniflow or longitudinal scavenging, in which the inlet and exhaust ports are located at the bottom and top of the cylinder, fulfills these requirements best; another layout often used is loop scavenging, where a kind of loop flow pattern is generated to scavenge the entire cylinder volume without shortcut flows to the exhaust ports.

After choosing a suitable scavenging model and under consideration of the equation of mass conservation

$$\frac{dm_{cyl}}{d\alpha} = \sum \frac{dm_{in}}{d\alpha} - \sum \frac{dm_{out}}{d\alpha} \qquad (3.45)$$

as well as the gas Eq. (3.36), again a system of equations is available which can be solved numerically for each crank angle iteration and which allows the calculation of the values of all gas exchange parameters.

Pipe elements

The exact simulation of the engine's cycle also necessitates the gas-dynamically correct simulation of the flow in the pipes. Under consideration of the laws of conservation for mass, momentum and energy, the following set of equations can be compiled to describe the flow in a pipe:

$$\frac{\partial \rho}{\partial t} = -\frac{\partial (\rho v)}{\partial x} - \rho v \frac{1}{A}\frac{dA}{dx}, \qquad (3.46)$$

$$\frac{\partial (\rho \cdot v)}{\partial t} = -\frac{\partial (\rho v^2 + p)}{\partial x} - \rho v^2 \frac{1}{A}\frac{\partial A}{\partial x} - \frac{F_{fr}}{V}, \qquad (3.47)$$

$$\frac{\partial E}{\partial t} = -\frac{\partial [v(E+p)]}{\partial x} - v(E+p)\frac{1}{A}\frac{dA}{dx} + \frac{q_W}{V}, \qquad (3.48)$$

where $E = \rho c_v T + \frac{1}{2}\rho c^2$ describes the energy content of the gas.

For the wall friction and wall heat losses contained in this set of equations, utilizing Reynold's analogy, the following equations can be used.

– Wall friction losses:

$$\frac{F_{fr}}{V} = \frac{\lambda_{fr}}{2D}\rho v|v|, \qquad (3.49)$$

where λ_{fr} describes the wall friction coefficient.
– Wall heat losses:

$$\frac{q_W}{V} = \frac{\lambda_{fr}}{2D}\rho |v| c_p (T_W - T) \qquad (3.50)$$

with T_W as manifold wall temperature.

The complete system of partial differential equations for the flow in pipes (Eqs. (3.46)–(3.48)) can now be solved using suitable methods, where the most common methods are characteristics

3.6 Modeling and computer-aided simulation of supercharged engines 43

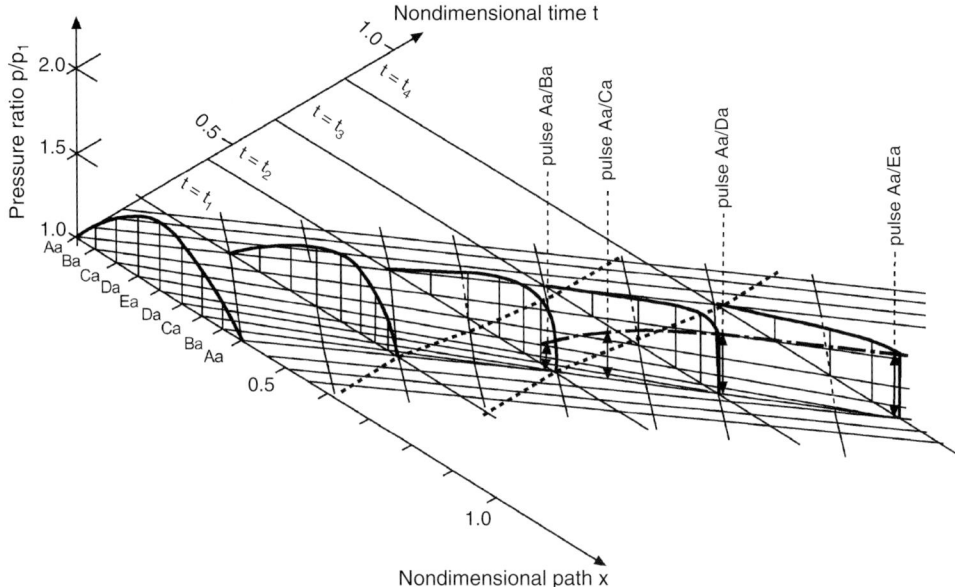

Fig. 3.16. Graphic view of a pressure wave in a manifold [109]

and finite-volume methods. The principles behind both methods will now be briefly discussed. For detailed studies it is referred to the literature [127, 128].

With the characteristics method, the solutions to this system of equations are calculated sequentially following the progression of the characteristics, which are the two Mach curves $v \pm a$. In the av finite-state diagram, the spacial and temporal changes in the pipe state can be determined, under consideration of the dependencies on sonic and flow speeds, iteratively at all grid points; the particle traces $dx/dt = v$ can be determined in the state diagram (Fig. 3.16). The basic difference between this solution algorithm and differences methods is the fact that with this method the grid points are not fixed from the outset for predetermined time and cell increments.

An especially advantageous solution algorithm based on the finite-volume method is the ENO scheme 1 [58, 131, 52]. It achieves at least the same accuracy as finite-differences methods of second order and at the same time has the same stability as methods of first order. It divides the manifold elements into one-dimensional finite-volume elements which exchange mass, enthalpy, and momentum flows over the element boundaries. In order to correctly determine the gas dynamic processes, however, the angle or time iterations Δt have to be adapted to the chosen cell sizes Δx, according to the CFL criterion (stability criterion according to Courant, Friedrichs, and Lewy):

$$\Delta t \leq \frac{\Delta x}{v + a}. \tag{3.51}$$

Starting with the status at the begin of each time iteration, and with the above mentioned flows over the cell boundaries in the temporal center of the time iteration, the result and thus the cell status at the end of the time iteration is calculated.

Plenum elements

For the modeling of a plenum element, basically the same equations can be used which describe the processes in a cylinder (Eq. (3.37)). Moreover, if the cylinder has a constant volume, the equation simplifies as follows:

$$\frac{d(m_F \cdot u)}{d\alpha} = -\sum \frac{dQ_W}{d\alpha} + \sum \frac{dm_{in}}{d\alpha} h_{in} - \sum \frac{dm_{out}}{d\alpha} h_{out}. \tag{3.52}$$

In special geometric designs of a plenum, momentum passages can occur in which the kinetic energy of an incoming mass flow only partially mixes inside the plenum. In such special cases, additional assumptions regarding the fraction of the passing kinetic energy must be made and the corresponding energy ratios in the above equation must be amended. However, for typical plenum layouts with several connections, the assumption of a complete mixing of the kinetic energy of the incoming flows is justified, so that for the overwhelming majority of plenums the thermodynamic behavior can adequately be described by Eq. (3.52). The wall heat losses Q_W have to allow for the conditions given by the plenum shape, surface geometry, and the flow pattern. After an estimate of the surface actively participating in the heat transfer, the corresponding wall heat transfer coefficients must be verified via measured data (comparison between measured and calculated temperatures).

Turbocharger

The turbocharger, as the central element of a supercharged engine, constitutes an extremely complex flow machine, both for the compressor and the turbine, as we have seen in previous sections. By means of mathematical models, which consider the losses in the compressor and turbine via isentropic efficiencies, it is relatively easy to describe the thermodynamics of a turbocharger operating at steady state.

$$P_C = P_T, \tag{3.53}$$

$$P_C = \dot{m}_C(h_2 - h_1), \tag{3.54}$$

$$P_T = \dot{m}_T \eta_{m,TC}(h_3 - h_4), \tag{3.55}$$

where P_C is the compressor power, P_T the turbine power, $\dot{m}_C$ and $\dot{m}_T$ the compressor or the turbine mass flow.

The enthalpy differences $h_2 - h_1$ and $h_3 - h_4$ can be calculated under consideration of the internal isentropic efficiencies of the compressor and the turbine (Eq. (3.3)). Together with the mechanical efficiency mentioned above, turbocharger total efficiency η_{TC} can be determined:

$$\eta_{TC} = \eta_{m,TC} \eta_{s-i,T} \eta_{s-i,C}. \tag{3.56}$$

The pressure upstream of the turbine is thermodynamically analogous to the resistance of a throttle. The temperature downstream of the throttle (turbine), however, has to be calculated based on Eq. (3.3), which in this case describes an expansion.

Furthermore, with changes of the charger speed, which is proportional to the angular velocity

3.6 Modeling and computer-aided simulation of supercharged engines

ω_{TC}, considerations must be made for rotor inertia I_{TC}, so that the following equation can be defined:

$$\frac{d\omega_{TC}}{dt} = \frac{1}{I_{TC}} \frac{P_T - P_C}{\omega_{TC}}. \qquad (3.57)$$

The efficiencies appearing in these equations, as well as the effective equivalent flow area for the turbine, can be derived from measured compressor and turbine maps. If such measurement data are not available, these data can be obtained either by scaling of similar maps or by empirical approximation methods which are based on the main geometric dimensions of turbo machines. Such numerical models were developed by various researchers, e.g., by Malobabic [93] for both compressors and turbines, as well as by Swain [136, 139] and Baines [13]. As can be seen in Fig. 3.17, these methods emanate from certain geometric boundary conditions. As a result, and considering empirical estimates for the losses, the maps can be drawn in good approximation (Figs. 3.18 and 3.19). If only incomplete measurement data exist, which often is the case especially with turbines (since the operational possibilities on a turbocharger test bench are

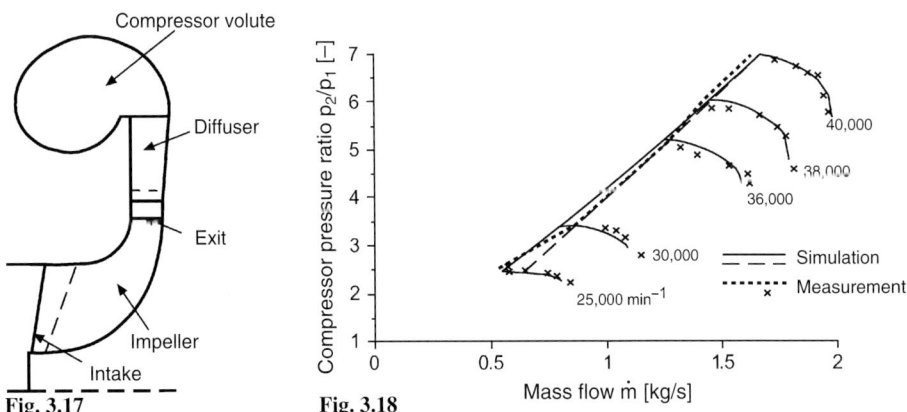

Fig. 3.17. Geometric parameters for the description of the compressor geometry, according to Swain [135]

Fig. 3.18. Comparison of measured and simulated speed curves of a compressor [135]

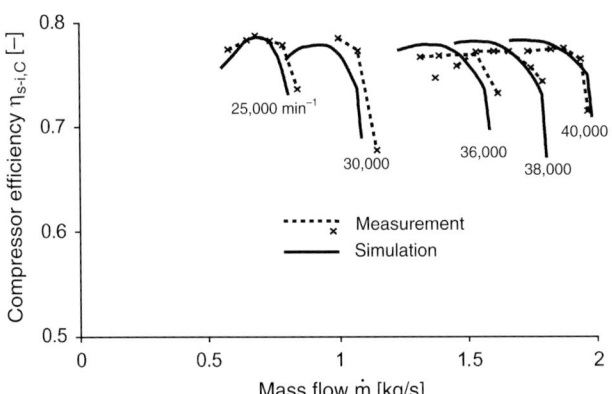

Fig. 3.19. Comparison of measured and simulated efficiency curves of a compressor [135]

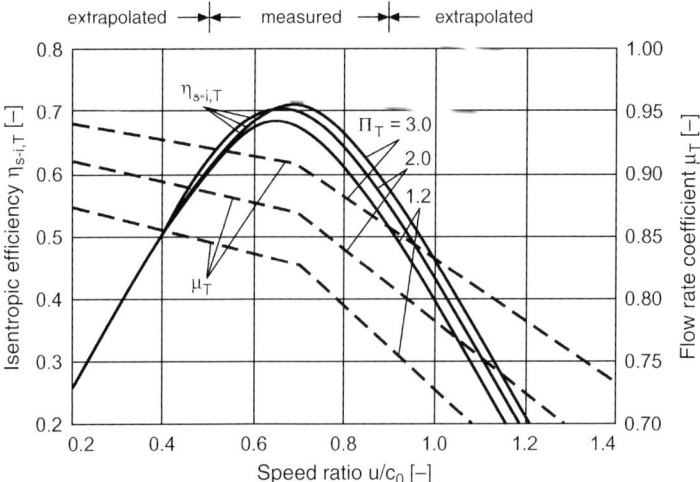

Fig. 3.20. Turbine map extrapolation method according to Bulaty [27]

limited), mathematical or phenomenologic-empirical extrapolation methods prove to be of value, of which the method formulated by Bulaty [27] has found wide acceptance (Fig. 3.20).

Displacement compressor

Basically, the modeling of a mechanically driven charger corresponds to a partial task of turbocharger simulation, i.e., only the compressor has to be modeled. Thus, e.g., the operating point of the charger can be freely chosen in order to define the design point of the compressor. In the rest of the map and engine operation, the charger operating points result from the either fixed or variable gear ratio between charger and internal combustion engine and the engine's swallowing capability. In the most extreme case, a complete uncoupling of engine and charger speed via an electric or hydraulic link could also be imagined. However, the losses associated with such energy conversions mostly limit their application possibilities in practice.

Thus the change of state as well, i.e., the temperature increase, in a mechanic charger can be described with the isentropic equation, as long as the pressure ratio and the corresponding isentropic efficiency are known (Eq. (2.14)).

The pressures mentioned, p_1 and p_2, establish themselves according to the flow restrictions in the intake system up- and downstream of the compressor and the engine, where the delivery characteristic of the compressor must be described on the basis of measured maps (Fig. 3.21). Lastly, the required power can be calculated by the following equation (also see Eq. (3.6)):

$$P_C = \dot{m}_C c_p T_1 \frac{1}{\eta_{\text{tot},C}} \left[\left(\frac{p_2}{p_1} \right)^{(\kappa-1)/\kappa} - 1 \right], \qquad (3.58)$$

where $\eta_{\text{tot},C}$ describes the total efficiency $\eta_{m,C} \eta_{s\text{-}i,C}$.

Charge air cooler

As will be shown in detail in Chap. 12, the charge air cooler represents a very important component of the entire supercharging system, provided that maximum power density and lowest brake specific fuel consumption as well as lowest emission levels are aimed at.

3.6 Modeling and computer-aided simulation of supercharged engines 47

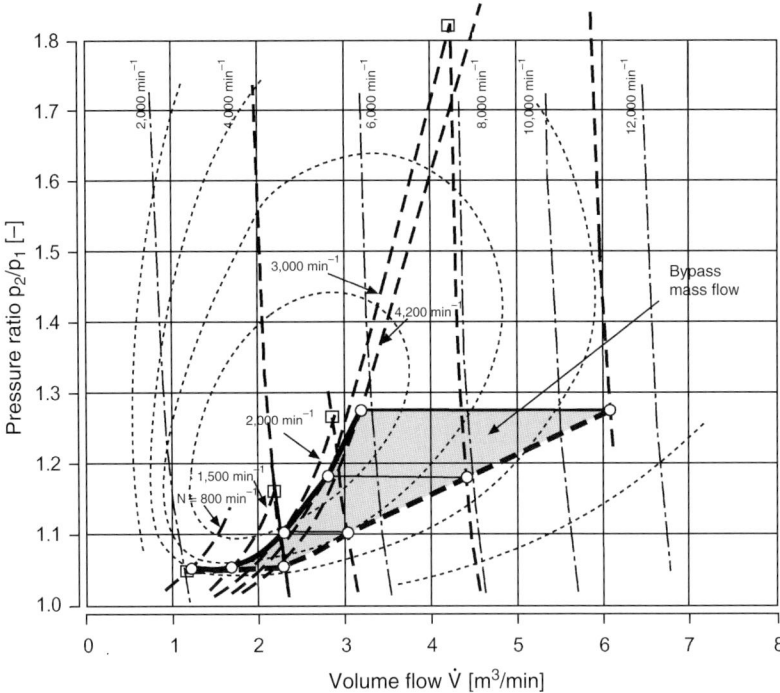

Fig. 3.21. Diagram of the flow characteristics of a 1 liter passenger car two-stroke diesel engine with blowby control, in a displacement compressor map. □, intersection points between engine and charger flow curves without blowby control; ○, intersection points between engine and charger flow curves with blowby control

Normally, charge air coolers are designed as conventional air-to-air heat exchangers, or in special cases as air-to-water heat exchangers.

To record the gasdynamic processes in a charge air cooler, i.e., the flow distribution into the individual sections of the matrix as well as the pulsations excited by the engine, all 3-D geometry parameters must be considered. For such a detailed analysis, it is advisable to simulate the charge air cooler by 3-D CFD methods embedded into a 1-D mathematical model (see Sect. 3.6.3).

For engine layouts as well as design of charge air coolers it has proven to be advantageous to describe the radiator by a model as sketched in Fig. 3.22.

In doing so it is decisive to correctly cover the pressure loss and the cooling effect of the radiator matrix; this can be achieved by means of tuning the heat transfer coefficient and the wall friction coefficient of the pipe element between the two plenums. It is also important to model the plenums as well as the gasdynamically active length between inlet and outlet in such a manner that the damping and the run time effects correspond to the actual circumstances. This can especially be of decisive importance if the gas in the intake system, and thus the charge air cooler, is excited to oscillation by the gas exchange process of the engine. Further modeling approaches, some simpler and others more in-depth, are described in the relevant literature, e.g., reference 24.

For the numeric simulation of transient engine operation, the inertia of the rotating mass of the engine has to be related to the actual load and engine torque or the corresponding power:

$$\frac{d\omega_E}{dt} = \frac{1}{I_E} \frac{P_E - P_{\text{load}}}{\omega_E}. \quad (3.59)$$

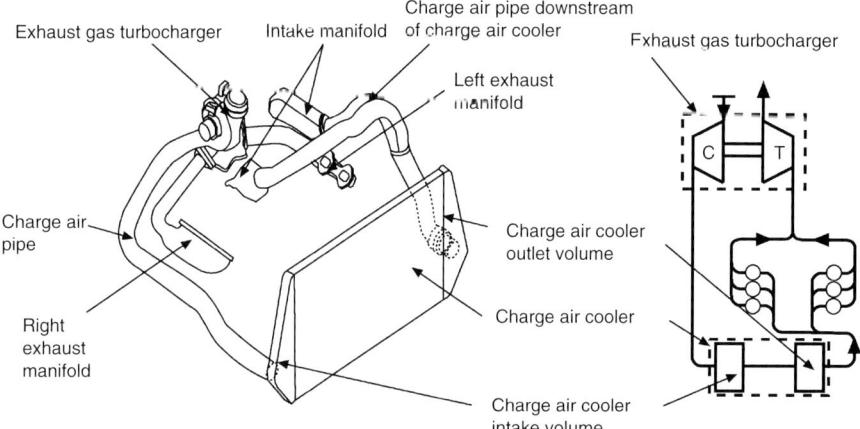

Fig. 3.22. Example for a charge air cooler substitute model for the modeling of a supercharged V6 engine with charge air cooling

Here, ω_E describes the angular speed of the internal combustion engine, I_E the equivalent moment of inertia of all rotating and translatory inertias, P_E the instantaneous engine power, and P_{load} the instantaneous power of the external load.

The transient engine processes can now also be described with the equations given above, where, however, the load dependency of many parameters, such as combustion data, fuel quantities, wall temperatures in the engine and especially in the exhaust system, must be considered. For this, the time-correct input of the heat capacities is also necessary in order to correctly simulate the lagged response of the turbine after a sudden load variation. The modeling of the manifolds has to consider the material properties and layout (cast iron, plastic, steel manifold, insulation, double wall manifolds, etc.).

Further subsystem elements often used in supercharged engines, such as **air filter, intake noise muffler, EGR (exhaust gas recirculation) radiator, exhaust gas catalyst,** and **muffler** may be individually configured from the components discussed above, i.e., pipe and plenum. Many commercially available software packages for thermodynamic cycle simulations also offer predefined modules for these elements.

3.6.3 Numeric 3-D simulation of flow processes

Parallel to the layout and optimization of the thermodynamic process of supercharged internal combustion engines using the 1-D simulation methods discussed in the last section, for best geometric design further detail optimizations of the gas-containing parts must be performed. This is possible by numeric 3-D CFD simulations. With them, objective analyses and assessments of designs and variants can be done.

For such an analysis, advanced engineering CAD drawings are necessary, on the basis of which the geometric boundary conditions can be exactly described.

Those 3-D design methods are especially advantageous which are based on so-called solids, i.e., bodily imaginary elements. On the basis of these elements, totally closed surface models can be derived. The calculation grids must now be fit into these surfaces. In the case of good surface quality, it is advantageous to use automated grid generators. For the latter tasks, a number of

3.6 Modeling and computer-aided simulation of supercharged engines 49

commercially available program packages exist (AVL-FAME, PRO-STAR).

Besides gas flows, fluid flows are also simulated by CFD methods, e.g., the coolant flow in the engine crankcase and the cylinder head. In general, the most important tasks of CFD simulation within the framework of the development of supercharged engines can be summarized (without tasks regarding application and production engineering such as engine compartment flow patterns or simulation of casting related processes) as follows:

- gas flows in the engine-related air conducting manifolds (both upstream and downstream of the compressor): minimization of pressure losses, acoustic phenomena, compressor oncoming flow, quality of the intake manifold flow, EGR feed to the intake manifold, EGR distribution in the intake plenum, mixture formation and wall agglomeration in the intake manifold
- gas flows in the engine-related exhaust manifolds: minimization of pressure losses in the exhaust manifold, turbine and waste-gate flow admission, catalyst flow admission, resonance mufflers for two-stroke engines
- internal engine processes: flow patterns during the compression and combustion processes, internal mixture formation in direct-injection diesel and direct-injection gasoline engines, propagation of the combustion zone and pollutant formation, pre- and swirl-chamber design for combustion processes with divided combustion chamber, two-stroke scavenging flow, flows in the fuel system
- coolant flows in fluid-cooled and air-cooled engines: optimization of flow ducts regarding heat transfer, minimization of pressure losses, elimination of dead water zones, elimination of cavitation

Several examples of such CFD analyses – especially for elements of supercharged engines – will be discussed in Sect. 5.5.

3.6.4 Numeric simulation of the supercharged engine in connection with the user system

The objective of any engine layout is the best fulfillment of as many requirements of the user system as possible. In the case of a passenger car, this would include, e.g., the acceleration capability, elasticity, maximum vehicle speed, gradeability, pollutant emissions, as well as smoothness and noise radiation.

With the methods shown in Sects. 3.6.2 and 3.6.3, an exact simulation of the engine-related processes is possible. However, reactions caused by the load (e.g., drivetrain as well as vehicle) on the internal combustion engine and the total environment (e.g., thermal system) cannot be comprehensively incorporated.

Such problems related to the total process thus have to be described by methods which reflect all relevant physical subprocesses (engine, supercharging system, drivetrain, engine control, thermal and electrical system, air conditioning of the passenger compartment, vehicle dynamics) with sufficient accuracy, without resulting in processing times for modeling, simulation, and evaluation which are not practically feasible.

Tools suited for this task are commercially available program packages (AVL-CRUISE, ADVISOR, GT-DRIVE, etc.) or the program system GPA (total process analysis) of the German Forschungsvereinigung Verbrennungskraftmaschinen (FVV; Research Consortium for Internal Combustion Engines, Frankfurt am Main). With these, the operational behavior of the complete system can be simulated and optimized. The possible applications extend from cold start analysis,

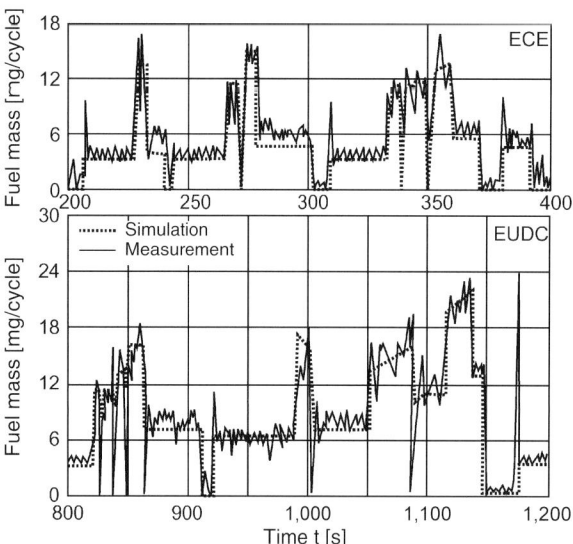

Fig. 3.23. Comparison of measured and simulated fuel consumption curves of a passenger vehicle in the NEDC

optimization of the transient operation (acceleration capability) to error simulations, with which the impacts of, e.g., incorrect sensor readings on engine control can be analyzed.

The program systems mentioned use various techniques to reduce the simulation times. As an example, GPA simulates and files an expanded engine map before the actual simulation of the complete system, which also covers all occurring transient operation situations. The time needed for the description of the engine behavior simultaneous to the cycle simulation is thereby greatly reduced. However, the acceleration of the computing time principally is linked to a certain reduction in accuracy.

It is therefore always necessary to identify the objectives of an analysis in order to choose, in each case, the tool with the proper accuracy and significance.

Figure 3.23 shows a comparison of the measured and the simulated fuel mass flows in the standardized NEDC. On the basis of such verified models, preoptimizations can be performed before development on a test bench or in a vehicle, and, e.g., the number of control strategies to be experimentally analyzed can be reduced.

4 Mechanical supercharging

4.1 Application areas for mechanical supercharging

If the precompressor (displacement or turbo compressor) is directly powered by the engine, it is classified as mechanical supercharging. As a general rule, a fixed gear ratio is sufficient for displacement compressors, while a variable gear ratio is necessary for most applications of turbo compressors.

Under the simplifying assumption of fixed gear ratios for both supercharging methods, slightly increasing pressure ratios will be obtained with increasing speed with the displacement compressor, while the pressure ratio curves show a parabolic gradient for a turbo compressor linked to the engine via transmission with fixed gear ratio, similar to a throttle curve. Depending on the specific application, the gear ratio must be chosen in such a way that either the desired power or the desired torque level at low speed is obtained. The following relationships can be established between engine torque T and engine speed n_E:

- **Constant speed operation**
  n_E = constant T = variable (e.g., generator operation)
  Displacement compressor and turbo compressor with fixed speed ratio are suited; the turbo compressor can be adapted very well.
- **Propeller operation**
  n_E = variable T increases parabolically (e.g., ship operation)
  If acceleration phases are not taken into account, the **turbo compressor** with fixed ratio is well suited for this application, due to the fact that its pressure characteristic is identical to the load curve.
- **Automotive operation**
  n_E = variable T = variable
  Torque backup at lower speed is desired. In such an application **only the displacement compressor with fixed ratio** provides acceptable torque curves. However, a variable gear ratio would here, too, enable a better match of the torque curve to the traction force hyperbola. In any case, engines with **flow chargers** can provide the desired torque curve **only with variable charger gear ratio** (e.g., CVT).

Since the volume of a mechanical charger increases about linearly with the flow rate, their application today is mostly limited to smaller automotive engines up to 3–4 liter displacement. The preferred application area is gasoline engines, where exhaust gas turbocharging is not yet in series production on a major scale, especially due to the high thermal load of the charger.

4.2 Energy balance for mechanical supercharging

To examine the energy balance of displacement compressors in collaboration with a reciprocating piston combustion engine, pV and TS diagrams are especially suited. Figure 4.1 shows the principle layout and Fig. 4.2 the pV and TS diagrams.

From the pV diagram, it can be well recognized that the compression work performed by the charger theoretically could be reclaimed to a considerable extent as positive gas exchange work. In practice, the efficiencies involved, during compression ($\eta_{ges,C}$ of about 60%) as well as during the gas exchange itself, prevent that. Only a recovery in the order of about 20–30% of the compressor driving power is attainable. Therefore, the disengagement of mechanically driven chargers in those load ranges where no boost pressure is needed is desirable or even necessary for efficiency reasons.

Further, in supercharged gasoline engines a deterioration of the high-pressure efficiency η_{HP} occurs, which is shown in Fig. 4.3 depending on charge air cooling, intake temperature, and the air-to-fuel ratio λ. Problems arise in that load area which needs only a partial boost pressure, as in the case of the quantity load control of a gasoline engine.

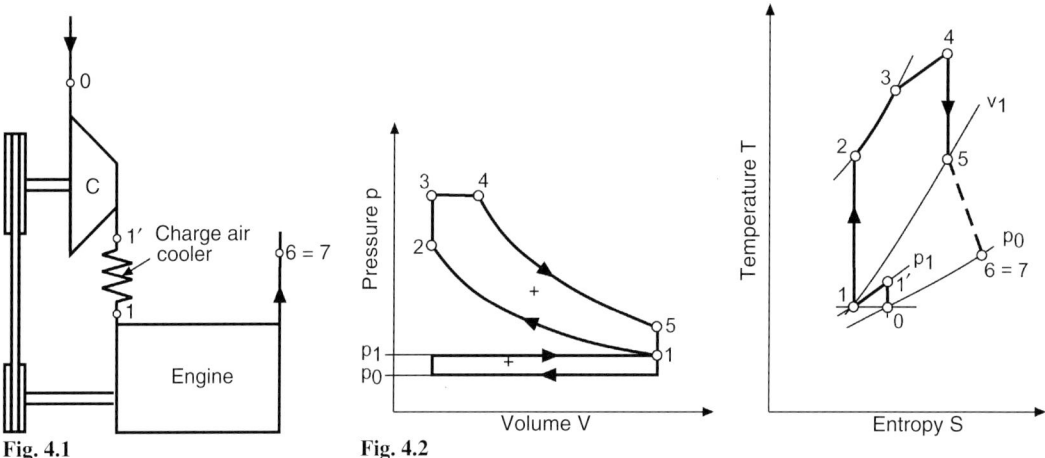

Fig. 4.1. Layout of a mechanically supercharged engine with charge air cooling

Fig. 4.2. pV and TS diagrams of a mechanically supercharged engine with charge air cooling

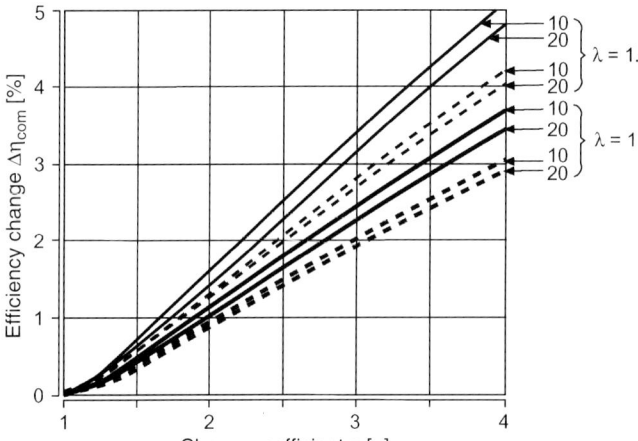

Fig. 4.3. Combustion efficiency deterioration in supercharged gasoline engines depending on air-to-fuel ratio λ, intake temperature, charge gradient, and charge air cooling. Dash line, with charge air cooling; solid line, without charge air cooling

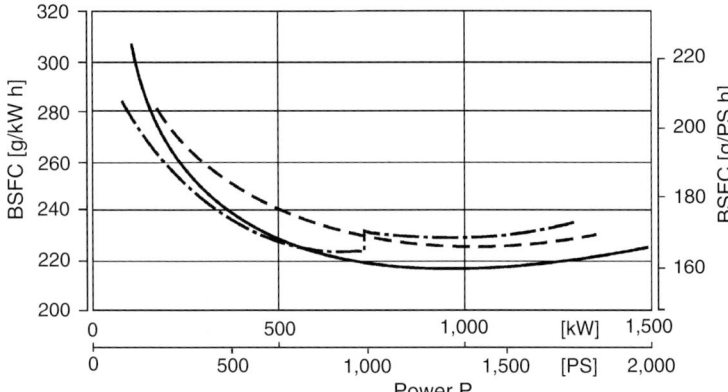

Fig. 4.4. Fuel consumption comparison between mechanically supercharged diesel engines with (dot and dash line) and without (dash line) charger disengagement at part load and in comparison to an exhaust gas turbocharged engine of equal power (solid line) [159]

Figure 4.4 shows a comparison of the part-load fuel consumption values of diesel engines of equal power, on the one hand, mechanically supercharged (but with and without charger disengagement), on the other, exhaust gas turbocharged. It is obvious that the control of the boost pressure of mechanically driven chargers, discussed in the next section, is of special importance.

4.3 Control possibilities for the delivery flow of mechanical superchargers

To estimate and evaluate the options for charge pressure control, necessary in most cases, the pressure–volume (mass) flow diagram of charger and engine already described in depth is useful. It is very reasonable to examine the control possibilities for four-stroke and two-stroke engines separately due to their highly different engine operating characteristics.

4.3.1 Four-stroke engines

Displacement compressors

If the delivery curves of the charger and the swallowing capacity of the engine are plotted together (Fig. 2.15), it can be seen that for displacement compressors, with the steep characteristic curves typical for both charger and engine, the difference in delivery mass between charger and engine increases only slightly with decreasing charge pressure. This means for the control system that only small amounts of the entire charge flows – as a function of load or speed – have to be manipulated, i.e., have to be either blown off or rerouted upstream of the charger via a bypass valve.

Bypass control is meaningful in the load area where a partial boost pressure is necessary, thus in the map area between Π_{max} and $\Pi = 1$. This boost pressure bypass control is necessary for gasoline engines with quantity control, since the mixture quantity aspirated by the engine is determined by the pressure upstream of the intake valve.

Also for the diesel engine a lowered part-load boost pressure is desirable, since the corresponding reduction of required charger driving power leads to a lower part-load fuel consumption.

Additionally, with this system layout (Fig. 4.5), air mass flow measurement is possible despite charger bypass, which is necessary in a gasoline engine with stoichiometric λ control and in a diesel engine with a controlled exhaust gas recirculation system.

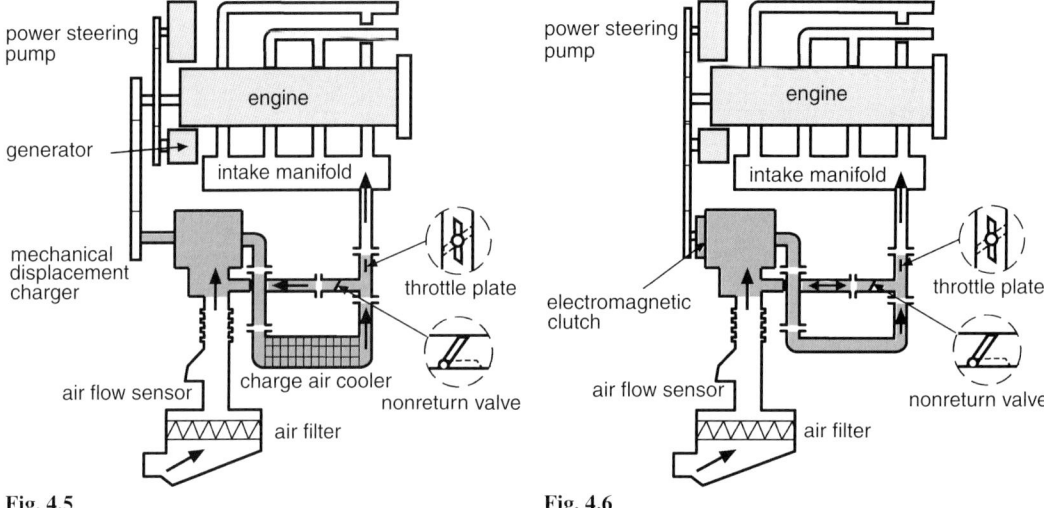

Fig. 4.5. Charger pressure control via compressor bypass layout for a permanently engaged charger [132]

Fig. 4.6. Charger pressure control via compressor bypass and charger disengagement [132]

The **disengagement** of mechanically powered chargers, whether equipped with a bypass setup or not, saves energy in the load range in which the engine can be operated as a naturally aspirated engine. Independent of the load control method of the engine, disengagement always makes sense since it avoids the charger's inner losses, reduces the operation time of the charger, and increases its durability. Figure 4.6 shows the basic principle layout. Charger disengagement is imperative when using chargers with internal compression, since otherwise unnecessary compression work will be expended. In this case, the nonreturn valve has to be opened by a controlled servo device or by the pumping work of the engine.

In a gasoline engine additionally the throttle plate, present in any event, can be used for load control in the complete load range.

Blowoff control, the most simple way to adapt the air quantity to the engine air requirement, should only be chosen in special cases, e.g., for emergency shutoff, since it is very inefficient.

Charger speed control is also only used in special cases, when the boost pressure of a displacement compressor has to be increased in the lower engine speed range. In this case, it is advantageous if the necessary gear ratio changes for the charger drive are moderate and, if necessary, may only be shifted stepwise.

Turbo compressors

Charger control for mechanically powered turbo compressors turns out to be totally different.

For a defined part-load boost pressure, arbitrarily chosen in the charger map, a variable charger speed control system has to be considered in any case. Otherwise, significant air quantities are to be blown by or blown off via a bypass layout. Controlled part-load boost pressure values can be obtained without waste of energy only with variable charger speed. Technical solutions for such a system are continuously variable belt or chain transmissions, as well as hydrodynamic transmissions (Foettinger) and hydrostatic drives. All these solutions can be controlled easily and are reliable. However, they suffer under the required very broad speed range and very high engine

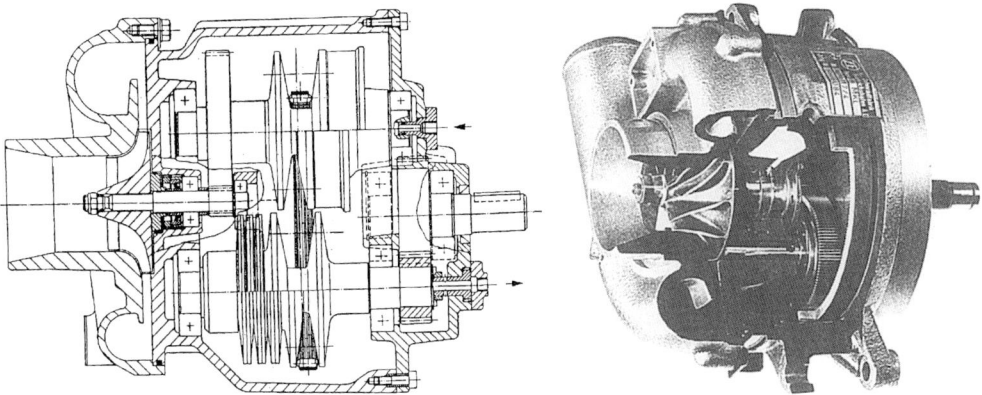

Fig. 4.7. Variable charger speed control of a mechanically powered turbo compressor [ZF]

speed gradients. In addition, all above mentioned technical solutions lead to significant system cost (Fig. 4.7).

Measures which influence the map width or the possible pressure ratio changes of turbo compressors will be discussed in detail in Chap. 5.

4.3.2 Two-stroke engines

The pressure–volume flow map of two-stroke engines with an approximately symmetric timing diagram corresponds to a delivery rate curve with starting point at zero delivery and dependence on the prevailing exhaust backpressure (Fig. 2.13). Significantly different swallowing capacity functions, closer to the four-stroke engine, occur in longitudinally scavenged two-stroke slow-speed diesel engines with asymmetric scavenge port timing (including variable exhaust closing).

From this behavior it follows that for such an application a mechanically driven turbo compressor with fixed speed ratio yields very good results, unless special operating conditions, such as fast load changes, accelerations, or speed changes of the engine play a major role. In contrast, displacement compressors with fixed gear ratio are not suitable. Adaptations for special conditions, e.g., high air-to-fuel ratio at idle speed, may necessitate a **variable speed control** of the charger for two-stroke engines as well. Mechanically driven chargers nowadays are of some importance for small high-speed engines, although such engines are barely in series production.

4.4 Designs and systematics of mechanically powered compressors

4.4.1 Displacement compressors

The piston compressor is still the classic example of a displacement compressor. In fact, its relevance today is limited to applications like scavenging pump for two-stroke engines, either by using the bottom side of the engine's piston (in small, inexpensive two-stroke engines as a crankcase scavenging pump) or in slow-speed cross-head engines as end compression stage and scavenging pump in combination with exhaust gas turbocharging.

For reasons of their installed size and weight, the so-called rotational piston chargers have gained more importance, since they can be operated at significantly higher speeds than the engine

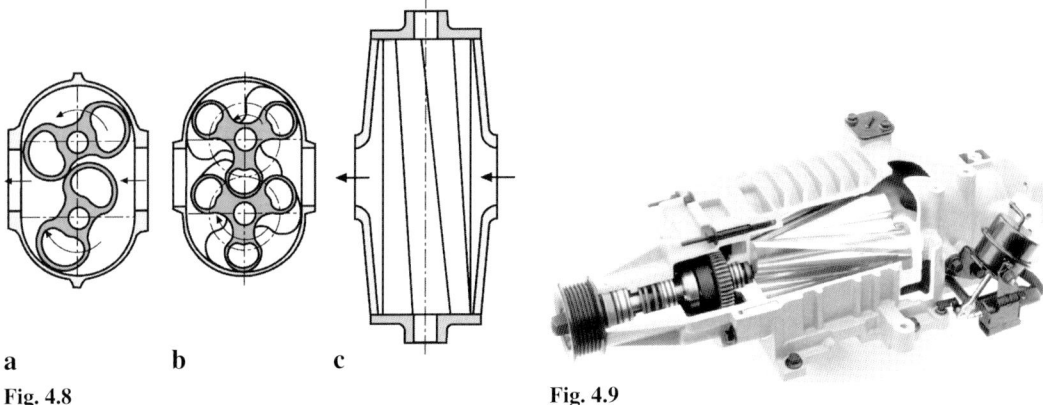

Fig. 4.8. Roots blower with two lobes (**a**) and three lobes (**b**); **c** top view [56]

Fig. 4.9. Eaton Roots blower

itself. Rotational piston chargers exist in very different designs. Systematic compilations of these were done by Felix Wankel, recently by G. Haider [56], and by other authors.

Within the framework of this book we will only take a closer look at those few types of chargers which are either in series production or feature special advantages for future applications.

Roots blower

Today, the Roots blower is the displacement compressor produced in largest quantities. Major manufacturers are Eaton and Ogura (Figs. 4.8 and 4.9). This design consists of a pair of double- or triple-lobed rotational pistons arranged on two separate shafts. Only an insignificant internal compression takes place. The dead space of the charger is large. Advantages are the, for a displacement compressor, simple design associated with comparably low manufacturing cost, sufficient durability with efficiencies staying constant due to contactless sealing of the working plenums, and the relatively small installed size due to high possible charger speeds.

Since the Roots blower has practically no internal compression, and in addition the air delivery occurs with large fluctuations due to design limitations, the control of boost pressure pulsations and the charger's noise emissions are its main problem areas. They can only be satisfactorily solved by special geometric design of the inlet and outlet ports as well as appropriate twisting of the rotational pistons.

Another problem is the low achievable boost pressure at low engine speeds, due to the gap losses between rotational piston and housing, which can only be sealed contactless. A variable charger drive gear ratio, as minimum changeable stepwise, is therefore under closer consideration by some designers.

In the future, the major applications will likely be in passenger car gasoline engines, with advantages such as an arrangement of the charger on the "cold" intake side of the engine (no additional measures on the "hot" exhaust manifold side are necessary), an instantaneous boost pressure buildup, the wide usable map, and only minor requirements regarding the realizable pressure ratio providing a good basis for series application.

It is used in increasing quantities, e.g., by Mercedes Benz in their 1.8 and 2.0 liter compressor engines for the C and E class and some SLK models.

4.4 Designs and systematics of mechanically powered compressors   57

Spiral charger

For several years, the spiral charger was in series production by Volkswagen (VW), under the name of G-Charger, to provide high-end power variants for some models. Manufacturing problems such as cost and performance uniformity led to its demise.

Spiral chargers are working according to the principle of interleaved scrolling spirals. They are eccentrically arranged in such a way that two spiral segments pump from outside inlets into two inner outlets (Fig. 4.10). During this process, the internal volume is reduced, which results in an internal compression. The advantages of this design are its small moment of inertia (about one-tenth to one-twentieth of that of rotational piston chargers), low noise emission level, and low weight.

Disadvantages are its complicated production and sealing problems both between the very long spiral delivery elements and the housing and in the outlet area. ABB Turbo Systems was attempting to reintroduce this charger type into the market under the name Ecodyno. The main application should again be the small high-power gasoline engine, due to the advantages mentioned above [132].

Wankel-2/3 (Ro) and -3/4 (Pierburg) charger

The companies KKK (today 3K Warner) and Pierburg have developed 2- and 3-lobed rotational piston chargers following patents by Felix Wankel (Figs. 4.11 and 4.12) and offered them as prototypes. These are both two-concentric-shaft rotational piston chargers with three or four work

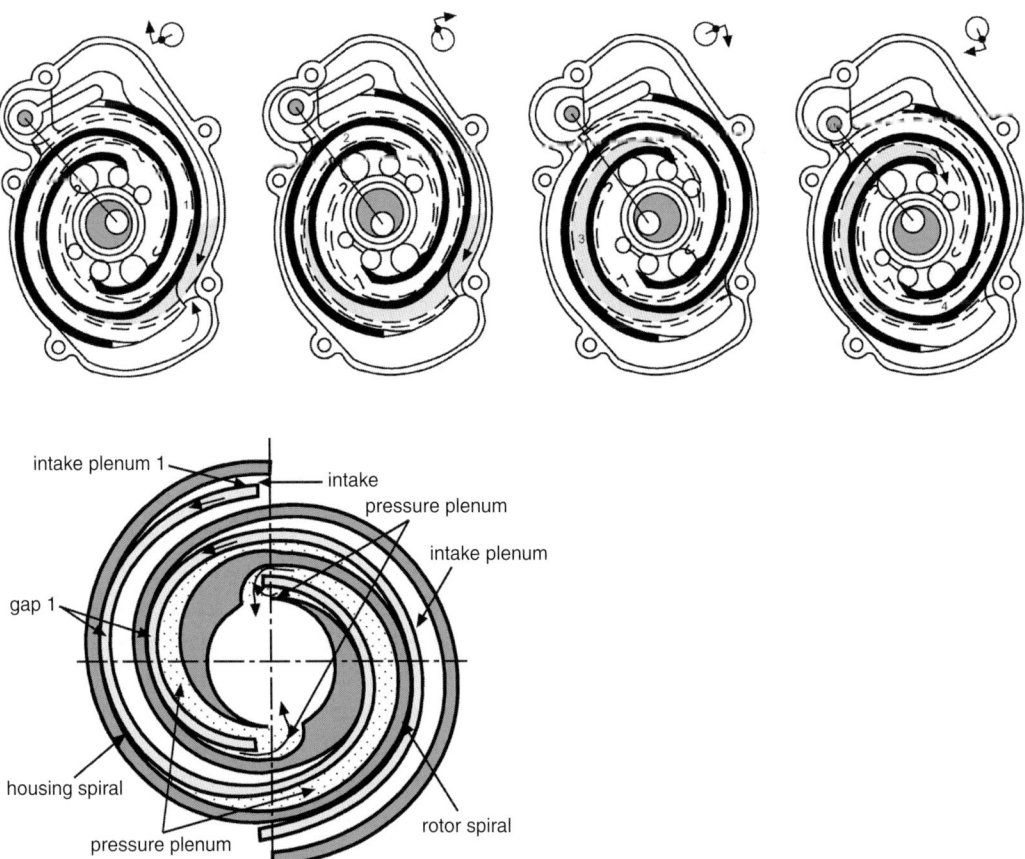

Fig. 4.10. Principle drawing showing the operation of a spiral charger [131]

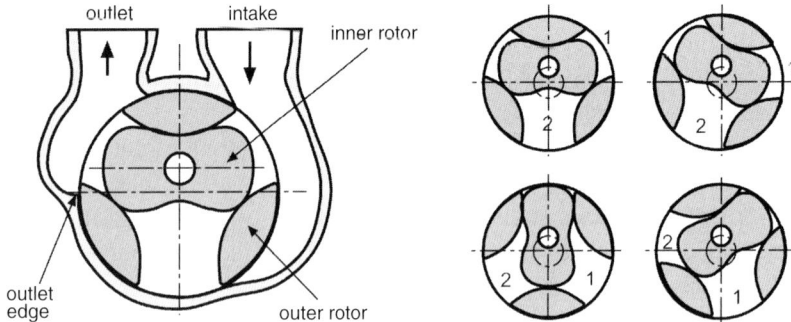

Fig. 4.11. Wankel-2/3 (Ro) charger [56]

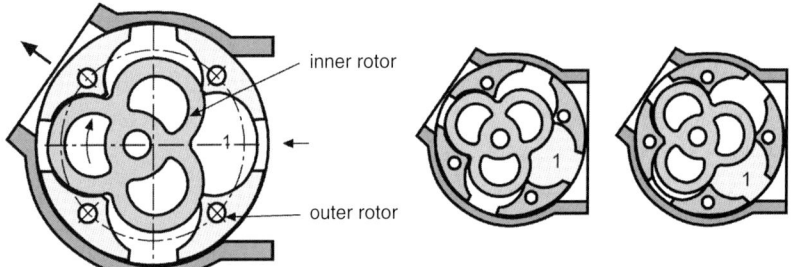

Fig. 4.12. Wankel-3/4 (Pierburg) charger [56]

plenums. In KKK's Ro charger, the two-lobed inner rotor dips into three corresponding recesses in the outer rotor. The speed of the inner rotor therefore is 1.5-fold of that of the outer rotor. The inner rotor is powered, and it drives the outer rotor via a pinion and an internally geared wheel.

In Pierburg's DK compressor, a three-lobed inner rotor dips into four recesses in the outer rotor. The inner rotor's speed is therefore 1/3 higher than that of the outer rotor.

Application areas would also be small supercharged gasoline engines. However, up to now these charger types have not been used in series production.

Lysholm screw-type compressor

Advantages of the screw-type compressor first introduced by Lysholm in the 1930s are especially the high internal compression as well as the high pressure ratios and efficiencies attainable. It is a two-separate-shaft rotational piston charger with main and secondary rotors (Fig. 4.13). Here, the main rotor has four, the secondary rotor six cogs. The main rotor runs at 1.5 times the speed of the secondary rotor. Especially the gap losses are critical, due to the high pressure ratios attainable. Production of the highly twisted cog profiles is very complex. The moment of inertia of the rotor is higher than that of a Roots blower. On the other hand, it enables an especially even delivery curve, and high charger speeds can be obtained, resulting in small charger dimensions.

Lysholm compressors are produced by Svenska Rotor Maskiner in Sweden and by IHI in Japan. They were briefly used for series application at Mazda and have been used by DaimlerChrysler in their AMG C32 model (Fig. 4.14).

Due to the high efficiencies and pressure ratios, the main application areas in the future for screw-type compressors will mostly be modern high-performance engines for vehicles or boats.

4.4 Designs and systematics of mechanically powered compressors

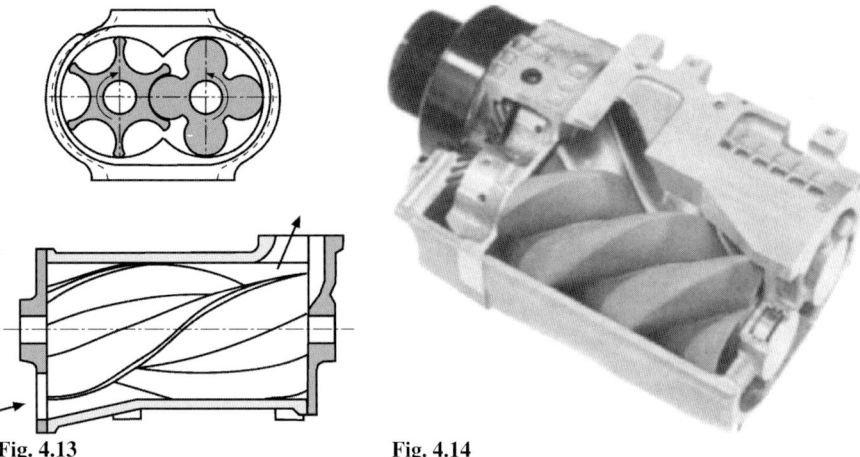

Fig. 4.13. Principle diagram of a screw-type (Lysholm) compressor [56]

Fig. 4.14. Screw-type compressor of the AMG C32 3.2 liter engine by DaimlerChrysler

4.4.2 Turbo compressors

The design of a turbo compressor does not have to be discussed in detail at this point since it does not differ from the turbocharger compressor. Instead of the turbine, only a corresponding drive, e.g., a mechanical connection to the crankshaft via CVT transmission (see Fig. 4.7) or an electric motor in a mechanically powered turbo compressor has to be incorporated. The system shown in Fig. 13.7 is used in the United States under the tradename Turbopac in small quantities for retrofitting commercial vehicle diesel engines.

5 Exhaust gas turbocharging

5.1 Objectives and applications for exhaust gas turbocharging

The clear objective of exhaust gas turbocharging is the increase in power density of reciprocating piston internal combustion engines by means of precompressing the work medium, i.e., air. It utilizes the exhaust gas energy which otherwise – due to the geometrically given expansion ratio of the crank mechanism – would be lost at the end of the high-pressure cycle. Simultaneously, the boundary conditions for combustion and the high-pressure cycle can be improved so that their control and emission level can be optimized.

Therefore, the main application areas for exhaust gas turbocharging are those in which high engine power density has to be obtained in combination with minimized emission and fuel consumption values. Thus, exhaust gas turbocharging will always be preferred, if it can be realized technically and at an acceptable cost.

5.2 Basic fluid mechanics of turbocharger components

This section will primarily discuss the basic fluid mechanics necessary for understanding supercharging equipment with flow compressors and turbines. It will not discuss the problems and methods associated with their layout and optimization. For that, we refer to the relevant literature [42, 43, 81, 90].

5.2.1 Energy transfer in turbo machines

Compression

In flow compressors, the pressure increase of the work medium occurs in several phases proceeding nearly simultaneously.

On the one hand, by adding external mechanical energy, the medium is forced into a vectored speed, i.e., kinetic energy (change of state from 1 to 2) in the compressor impeller.

This is then changed into pressure energy (change of state from 2 to 3), partially by deceleration of the medium in the divergent blade channels of the compressor impeller itself, and partially in a downstream static diffuser.

The addition of energy and the pressure increase (in the decelerated flow) can be described using the first law of thermodynamics for open systems (also see Eq. (2.15), without consideration of the influence of geodetic altitude):

$$h_1 + c_1^2/2 = h_2 + c_2^2/2 + w_t + q_{add}, \qquad (5.1)$$

where h describes the enthalpy, w_t the technical work added (or subtracted) from outside, and q_{add} the heat added (or subtracted) from outside.

5.2 Basic fluid mechanics of turbocharger components

Under the assumption of an adiabatic system (1–2), the following applies for the addition of kinetic energy w_t:

$$h_1 + c_1^2/2 = h_2 + c_2^2/2 + w_t, \tag{5.2}$$

$$w_t = \frac{c_1^2 - c_2^2}{2} + (h_1 - h_2). \tag{5.3}$$

For the pressure rise by flow deceleration (2–3) the following applies:

$$h_2 + c_2^2/2 = h_3 + c_3^2/2 \tag{5.4}$$

and with $h = u + p/\rho$

$$\frac{p_3}{\rho_3} - \frac{p_2}{\rho_2} = \frac{c_2^2 - c_3^2}{2} + u_2 - u_3. \tag{5.5}$$

Expansion

The desired gain in technical work is also obtained in processes occurring nearly simultaneously.

On the one hand, the pressure energy of the medium is partially changed into kinetic energy (change of state from 1 to 2) in the converging blade channels or volute.

This, as well as the remaining pressure energy, is now converted in the rotor into mechanical work (change of state from 2 to 3) via flow deflection and further pressure reduction (actio et reactio).

These changes of state can again be described using the first law of thermodynamics for open systems (Eq. (5.1)).

For the conversion of pressure energy into kinetic energy (1–2) the following applies:

$$c_1^2 - c_2^2 = 2\left(\frac{p_2}{\rho_2} - \frac{p_1}{\rho_1} + u_2 - u_1\right). \tag{5.6}$$

For the conversion of the kinetic energy and the remaining pressure energy (enthalpy) into mechanical work (2–3) the following applies:

$$w_t = \frac{c_2^2 - c_3^2}{2} + h_2 - h_3. \tag{5.7}$$

Generally, the reverse process – accelerated flow – is easier to comprehend. Here, in accordance with the law of energy conservation, pressure is converted into velocity (conversion of potential pressure energy into dynamic flow energy).

The discussion of the turbine and its special properties will be continued in depth in Sect. 5.4.2.

5.2.2 Compressors

Axial compressor

Since the pressure increase via deceleration (Bernoulli) can best be illustrated using an axial compressor stage as example, we will use this layout of a flow engine to describe the relevant processes leading to a pressure gain using energy input (Fig. 5.1).

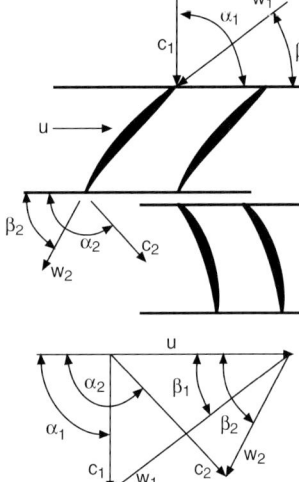

Fig. 5.1. Flow process with relevant velocity triangles for an axial compressor stage [120]

As can be seen, the blade pitch for both rotor and diffuser is such that the inlet angles are smaller in comparison with the corresponding outlet angles. Therefore, the areas measured vertical to the blade profiles must increase (compressor!) and the relative flow velocity w must decrease in the channels formed by the blades, i.e., $w_1 > w_2$.

Since the compressor rotor is powered, and thus energy is added, the absolute velocity c of the medium to be compressed nevertheless increases: $c_2 > c_1$. This energy is utilized in the diffuser for an additional pressure increase, by deceleration of the medium. The ratio between the pressure increase in the rotor and the total pressure gain in the stage is called reaction rate r. It is exactly defined as the ratio between the enthalpy conversion in the rotor and the total enthalpy conversion of the compressor:

$$r = \frac{\Delta h_{s,\text{C-Rot}}}{\Delta h_{s,\text{C}}}. \tag{5.8}$$

Characteristic for an axial compressor is that its diameter is nearly constant. Thus, in the optimum layout, several compressor stages are compiled into a multistage medium-pressure or high-pressure compressor (e.g., for gas turbines).

In an axial compressor, for pressure generation no change in the rotor diameter, i.e., no increased outlet diameter, is needed. Therefore, its inlet diameter is the largest diameter of the entire compressor. They are therefore predestined for large air quantities at a given outer diameter (jet aircraft). However, to generate higher pressures, they mostly need several stages since the pressure levels obtainable per stage are far lower than those of a radial compressor.

Radial compressor

The pressure increase, per stage, in a radial compressor strongly depends on the blade shape (Fig. 5.2; left, rearward bent; right, straight blades), and additionally on the ratio between the inlet and the outlet diameter of the compressor impeller. Its total pressure generation occurs in three steps.

5.2 Basic fluid mechanics of turbocharger components

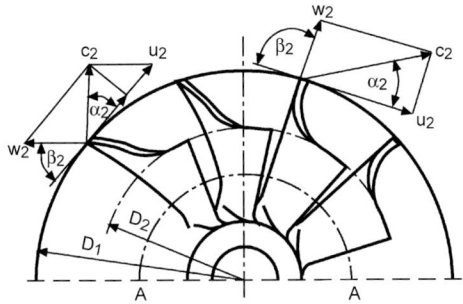

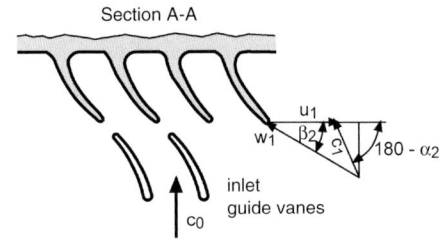

Fig. 5.2. Radial compressor with differing blade designs (straight and rearward bent), inlet guide vanes, corresponding speed triangles

1. Pressure increase in the centrifugal field (outlet diameter larger than inlet diameter):

$$\Delta p \sim u_2^2 - u_1^2, \tag{5.9}$$

   i.e., the enthalpy increase is proportional to the squares of the circumferential speed differences.
2. Deceleration of the relative medium velocity w in the impeller analogous to the increase in flow areas:

$$\Delta p \sim w_1^2 - w_2^2. \tag{5.10}$$

3. Pressure increase in the outlet diffuser:

$$\Delta p \sim c_2^2 - c_3^2. \tag{5.11}$$

Here, in a blade or disc diffuser, downstream of the impeller, the medium is decelerated from the absolute outlet velocity at the radial impeller c_2 to the outlet velocity of the compressor c_3.

Therefore, the total pressure increase and enthalpy increase in a radial compressor corresponds to

$$\Delta p_{\text{tot}} \sim (u_2^2 - u_1^2) + (w_1^2 - w_2^2) + (c_2^2 - c_3^2). \tag{5.12}$$

Due to their additional significant pressure increase in the centrifugal field, radial compressors are predestined for high pressure ratios in a single stage at comparably low flow rates. With that they are especially well suited for application as exhaust gas turbochargers, mostly designed in a single-stage layout.

The characteristics of radial compressors and their control possibilities will now be discussed on the basis of the compressor map.

In the pressure–volume flow diagram, the surge limit was defined as the border to the instable region of small flow rates and higher pressures. The choke limit is defined as the flow rate limit at maximum compressor speed.

At the **surge limit**, the flow in the compressor impeller stalls, resulting in pressure waves in the charger as well as in the charge air manifold upstream of the compressor, the so-called pumping. There are several ways to avoid this stalling and thus pumping, which all must aim to design the inlet into the compressor impeller free of wrong intake (admissions) angles compared to the blade angle of the impeller. These technical possibilities will be discussed in more detail in Sect. 5.4.3.

The **choke limit** is characterized by the fact that the gas flow in the narrowest area of the compressor inlet reaches sonic speed. The volume flow cannot be increased even by increasing the compressor speed. Therefore, all curves of constant compressor speed leading to a higher than the critical pressure ratio Π (see Fig. 5.8) approach one maximum flow value at pressure ratio 1. However, in Sect. 5.4.3, a method of influencing this limit to a small degree will be described. Figure 5.3 shows the most important characteristics, including how the surge limit narrows the useable map area, the maximum charger speed, and the thus achievable maximum charge pressure, as well as the choke limit (sonic speed at the compressor inlet).

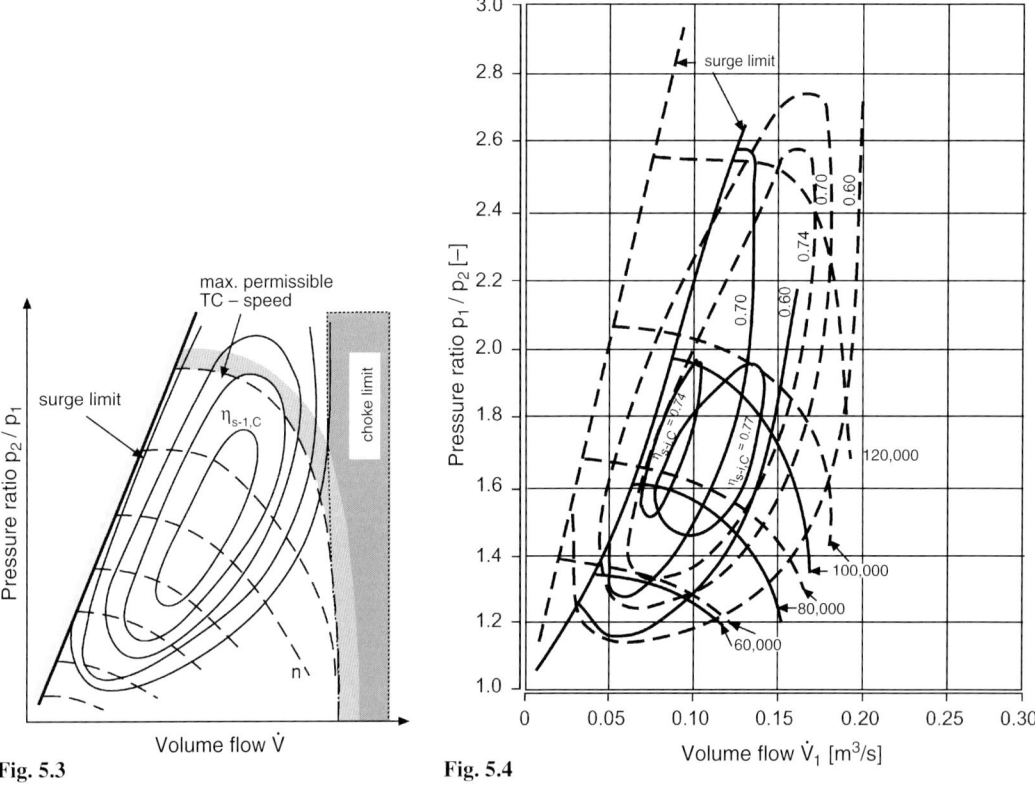

Fig. 5.3.   Fig. 5.4

Fig. 5.3. Principle map limitations of radial compressors: surge, speed and choke limit

Fig. 5.4. Comparison of two compressor maps; with straight-ending blades (solid lines) and with rearward bent blades (dash lines) [KKK, now 3K-Warner]

5.2 Basic fluid mechanics of turbocharger components

Because of the significant increase in the strength of compressor impeller materials, today high-performance compressors can be designed with blade ends subjected to high stresses which are caused by heavy bending instead of pure tension forces. This allows the use of backward bent compressor blades.

With these, at a given impeller diameter the channel length, i.e., the length between impeller inlet and outlet, is increased, resulting in an increased pressure gain in the impeller due to the associated decrease in relative speed of the medium in the blade channel. As a result of this, the efficiencies are better. Further higher pressure ratios as well as wider maps result from an increased insensibility of the channel flow. Figure 5.4 shows a comparison of a compressor map with straight-ending blades and one with backward bent blades. Today, for high-volume production chargers used in passenger cars and trucks, backward bent blades are state of the art. For cost reasons, swirl restrictors and/or vaned diffusers are used only in special applications or for expensive engines.

5.2.3 Turbines

Axial turbine

Similar to axial compressors, the energy gain by pressure and enthalpy reduction can best be explained for an axial turbine stage (Fig. 5.5). Therefore, we will more closely describe the relevant processes here also using this type of turbine.

From the pitch of the guide vanes it can be seen that, starting with a rectangular intake flow profile of the gas at a velocity c_0, the flow is accelerated to velocity c_1 due to the flat outlet angle β_1. For a given circumferential speed u of the turbine, this evolves into the inlet angle of the relative flow w_1 into the turbine rotor. While further accelerating the flow to w_2 and c_2 in the rotor, energy is transferred to the rotor. Analogous to the compressor, the ratio between the enthalpy decrease in the turbine rotor and the entire turbine is again called reaction ratio r.

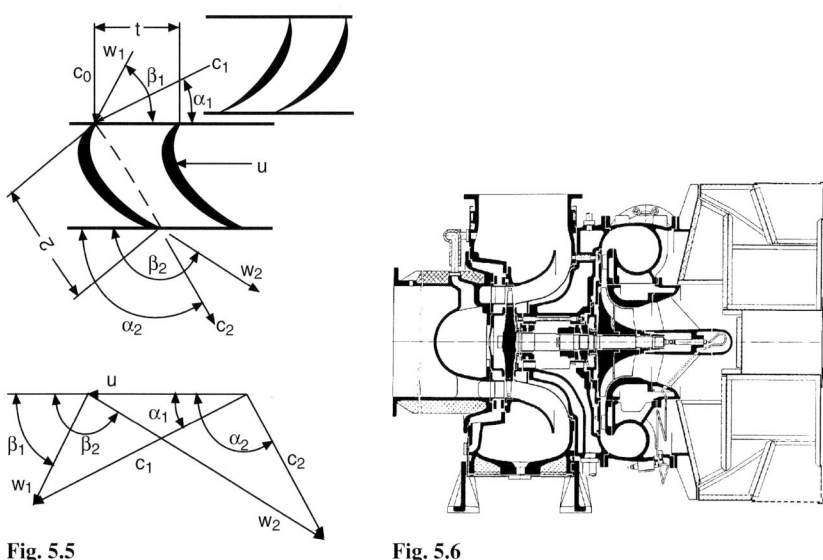

Fig. 5.5. Axial turbine stage with relevant velocity triangles [121]

Fig. 5.6. Single-stage axial turbine of a large exhaust gas turbocharger [MAN]

If the total enthalpy is converted into velocity in the stator, i.e., with a reaction rate of 0, such configuration is called a straight action turbine.

For axial turbines as well, several stages are possible. They are used for aircraft or stationary gas turbine applications. For large exhaust gas turbochargers, single-stage axial turbines are state of the art, for efficiency and intake flow reasons (Fig. 5.6).

Radial turbine

In the radial turbine, analogous to the radial compressor, the energy conversion occurs also in steps.

At first, the exhaust gases are accelerated in the mostly bladeless circular inlet nozzle as part of the volute according to

$$\Delta p \sim c_2^2 - c_1^2 \qquad (5.13)$$

The conversion of this momentum of the gas flow – together with a further pressure decrease in the rotor – then results in a corresponding gain of mechanical energy due to
- the pressure reduction in the rotor caused by the increase of the relative velocity w:

$$\Delta p \sim w_2^2 - w_1^2 \qquad (5.14)$$

- and the conversion of the circumferential velocity difference u:

$$\Delta p \sim u_1^2 - u_2^2. \qquad (5.15)$$

Figure 5.7 shows the velocity triangles of a radial turbine.

Special features of the turbine and its pressure–volume flow map

Here the characteristics of the operating behavior of a turbine will be discussed by means of the appropriate pressure–volume flow map. This has not yet been discussed, and it has to be

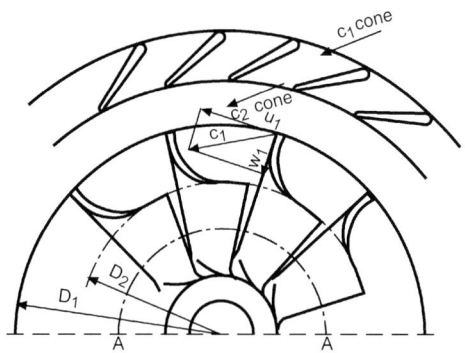

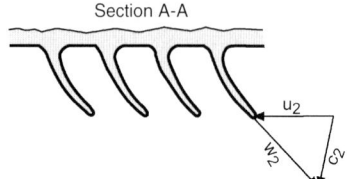

Fig. 5.7. Function and velocity triangles of a radial turbine [159]

5.2 Basic fluid mechanics of turbocharger components

kept in mind that the boundary conditions for the turbine are quite different from those for a compressor.

- The volume or mass flow through the turbine is predetermined by the engine. Even more important, the pressure downstream of the turbine approximates ambient pressure, without flow limitation by a downstream volume pump, which the engine is on the compressor side.
- Moreover, varying exhaust gas temperatures occur depending on load and speed of the engine, which influence the volume flow through the turbine.
- Finally, the compressibility of the exhaust gases must be taken into consideration.

Thus, considering the flow conditions given in an exhaust gas turbocharger, characteristic swallowing lines can be identified for a turbine which in first approximation correspond to those of an adequate opening or nozzle.

The flow velocity downstream of the cylinder of a reciprocating piston engine – for the sake of simplicity here assumed without any speed in the cylinder itself – results from the existing enthalpy difference in the exhaust gas upstream and downstream of this nozzle:

$$c_4^2/2 = h_3 - h_4 \rightarrow c_4 = \sqrt{2(h_3 - h_4)}. \tag{5.16}$$

For ideal gases – assumed here – the following applies:

$$h_3 - h_4 = c_p(T_3 - T_4) \tag{5.17}$$

and

$$\frac{T_3}{T_4} = \left(\frac{p_4}{p_3}\right)^{(\kappa-1)/\kappa} \tag{5.18}$$

as well as

$$T_3 = \frac{p_3}{\rho_3 R} \tag{5.19}$$

$$\frac{c_p}{R} = \frac{\kappa}{\kappa - 1}. \tag{5.20}$$

Inserting these terms into Eq. (5.16), the outlet velocity c_4 can be calculated as a function only of the turbine pressure ratio and the state of the gas upstream of the turbine, 3, as

$$c_4 = \sqrt{2\frac{\kappa}{\kappa - 1}\frac{p_3}{\rho_3}\left[1 - \left(\frac{p_4}{p_3}\right)^{(\kappa-1)/\kappa}\right]}. \tag{5.21}$$

The mass flowing through the turbine or the equivalent area (nozzle) is

$$\dot{m}_T = A_{T,\text{eff}} \rho_3 c_4 \tag{5.22}$$

and with $\rho_4/\rho_3 = (p_4/p_3)^{1/\kappa}$ it is

$$\dot{m}_T = A_{T,\text{eff}} \psi \sqrt{2 p_3 \rho_3}, \tag{5.23}$$

where ψ is the mass flow function already discussed, which only depends on the turbine pressure ratio and the upstream gas conditions (see Eq. (2.18)).

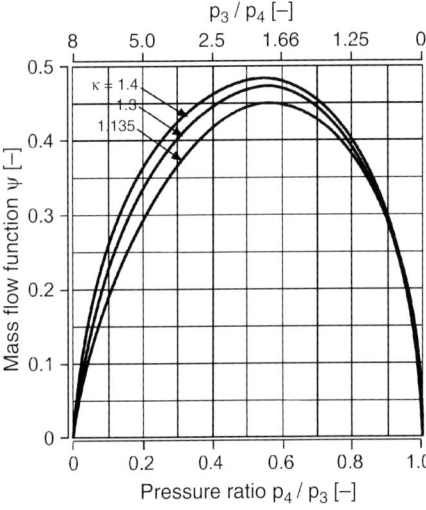

Fig. 5.8. Mass flow function ψ for $\kappa = 1.135, 1.3, 1.4$

This equation, shown in Fig. 5.8 for three different κ values, has two zero-crossings, at $p_4/p_3 = 0$ and 1. In between it shows a maximum at the so-called critical pressure ratio, which only depends on the gas condition, which is specified by the adiabatic exponent κ. If the pressure ratio at the turbine is kept constant, and thus ψ is constant, the turbine volume flow only depends on the initial state of the gas upstream of the turbine. With $p_3 v_3 = RT_3$, we get

$$\dot{m}_T = A_{T,\text{eff}} \psi p_3 \sqrt{\frac{2}{RT_3}} = A_{T,\text{eff}} \psi \frac{p_3}{\sqrt{T_3}} = \text{const.} \tag{5.24}$$

From this it follows that at constant pressure p_3 upstream of the turbine, the gas mass flowing through the turbine decreases in relation to $1/\sqrt{T_3}$. At constant temperature T_3, the flow rate is directly proportional to the pressure p_3. With this, pressure and temperature can be eliminated as parameters in the turbine map to be developed, by "standardizing" the turbine flow rate $\dot{m}$ with

$$\dot{m}_T = \dot{m}_T^* \cdot \frac{p_3}{p_0} \cdot \frac{1}{\sqrt{T_3/T_0}} \hat{=} \dot{m}_T^* \cdot \frac{p_3}{p_0} \sqrt{T_0/T_3}.$$

With this conclusion we can derive the **turbine map commonly used today**. The turbine characteristic (with fixed geometry) is shown **as turbine expansion and pressure ratio against the flow rate reduced by** $p_3/\sqrt{T_3}$. A flow curve, the so-called swallowing capacity function of the turbine, results, which approximates the flow characteristic of an equivalent nozzle area.

Therefore, in an exhaust gas turbocharger the resulting exhaust backpressure and thus the achievable charge pressure, under consideration of the efficiencies, only depend on the **turbine housing area** chosen, provided that it represents the flow-limiting nozzle area. This can be seen in Fig. 5.9.

Turbines with variable turbine geometry (VTG charger), which will be discussed later in detail, have a very wide turbine map, comparable to a compressor map, due to their various positions of the blades in the turbine inlet guide ring as an additional parameter.

5.2 Basic fluid mechanics of turbocharger components

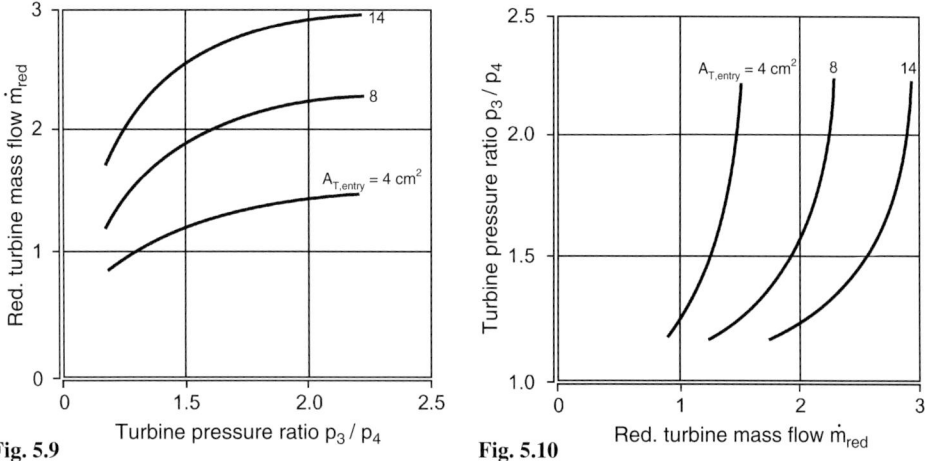

Fig. 5.9. Relationship between exhaust gas backpressure and turbine neck cross section area $A_{T,entry}$ (previous way of mapping a turbine)

Fig. 5.10. Turbine pressure ratio–volume flow map, new way of map plotting, similar to compressor map

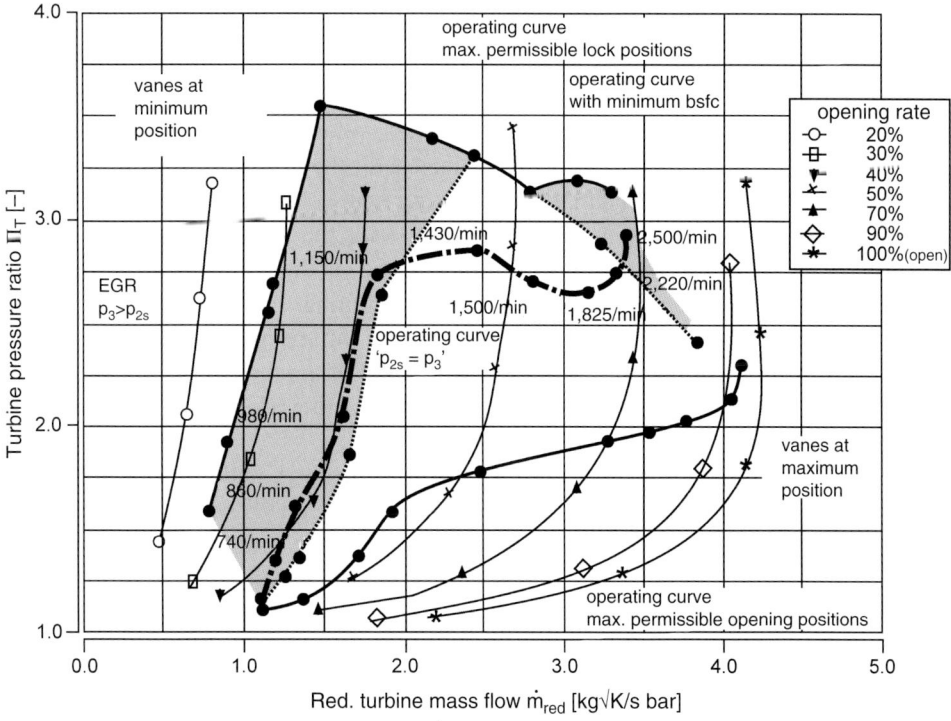

Fig. 5.11. VTG turbine swallowing lines for full-load and part-load in new way of mapping [DC]

Especially for these, plotting the map similar to the compressor map, i.e., **"turbine pressure ratio against the reduced turbine volume flow"**, results in a very useful way of plotting (Fig. 5.10). The reduced turbine volume flow is that volume flow which the turbine actually handles or must process – at given pressure and temperature conditions. This kind of mapping was also proposed by Watson and Janota [147].

With this way of plotting the map

- we can chart turbine work curves, e.g., for full-load and part-loads, and
- we obtain the desired similarity with the compressor map, to support the easier understanding and descriptiveness of such diagrams, as Fig. 5.11 shows.

For these reasons, the presentation of the turbine map as proposed by Watson is very much recommended.

Characterizing the turbine with an equivalent area or nozzle, and inserting ψ_{max}, which results in the maximum possible speed at the outlet area, i.e., sonic speed, we get

$$\dot{m}_T = A_{T,eff} \psi_{max} \sqrt{2 p_3 \rho_3}. \tag{5.25}$$

It can be easily seen that the turbine flow only depends on the gas condition upstream of the turbine and no longer on the backpressure. In other words, expansion ratios beyond the "critical" ratio, i.e., greater than 1.8–2.0, seem not to be practical since then the exhaust gas energy can no longer be utilized completely.

Therefore, from now on we will abandon the hypothesis of the **"equivalent nozzle"**, due to the fact that only the particular **relative speed** in the flow-conducting parts of the turbine is **"critical for sonic speed"**.

In a radial turbine rotor the narrowest area is always at the turbine outlet. Since the turbine rotor operates at considerable circumferential speed, at an exhaust gas temperature of, e.g., 620 °C, an oncoming flow to the rotor accelerated to nearly sonic speed at $c_1 = 550$ m/s, and at a circumferential speed of $u_1 = 400$ m/s, the relative intake velocity into the rotor, w_1, is only about 290 m/s (Fig. 5.12). From this relative intake velocity w_1, the medium can now be further accelerated in the rotor to the sonic speed in the medium as outlet velocity w_2.

Starting from a reasonable expansion ratio of about 1.8 for the acceleration in the nozzle, and further accelerating in the rotor to about the sonic speed of the medium, which is about 580 m/s, an additional expansion ratio of 1.6 is obtained, resulting in an "apparent expansion ratio" of about 3.5, which can be utilized in a single-stage turbine without pressure losses. Additionally, taking into account pressure losses due to friction of the medium in the inlet guide vanes and rotor, a maximum **expansion ratio of about 4** can be achieved in a **single-stage radial turbine**.

In axial turbines it is possible to obtain supersonic speeds in the inlet guide vanes and to utilize these velocities in an action turbine (Laval turbine). In practice, the conditions in the turbine are even more complex, since with pulse turbocharging, transient pressure conditions exist during the exhaust strokes of the cylinders of the engine (Fig. 5.13).

Thus we have to note that the rather simple mapping of the turbine operational characteristics

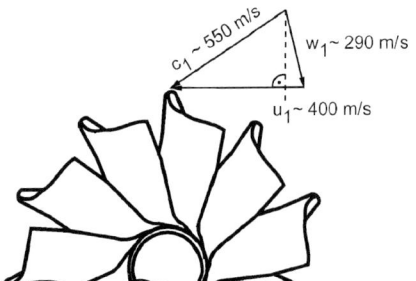

Fig. 5.12. Absolute and relative intake speeds at the turbine rotor

5.2 Basic fluid mechanics of turbocharger components

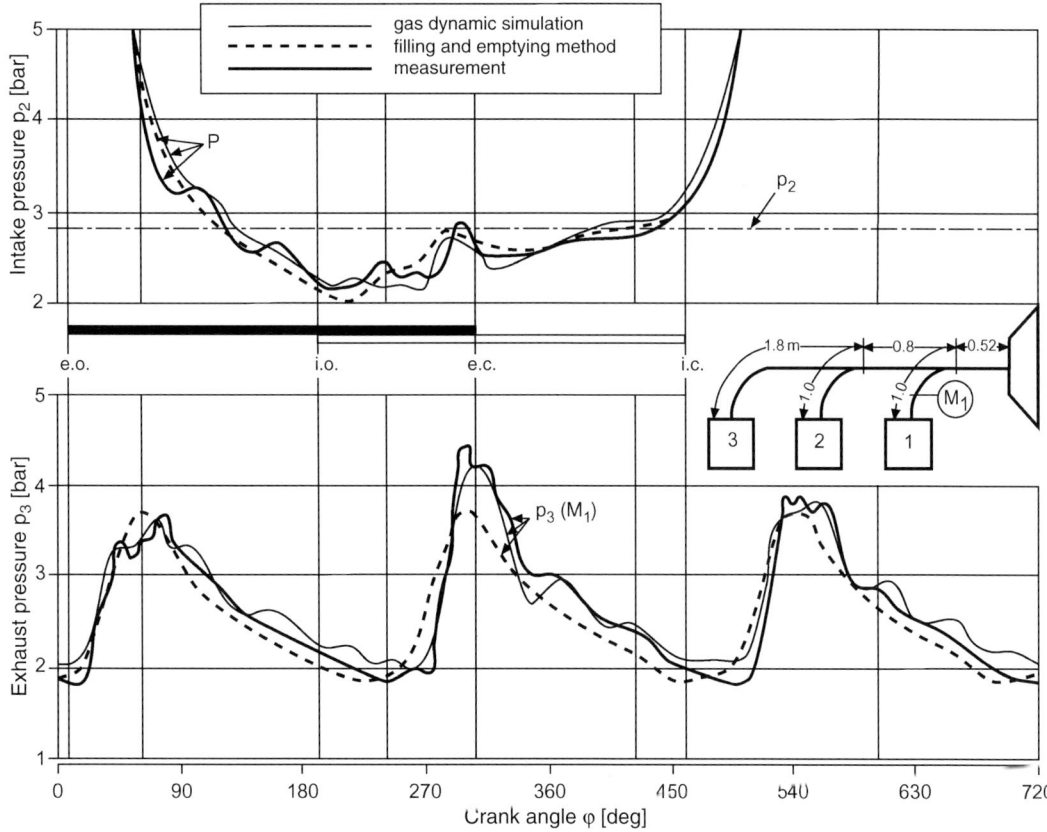

Fig. 5.13. Pressure conditions upstream of the turbine during the exhaust stroke of the cylinders [159]. e.o., exhaust opens; i.o., inlet opens; e.c., exhaust closes; i.c., inlet closes

with this turbine swallowing capacity function enables just a rough description of its mean operation behavior and is therefore only of very limited value for detailed turbine layout as required today. Both in stationary engine operation and, especially, during transient engine processes, the turbine is operated under conditions which can no longer be characterized with sufficient accuracy by such a simple function.

Although the mean turbine speed changes only slightly, the instantaneous turbine speed as well as operating points significantly deviate from the corresponding average values because of the widely varying pressures and mass or volume flows. For the map diagram this means that the corresponding operating points are no longer located on this mean swallowing capacity function. Consequently, the map must be shown in an extended operating area (Fig. 5.14).

The measurement of such maps is very complex and must be performed at special turbocharger test benches, on which the exhaust gas turbocharger cannot be operated in the steady-state operating points only, i.e., in power equilibrium with the compressor. Rather, it is necessary significantly to be able to freely adjust pressure, temperature, and mass flow conditions both on the compressor side (Fig. 5.15) as well as on the turbine side (Fig. 5.16).

Such extended maps – instead of the mean swallowing capacity functions for fixed-geometry turbines and a corresponding map for turbines with variable geometry – are of the utmost significance, especially for correct thermodynamic cycle simulations.

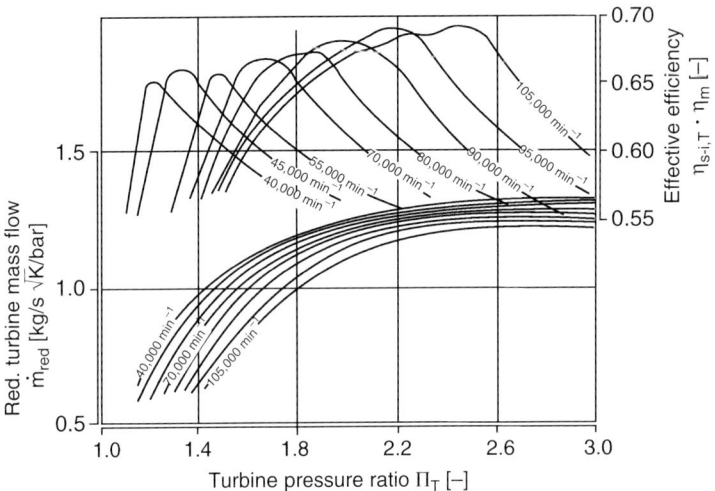

Fig. 5.14. Turbine map with extended operating area

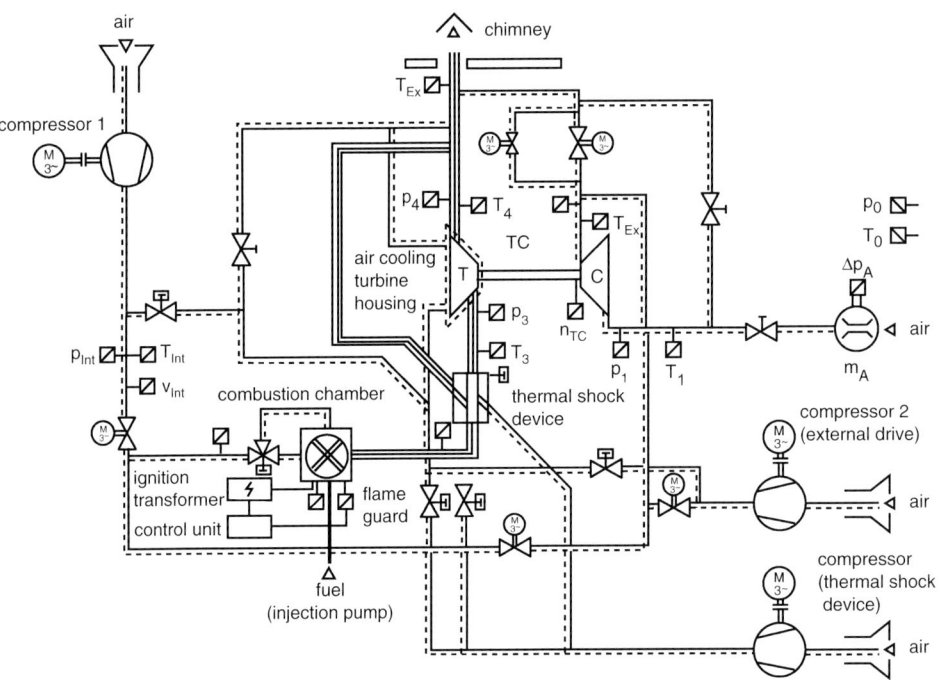

Fig. 5.15. Turbocharger test bench with freely adjustable pressure, temperature, and mass flow conditions [112]

Similar to those of compressors, and corresponding to the turbine layout, these maps show different characteristics:

- for axial turbines, the map width is very narrow (Fig. 5.17),
- for radial turbines, the map is significantly wider because of the varying centripetal forces at different speeds (Fig. 5.18).

5.2 Basic fluid mechanics of turbocharger components

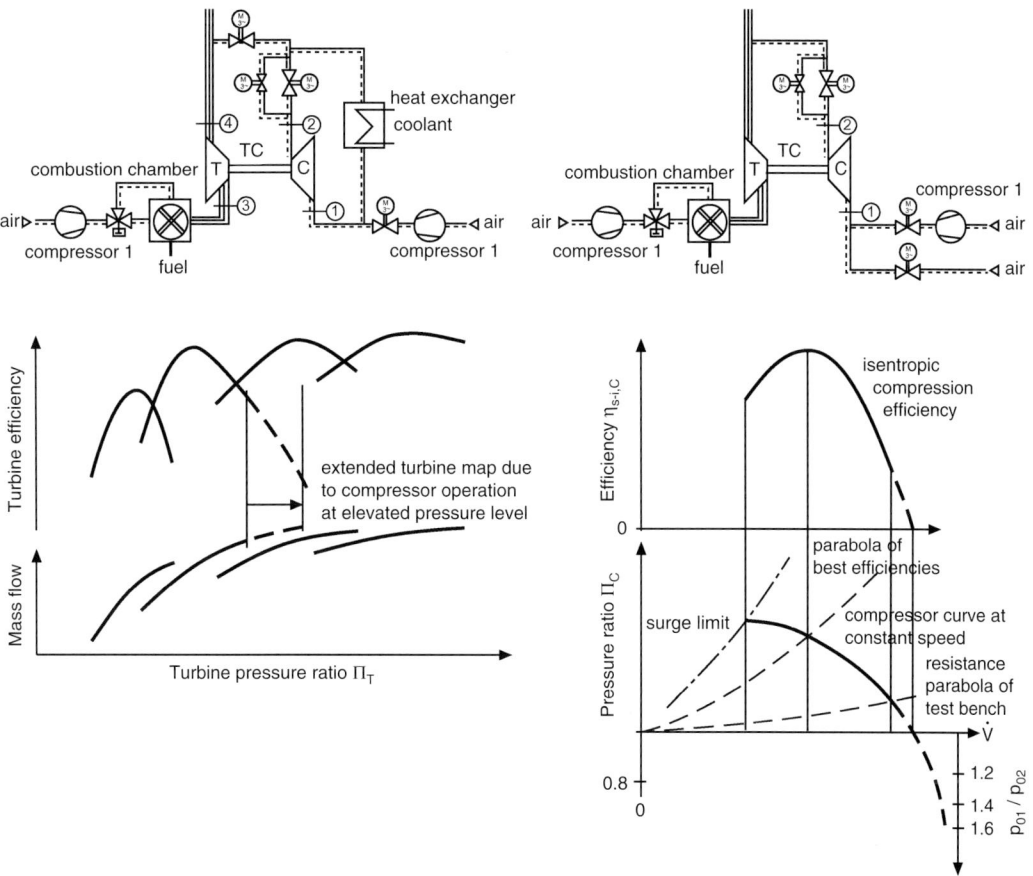

Fig. 5.16. Determination of extended compressor and turbine maps [112]

In summary, the characteristics of turbines are as follows.

Turbines, like compressors, can be described in **maps** which **show differing characteristics depending on their layout**.

At steady-state engine operation, under consideration of mean exhaust gas volume flows and turbine pressure ratios, the turbocharger operates only in a very narrow map area, the so-called turbine swallowing capacity function.

For cycle simulations – especially of transient engine operating conditions – the complete turbine map must be taken into account.

For a first coarse layout of exhaust gas turbochargers, the mean turbine swallowing capacity function may be used with sufficient accuracy.

Turbines with variable geometry can be treated like a band of fixed-geometry chargers. Their operating behavior can be described by a corresponding number of maps or mean swallowing capacity functions, each for a specific blade position.

The compilation of the mean swallowing capacity functions of a turbine with variable geometry results in an extended map, in which the engine operating points and thus the corresponding blade positions can be displayed.

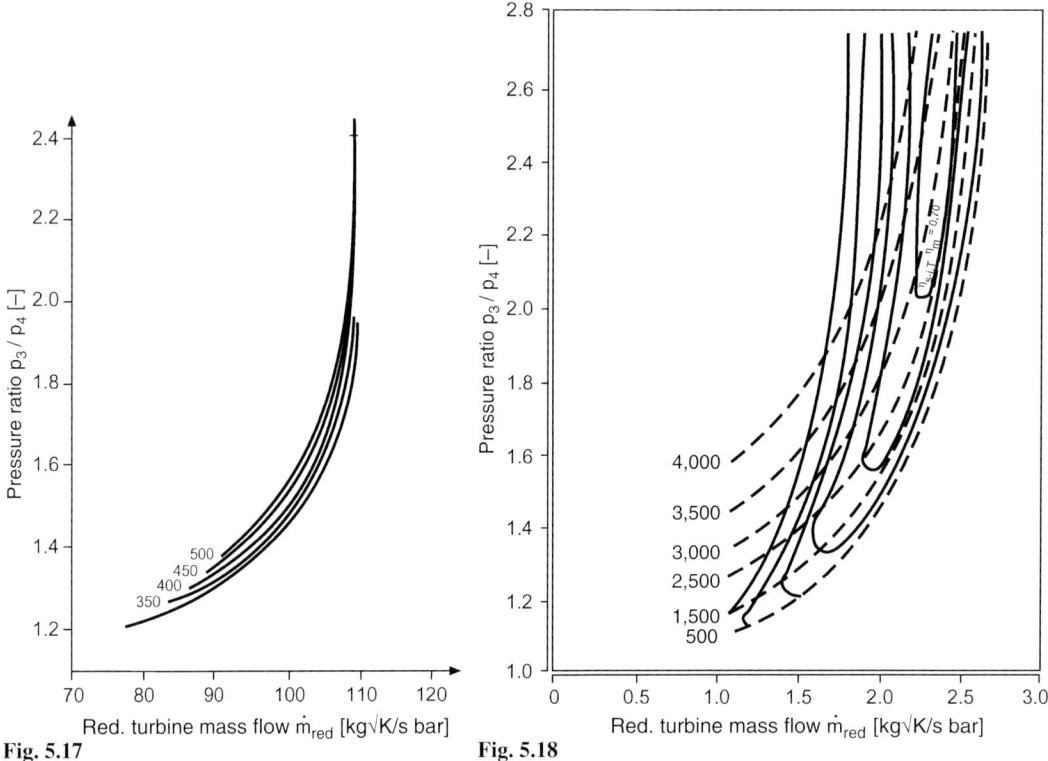

Fig. 5.17. Map of an axial turbine [147]

Fig. 5.18. Map of a radial turbine [147]

5.3 Energy balance of the charging system

Simply by coupling a flow compressor with a flow turbine on a common shaft and supplying this turbine with engine exhaust gas, exhaust gas turbocharging generates charge pressure without a mechanical connection to the engine. There exists a thermodynamic coupling. The turbocharger is freely spinning and the charger speed adjusts itself corresponding to the respective power equilibrium between compressor (P_C) and turbine (P_T). Thus, the attainable charge pressure is also subject to these equilibrium conditions.

$$P_C + P_T + P_{fr} = 0 \tag{5.26}$$

describes the power balance of compressor and turbine under consideration of a friction power loss P_{fr} of an actual charger (bearings, shaft seals, etc.), and

$$\dot{m}_T + \dot{m}_C + \dot{m}_F = 0 \tag{5.27}$$

describes the mass balance, where $\dot{m}_F$ is the added fuel mass. The equations for compressor and turbine power can be arranged as follows:

$$P_C = \frac{\dot{m}_C \Delta h_{s,C}}{\eta_{s-i,C} \eta_{m,C}}, \tag{5.28}$$

$$P_T = \dot{m}_T \Delta h_{s,T} \eta_{s-i,T} \eta_{m,T}, \tag{5.29}$$

where $\eta_{\text{s-i,C}}$ describes the isentropic compressor efficiency, $\eta_{\text{s-i,T}}$ the isentropic turbine efficiency, $\eta_{\text{m,C}}$ the mechanical compressor efficiency, and $\eta_{\text{m,T}}$ the mechanical turbine efficiency. $\Delta h_{\text{s,C}}$ describes the isentropic enthalpy increase in the compressor and, correspondingly, $\Delta h_{\text{s,T}}$ the isentropic enthalpy decrease in the turbine.

With the corresponding enthalpy changes in the compressor and the turbine we get

$$\Delta h_{\text{s,C}} = R_A T_1 \frac{\kappa_A}{\kappa_A - 1} \left[\left(\frac{p_2}{p_1} \right)^{(\kappa_A - 1)/\kappa_A} - 1 \right], \tag{5.30}$$

$$\Delta h_{\text{s,T}} = R_{\text{Ex}} T_3 \frac{\kappa_{\text{Ex}}}{\kappa_{\text{Ex}} - 1} \left[1 - \left(\frac{p_4}{p_3} \right)^{(\kappa_{\text{Ex}} - 1)/\kappa_{\text{Ex}}} \right]. \tag{5.31}$$

Inserting these into the balance Eq. (5.26) and performing some simplifications results in the following, so-called **main turbocharger equation**:

$$\Pi_C = \left\{ 1 + \frac{m_T}{m_C} K_1 \frac{T_3}{T_1} \eta_{\text{TC}} \left[1 - \left(\frac{p_4}{p_3} \right)^{(\kappa_{\text{Ex}} - 1)/\kappa_{\text{Ex}}} \right] \right\}^{\kappa_A/(\kappa_A - 1)}, \tag{5.32}$$

where

$$K_1 = \frac{R_{\text{Ex}}}{R_A} \frac{\kappa_A - 1}{\kappa_A} \frac{\kappa_{\text{Ex}}}{\kappa_{\text{Ex}} - 1}.$$

This main equation indicates which gas and physical conditions of the engine influence the attainable charge pressure:

$$\Pi_C = f\left(\frac{T_3}{T_1}; \eta_{\text{TC}}; \frac{p_4}{p_3} \right). \tag{5.33}$$

A few additional, general turbocharger characteristics which can be seen in the pressure–volume flow map should be mentioned.

The **charger speed** and thus the charge pressure is **not related to the engine speed**. It increases with increasing turbine power, i.e., with increasing exhaust gas flow rate and increasing exhaust gas temperature (energy supply to the turbine).

In an exhaust gas turbocharger, a **change in charge pressure** can be achieved only by a **change in charger speed**. This means that for any increase of the charge pressure, first the charger has to be accelerated via additional power to be generated by the turbine.

5.4 Matching of the turbocharger

5.4.1 Possibilities for the use of exhaust energy and the resulting exhaust system design

As mentioned in Sect. 3.2.3, exhaust gas turbocharging is the preferred method for utilizing the remaining energy in the cylinder charge at the end of the expansion stroke.

The most commonly used methods are constant-pressure turbocharging and pulse turbocharging, combined with a corresponding layout of the exhaust system.

Constant-pressure turbocharging

Figure 5.19 shows the *pV* diagram for the high-pressure cycle, as well as for the compressor and turbine work for this application. Due to the approximately constant exhaust gas backpressure, it exhibits the simplest thermodynamic conditions and is therefore especially suited for a discussion of the basic relations.

For this type of exhaust gas turbocharging, a correspondingly dimensioned plenum is located between the exhaust ports of the individual cylinders, designed to dampen exhaust pressure pulses occurring at "outlet valve opening". Thus the turbine will be admitted with exhaust gas pressure and temperature as constant as possible, i.e., constant energy flow. However, in this case the area 4–5–1, which contains the dynamic energy fraction, obviously cannot be utilized. On the other hand, due to the time-constant exhaust gas mass flow ($\dot{m} \approx$ const.), the turbine swallowing capacity can be comparably small, and good turbine efficiencies can be achieved.

If in first approximation the charge pressure p_2 is set equal to the pressure in the exhaust system ($p_2 = p_3$), the engine is always operated at a higher pressure level. The power required to drive the compressor is generated by the turbine. However, since the exhaust gas temperature at "outlet valve opening" is much higher than the compressor intake temperature, a larger volume is expanded in the turbine ($V \sim T$). Theoretically – at identical pressure drop as in the compressor ($p_3/p_4 = p_2/p_1$) – a larger turbine power could be generated than the compressor needs. Conversely, this means that, depending on the efficiency of the charger, p_2 will be higher than p_3, thus creating a so-called positive scavenging pressure gradient during the valve overlap phase [101].

More details can be obtained from the *hs* diagram of this process (Fig. 5.20). Here, as a complement to the *pV* diagram, the thermodynamic conditions are shown for a supercritical pressure

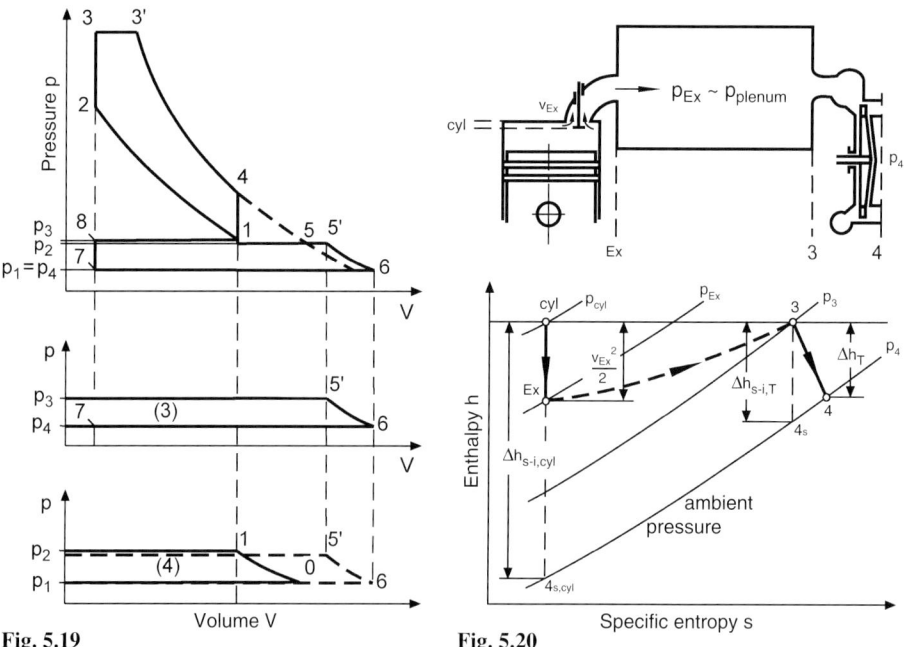

Fig. 5.19. *pV* diagram for four-stroke engines with constant-pressure turbocharging

Fig. 5.20. Principle schematic and *hs* diagram for four-stroke engines with constant-pressure turbocharging

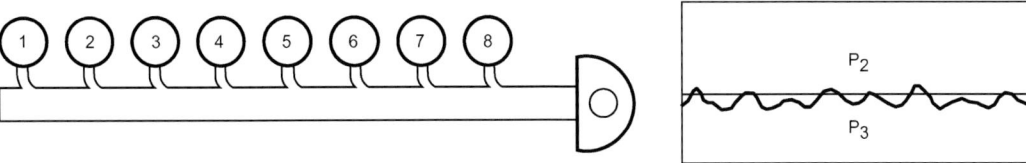

Fig. 5.21. Manifold layout and exhaust pressure traces for an 8-cylinder four-stroke engine with constant-pressure turbocharging [159]

ratio between cylinder pressure p_{cyl} and plenum pressure p_3 upstream of the turbine. The basis for this diagram are the assumptions of a heat-insulated exhaust system and a mean specific enthalpy h_0 for the total exhaust gas mass flowing out of the cylinder. An important fact shown in this diagram is that, due to the flow processes from the cylinder to the exhaust plenum, besides pressure losses and the loss of momentum, a significant increase in entropy occurs.

As a consequence, instead of an isentropic expansion to ambient pressure with a theoretically available enthalpy difference $\Delta h_{s-i,cyl}$, only the smaller enthalpy gradient $\Delta h_{s-i,T}$ is available for the turbine. Of this – due to turbine losses – only Δh_T can be utilized for the generation of compressor power. The inferior utilization of the exhaust gas enthalpy is, however, at least partially compensated for by better turbine efficiencies [101].

Figure 5.21 shows in principle how this can be achieved, on the one hand with respect to a certain layout and dimensioning of the exhaust manifold, on the other with given exhaust gas temperatures and turbocharger total efficiencies. Calculated exhaust gas charge pressure ratios, depending on charger efficiencies and exhaust gas temperatures, are shown in Fig. 5.22.

However, a serious **disadvantage** of constant-pressure turbocharging is the fact that with any change in the operating condition of the engine, the large exhaust plenum has to be brought to the new pressure and temperature level, which leads to significant problems under transient operating conditions.

The **advantages** of constant-pressure turbocharging are
– a simple exhaust system for multicylinder engines and
– low fuel consumption due to low gas exchange work.

Nowadays, the **major application** of constant-pressure turbocharging is in highly charged slow-speed engines in stationary use and with load patterns where transient operation is either modest or not relevant.

Pulse turbocharging

Pulse turbocharging, shown in Fig. 5.23 in the pV diagram, utilizes – additionally to the quasi-static energy (pressure and temperature) – the kinetic energy in the exhaust gas, present in the form of pressure waves from the blow down pulses. In this case, the admission into the turbine occurs with variable exhaust gas pressures and temperatures, i.e., under transient conditions.

The pV diagram could be interpreted in such a way that there would be no backpressure of the exhaust gas if the enthalpy h_4, present in the exhaust gas at "outlet valve opening", were completely converted into kinetic energy, i.e., exhaust flow velocity, and subsequently processed in an action turbine. In comparison to constant-pressure turbocharging this represents a gain, since an isentropic expansion is performed instead of the irreversible throttling to the turbine inlet pressure level in the plenum. However, the advantage of this process cannot be fully utilized, due to both the losses in the exhaust valve gap and the lower turbine efficiencies caused by transient gas admission to the turbine offside the turbine peak efficiency.

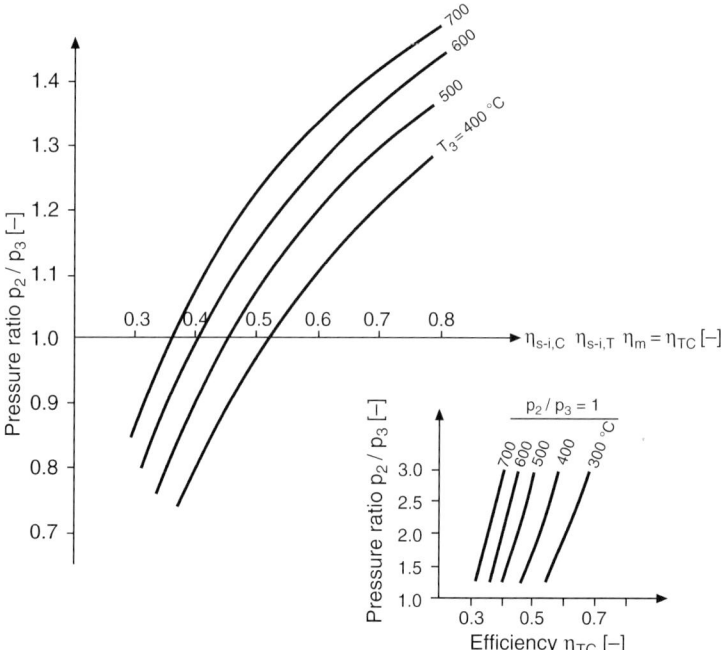

Fig. 5.22. Attainable exhaust gas charge pressure conditions depending on charger total efficiencies and exhaust gas temperatures [147]

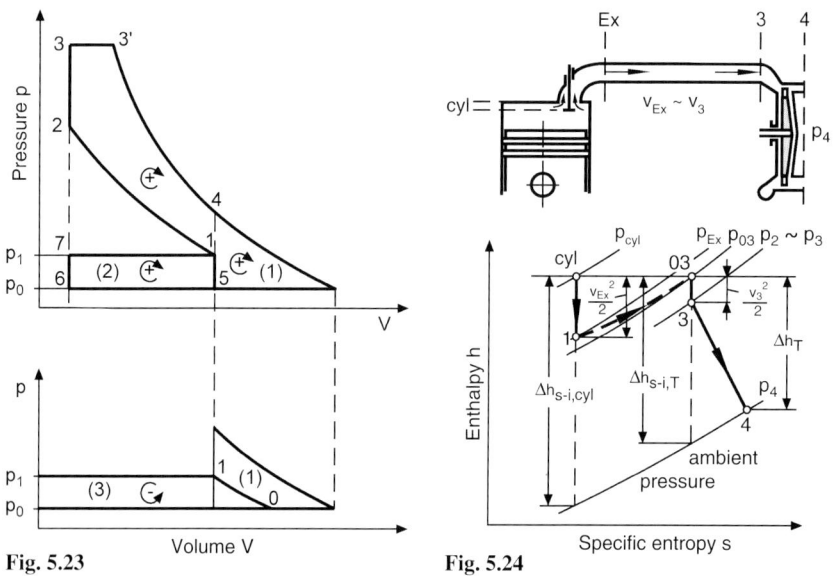

Fig. 5.23

Fig. 5.24

Fig. 5.23. *pV* diagram for four-stroke engines with pulse turbocharging

Fig. 5.24. Principle schematic and *hs* diagram for four-stroke engines with pulse turbocharging [110]

An examination of the processes in the *hs* diagram (Fig. 5.24) shows more details. Due to the supercritical pressure ratio between cylinder and exhaust manifold p_{cyl}/p_3, the exhaust flow at the valve gap reaches sonic speed and the pressure drops to p_{Ex}. Thus, with reduced throttling losses, an increased fraction Δh_T of the theoretical enthalpy gradient Δh_{s-i} can be utilized. The inferior turbine efficiencies, caused by irregular gas admission, are overcompensated by the higher energy content of the flow pulses [101].

Advantages for pulse turbocharging in comparison to constant-pressure turbocharging can be achieved
- in its thermodynamic behavior – however, these advantages decrease with increasing charge rate (Fig. 5.25);
- **especially in transient engine operation**, due to significantly improved acceleration behavior of the turbocharger and consequently the entire engine.

Nowadays, the **major application** area for pulse turbocharging is in engines with mostly transient operation, i.e., primarily in automobile engines.

Exhaust system

To incorporate the described exhaust gas turbocharging versions into engine design, a corresponding layout and dimensioning of the exhaust system is necessary. The goal of its layout is to improve the utilization of the exhaust gas energy as much as possible. In the case of constant-pressure turbocharging, main objective is a pressure recovery as good as possible. In the case of pulse turbocharging, main objective is the best possible conversion of the pressure pulses into kinetic energy with minimum losses.

The first exhaust gas turbocharged engines were designed with a common exhaust manifold, i.e., with constant-pressure turbocharging. A breakthrough in design was achieved with the implementation of Buechi's patent (DRP 568855) by splitting up manifolds and combining certain cylinders, which is of major importance for automotive applications. According to this patent, exhaust manifold and intake area into the exhaust gas turbine must be designed and the valve timing chosen such that at the beginning of the exhaust process the pressure in the exhaust manifold after the opening of the exhaust valve (blow down pulse) is higher than the pressure in the intake manifold, i.e., higher than the charge pressure. However, towards the end of the exhaust process, the exhaust manifold pressure must fall below the charge pressure.

In four-stroke engines this results in a total exhaust valve opening period of 260 to 300° crank angle (CA) before the next cylinder may exhaust into the same exhaust port. In practice, this interval can be somewhat shorter, due to wave propagation times and the delay of the blow down pulse. Ideal conditions for pulse turbocharging are therefore obtained using a firing distance of 240° CA for four-stroke engines, and of 120° CA for two-stroke engines, within one manifold branch.

For the four- and six-cylinder engines with pulse turbocharging, today predominantly used in automobile engines, this requires a twin-flow exhaust gas manifold arrangement. For a nine-cylinder engine, common in shipbuilding, a triple-flow arrangement accordingly would be necessary.

For **multipulse layouts**, the rule for minimum firing distance discussed above results in the following layout variants:
- four-cylinder engine, twin-flow layout with a turbine housing divided into two branches;
- eight-cylinder engine with one turbocharger, quadruple-flow layout with a turbine housing divided into four branches;
- eight-cylinder engine with two turbochargers, quadruple-flow layout with turbine housings divided into two branches;
- five-cylinder engine with symmetric firing order, triple-flow layout.

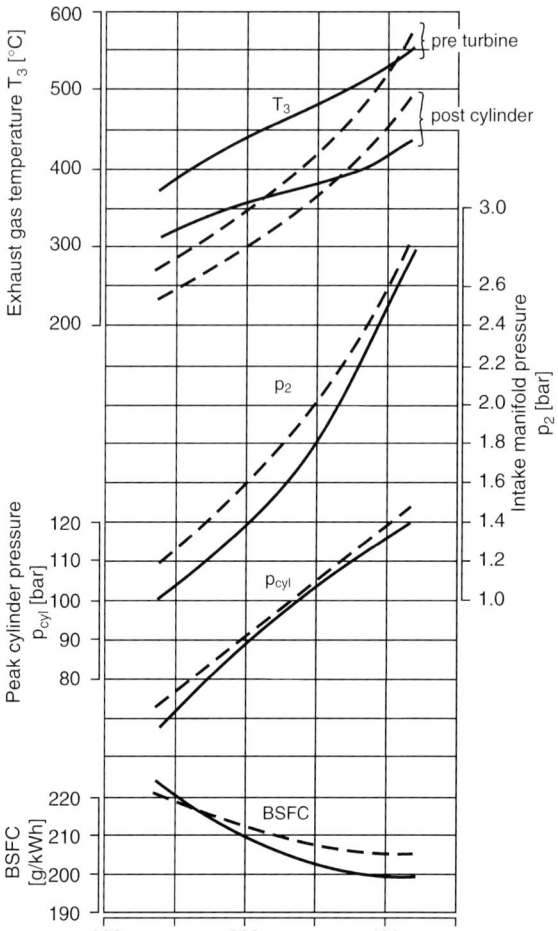

Fig. 5.25. Comparison between constant-pressure (solid line) and pulse turbocharging (dash line) of a medium-speed diesel engine [159]

Figure 5.26 shows examples for manifold arrangements and cylinder combinations for four-stroke inline engines, as well as the associated exhaust gas and charge pressure curves.

Besides combining the manifolds of certain cylinders, there is an additional method of utilizing the gas dynamics of the exhaust process which avoids the disadvantages of pulse pressure turbocharging, e.g., its inferior turbine efficiency.

This method is the **pulse converter**. The pulse converter also uses narrow exhaust gas manifolds which are combined exactly as for pulse turbocharging. However, here they are not channeled into separate turbine branches but are combined in the pulse converter.

In the pulse converter, the pressure energy present in the particular outlet flow is converted into kinetic energy by narrowing the manifold area and thus accelerating the velocity of the particular exhaust gas mass flow. Thus, the pressure differences between the individual manifold lines are reduced. In this way, a kind of injector effect is achieved which prevents the return of the pressure waves into the other manifold branches and thus interference with the scavenging process. Downstream of the pulse converter, the kinetic energy is exchanged among the pulses of the individual cylinders, and can be regained into pressure energy in a subsequent diffuser (Fig. 5.26).

5.4 Matching of the turbocharger

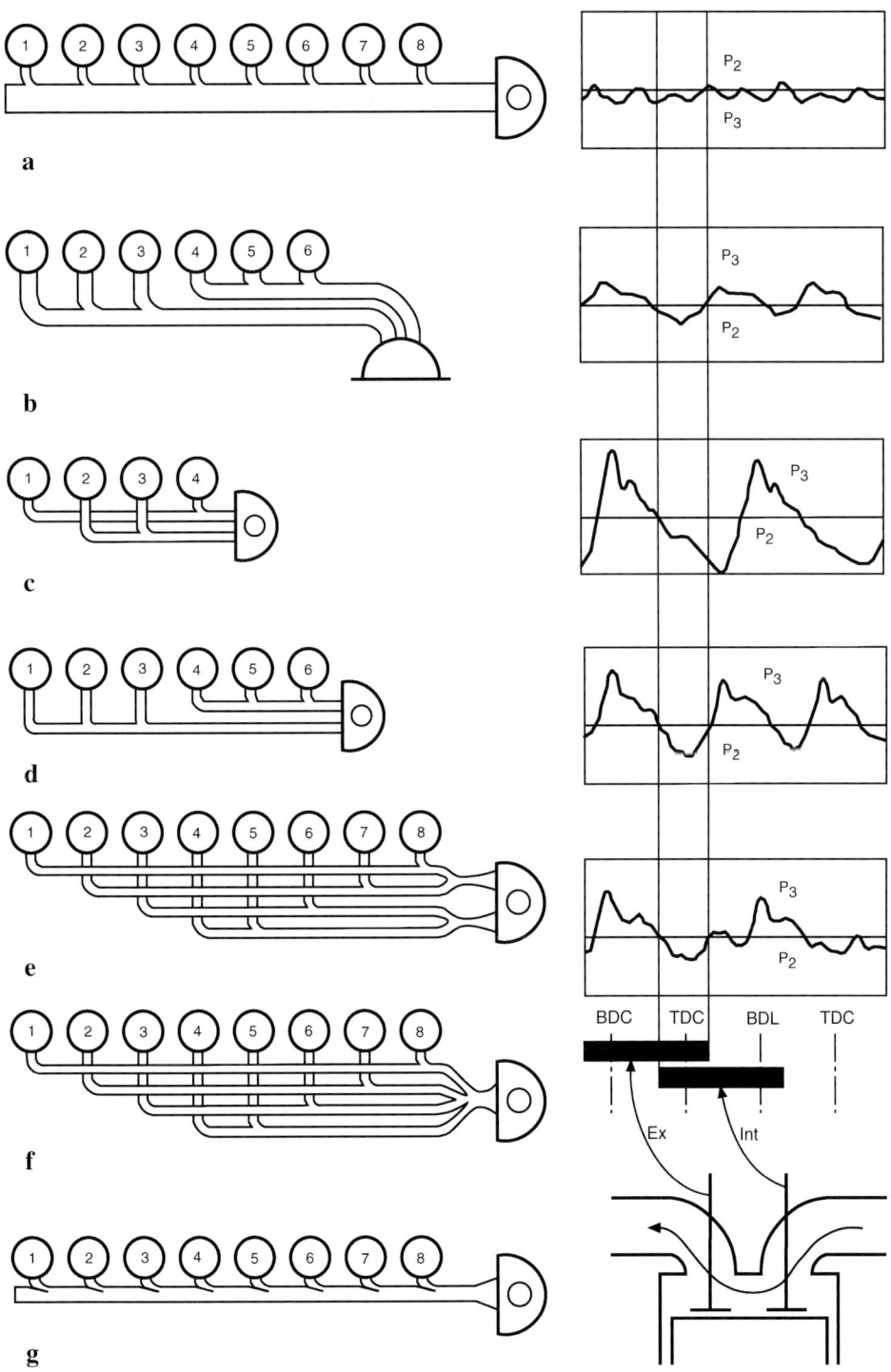

Fig. 5.26. Manifold and cylinder combinations, with their achievable charge pressure and exhaust pressure curves, for four-stroke inline engines: **a** constant-pressure, **b** Buechi 1925, **c** 2-combined pulse, **d** 3-combined pulse, **e** dual-flow pulse converter, **f** multipulse converter, **g** modular pulse converter [159]

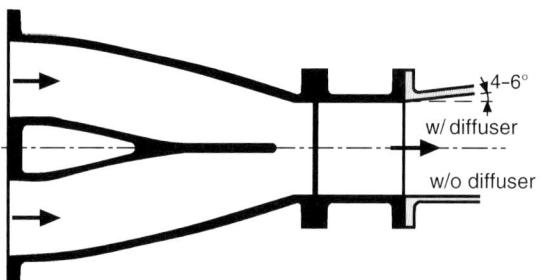

Fig. 5.27. Layout of a pulse converter with and without a subsequent diffuser

For efficiency reasons, in most cases this pressure recovery is not utilized, but the turbine is designed more as an action turbine. The improved efficiency results from a more uniform flow admission into the turbine. Figure 5.27 shows such a pulse converter, both with and without a diffuser for pressure recovery.

5.4.2 Turbine design and control

Design via nomograms

In the past, for the selection of the turbine rotor and its dimensions the following rough approximation via nomograms was used: The particular volume flow $\dot{V}$ is determined from the pressure–volume flow map of the engine to be supercharged (for new designs one can take the displacement and the desired speed range), for both the lowest and the highest full-load speed, possibly also for one or two intermediate engine speeds, and under estimation of the necessary charge pressure. These values are then used as entry into the nomograms shown in Fig. 5.28.

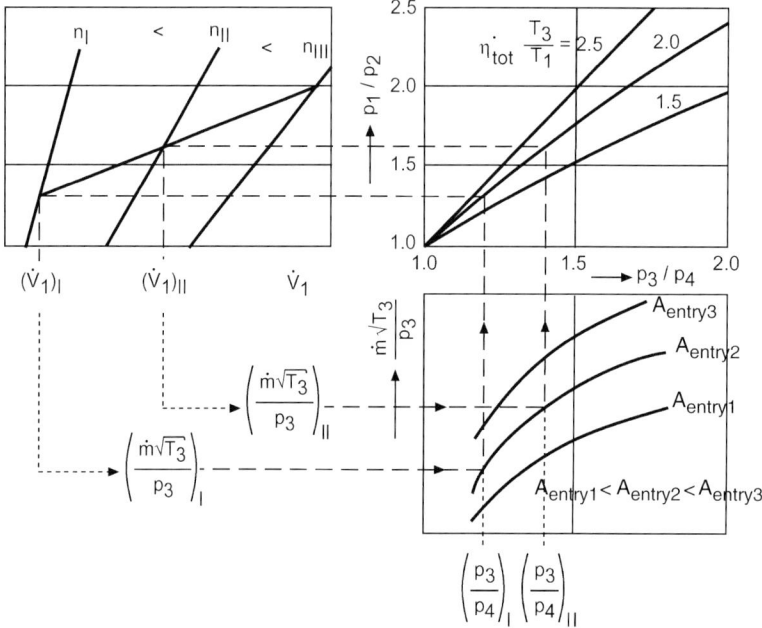

Fig. 5.28. Nomograms for turbocharger preselection [KKK, now 3K-Warner]

5.4 Matching of the turbocharger

The charge pressure conditions for the selected speeds obtained this way can now be correlated with the initial estimates and corrected. It is obvious that such a rough, iterative process can no longer satisfy today's demands for the layout of the supercharging system for a new engine. This is especially true if, e.g., in the case of pulse turbocharging, the exhaust manifolds have to be dimensioned and combined under consideration of gas dynamics. Accordingly, nowadays numeric simulation processes are utilized for the layout of supercharging systems.

Turbine selection via actual thermodynamic process simulation and maps
The physical basics and the corresponding mathematical models for such thermodynamic cycle simulation programs were presented in Sect. 3.6 in detail.

For an actual exhaust gas turbocharger layout, first the complete engine is modeled and then the engine air requirement at the turbine design point is determined from the desired air-to-fuel ratio (this will differ depending on the combustion process). In doing so, values for volumetric efficiency and specific fuel consumption are taken from similar engines and adjusted to the engine displacement. Additionally, a first estimation can be made for the necessary compressor pressure ratio (if need be, considering charge air cooler pressure losses).

With this value, a cycle simulation can be started in the design point, using estimated turbocharger efficiencies. Considering the necessary power equilibrium between compressor and turbine, the equivalent turbine area is adapted in iterative steps until the compressor reaches the desired pressure ratio. Once this is done for several operating points in the map of the engine to be adapted, the exhaust gas turbocharger dimensions can be scaled – again starting with actual charger data from a similar engine, and then entering geometric changes of both compressor and turbine dimensions (Fig. 5.29) – i.e., its main dimensions can be specified, which will already be very close to the actual design. For three specific cases taken from Fig. 5.29, Fig. 5.30 shows the resulting charger performance data for specific scaling coefficients.

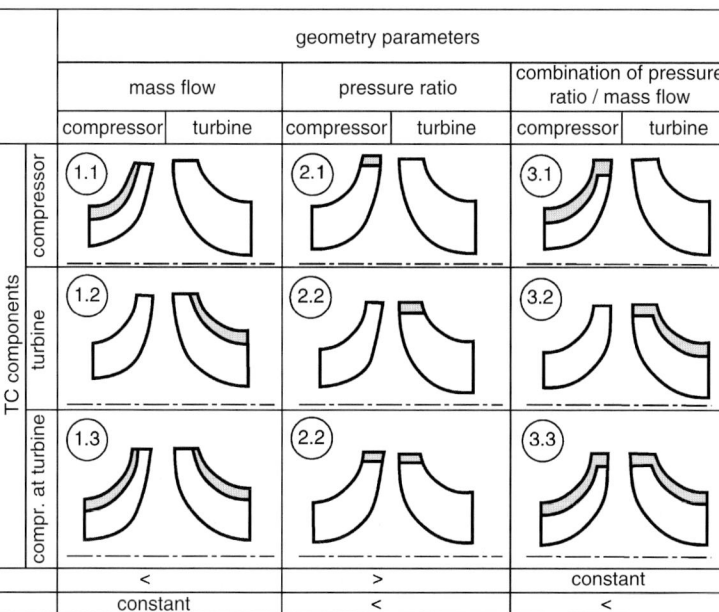

Fig. 5.29. Possible change of the charger dimensions while scaling [DC]

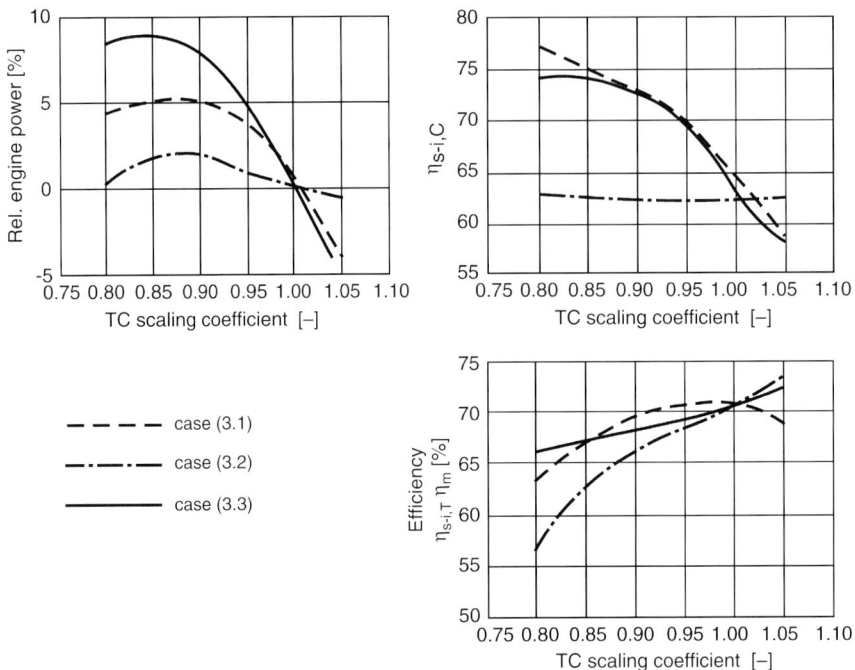

Fig. 5.30. Changes of engine and charger characteristics depending on scaling coefficient (cases according to Fig. 5.29)

Possibilities for matching the turbine

In today's exhaust gas turbochargers of high-volume series production, usually a radial turbine is applied. Therefore its free parameters are similar to those of a radial compressor. Figure 5.31 shows the most important major dimensions of a single-flow radial turbine.

For the first rough selection of the housing, the characteristic housing area A_{Volute} at the inlet to the volute is of great importance for the achievable turbine power.

Additionally, the so-called A/R ratio is used for characterization of the housing identification. A/R describes the ratio between the turbine entry cross-sectional area A (cm^2) at the transition from the turbine inlet area into the volute and the radius R (cm), which is defined as the distance from the center of the shaft to a theoretical mean flow path in the flow channel, by which the mass flow is halved (see Fig. 5.31 with A_S/R_S). A/R therefore is a measure for the flow capacity of the turbine housing. In divided turbine housings (e.g., twin-flow housing), A is the sum of both channel areas. A/R has to be considered together with the so-called trim of the turbine.

Trim denotes the tuning of the contour of a turbine rotor for a specified flow range. The trim and the A/R ratio together fully characterize the swallowing capacity of the turbine for a constant rotor diameter (Fig. 5.32). Numerically, trim is defined as

$$T = (d/D)^2 \cdot 100, \quad (5.34)$$

i.e., the ratio between the squares of two diameters, rotor outer diameter d and turbine rotor gas outlet diameter D.

The **compressor impeller-to-turbine rotor diameter ratio** D_C/D_T is another main parameter identifying the behavior of exhaust gas turbochargers. The turbine rotor diameter is chosen in such a way that for a specified compressor performance the turbine operates at best efficiency.

5.4 Matching of the turbocharger

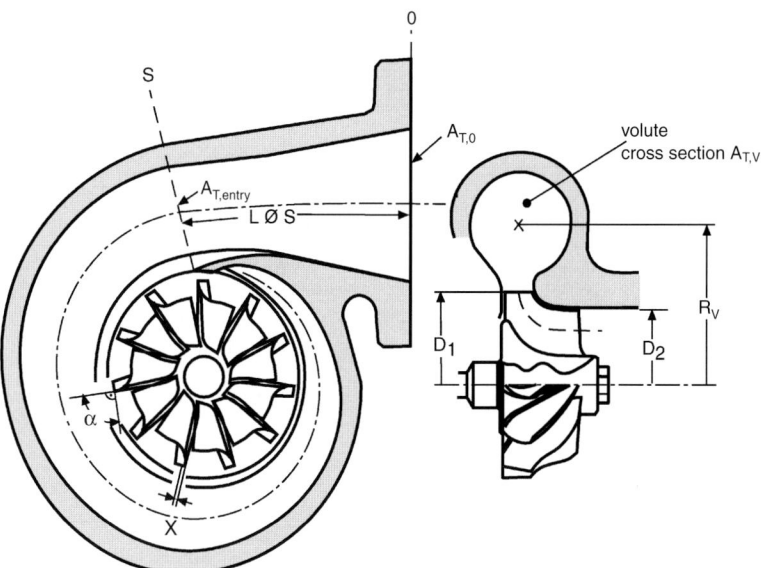

Fig. 5.31. Main dimensions of a single-flow radial turbine, with definitions of A and R. α, inlet angle; X, thickness of turbine blade

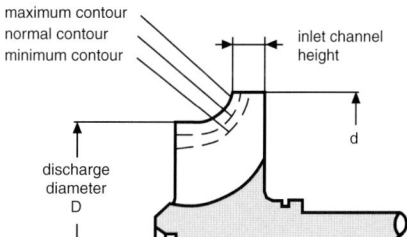

Fig. 5.32. Trim of a turbine rotor contour

This can be done by plotting the turbine efficiencies against the so-called turbine blade speed ratio, defined as the ratio between the circumferential rotor speed u and the theoretical gas expansion velocity c_0 which would be achieved if the exhaust gas were expanding without losses, in a nozzle, from the turbine inlet pressure p_3 to the static turbine outlet pressure $p_{4\text{stat}}$.

Figure 5.33 shows the efficiency behavior of a radial turbine in such a diagram, i.e., it shows the isentropic turbine efficiency depending on the blade speed ratio S: $S = u/c_0$. In this special case, c_0 describes a particle velocity and the letter u was chosen to be comparable with the internationally used notation of S.

To obtain performance equilibrium between compressor and turbine, the following relationship between the compressor-impeller and turbine-rotor diameters of a radial turbine applies:

$$\frac{D_C}{D_T} = \frac{1}{u/c_0}\sqrt{\frac{\eta_T}{2}m}, \tag{5.35}$$

where D_C is the compressor outlet diameter, D_T the turbine inlet diameter, u/c_0 the turbine blade speed ratio mentioned above, η_T the turbine efficiency ($\eta_{\text{s-i,T}}\eta_m$) and m is the slip factor of the compressor (0.8–0.9).

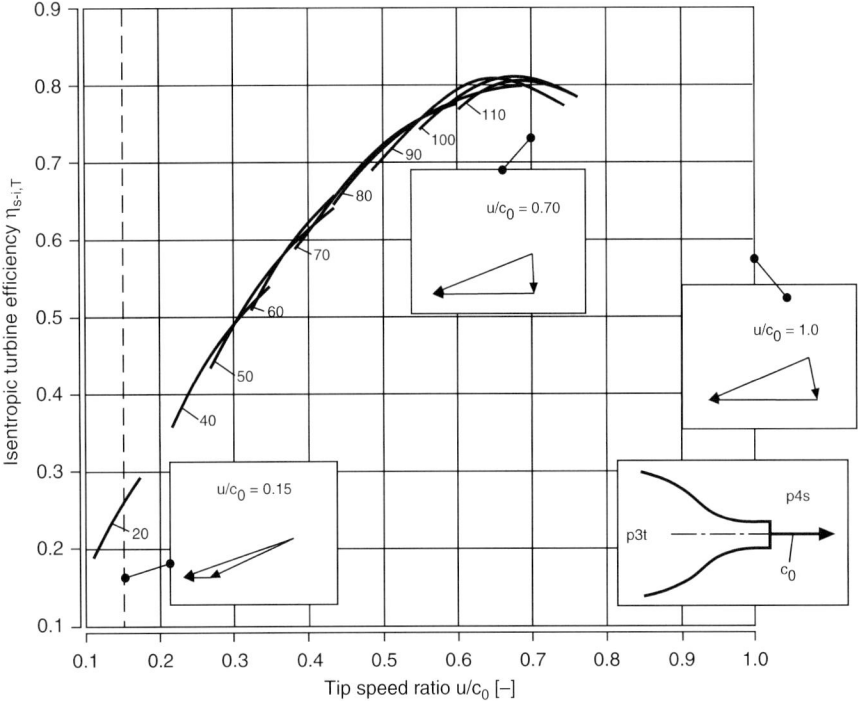

Fig. 5.33. Efficiency behavior of the radial turbine depending on the blade speed ratio u/c_0 [DC]

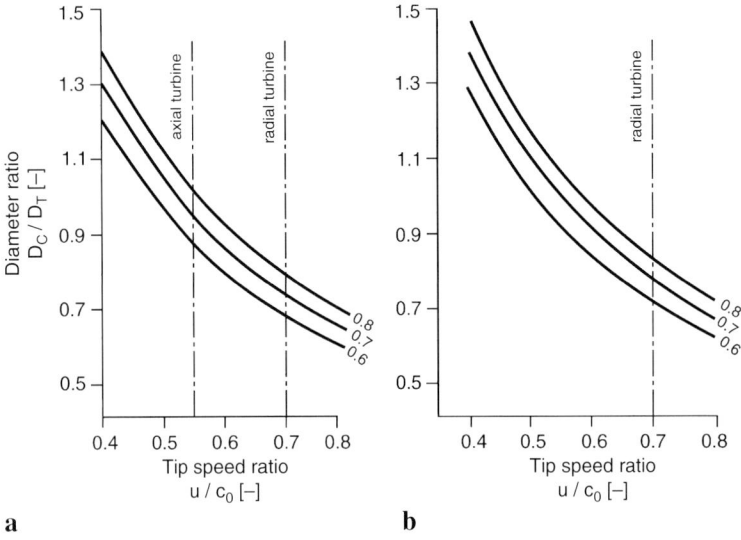

Fig. 5.34. Relationship between compressor and turbine rotor diameters for the slip factors 0.8 (**a**) and 0.9 (**b**), with characteristic operating ranges of axial and radial turbines

For slip factors of 0.8 and 0.9, Fig. 5.34 shows these relations in characteristic operating ranges of axial and radial turbines. Additionally, Table 5.1 lists some values of D_C/D_T for series production exhaust gas turbocharger combinations for truck engines, with and without waste gate and with VTG charger.

5.4 Matching of the turbocharger

Table 5.1. Relationship between compressor and turbine rotor sizes for various charger types

Truck engine and charger type	D_C/D_T
Diesel engine with fixed-geometry charger	1.15
Diesel engine with fixed-geometry charger and waste gate	1.07
Diesel engine with VTG charger	0.98

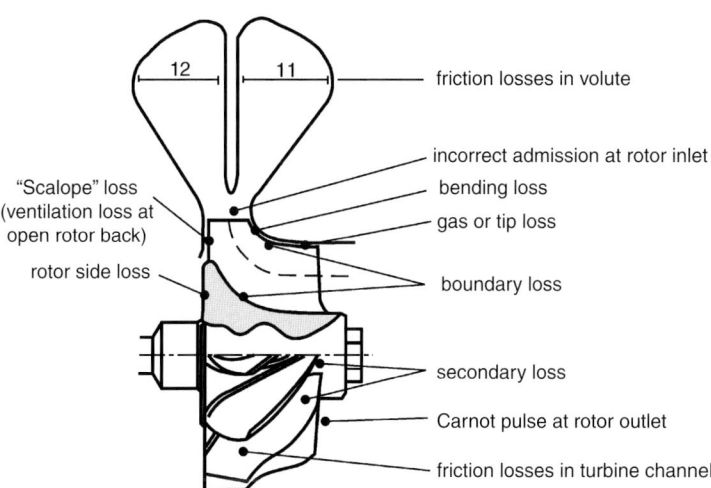

Fig. 5.35. Flow losses of a turbine

Figure 5.35 shows the **flow losses** occurring in a turbine. For smaller exhaust gas turbocharger turbines, the gap between rotor and housing represents a noticeable fraction of the total losses. For turbines operated in a wide operating range, the flow angles at the turbine inlet can significantly deviate from the optimum values, resulting in admission losses and flow separation losses in the turbine rotor (Fig. 5.36).

In smaller turbines, which mostly are equipped with waste gates and therefore have smaller turbine housing areas, friction losses also have a noticeable effect. Especially if the kinetic energy (present in the form of pressure waves) in the exhaust flow is of importance, for particular engine designs this can be achieved most effectively if a so-called twin-flow turbine is utilized. In such a turbine, the housing is separated into two symmetric inlet volutes, creating a flow division (Fig. 5.37a). In contrast, double-flow housings (Fig. 5.37b) are only used for special applications.

Layout and calculation of such a system, which has extremely effective flow dynamics, requires knowledge of the flow characteristics of **twin-flow housings** under transient operating conditions. This has to be obtained using sophisticated measurement techniques, and the recent emphasis on further improved efficiency of modern turbocharged engines has led to increasing interest in this type of turbochargers. Figure 5.38 shows a comparison between the flow characteristics of double-flow and twin-flow turbine housings for the case of nonsymmetric admission. Nowadays, better measurement techniques, as well as more precise simulation software tools are available

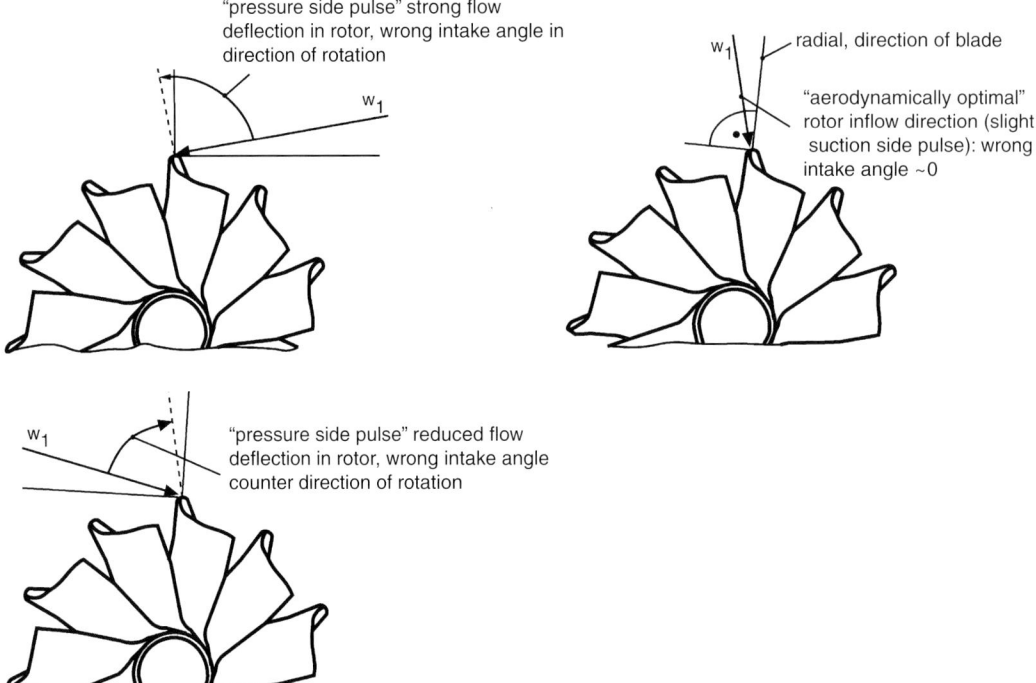

Fig. 5.36. Incorrect-admission and deflection losses in a turbine rotor

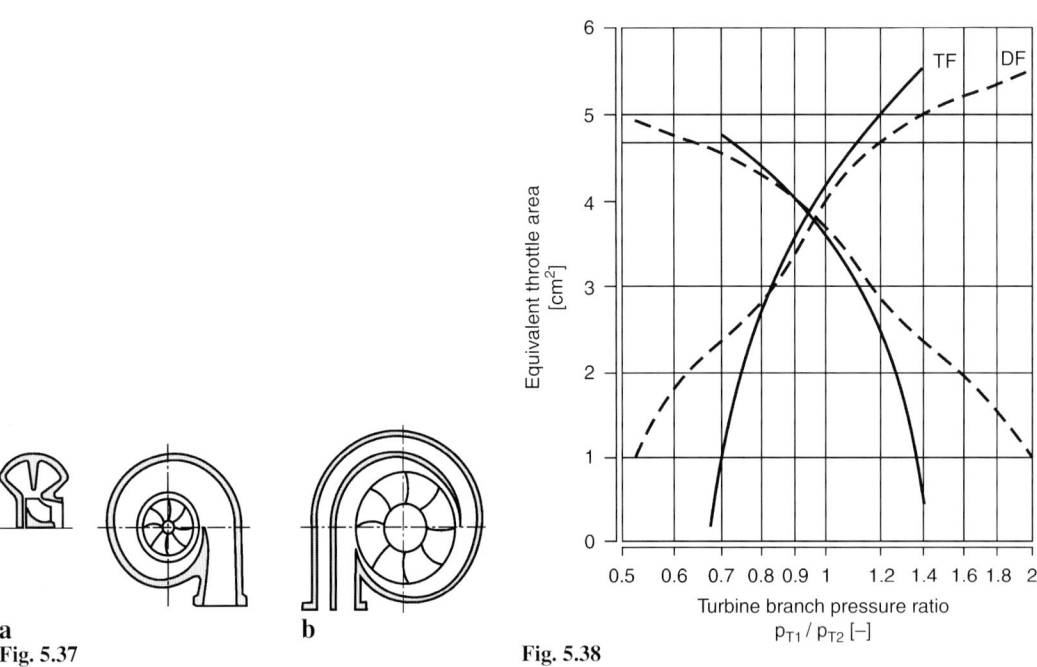

Fig. 5.37. **a** Twin-flow turbine housing, **b** double-flow housing

Fig. 5.38. Comparison of the flow characteristics of double-flow turbine housings (DF) and twin-flow housings (TF) at nonsymmetric admission [159]

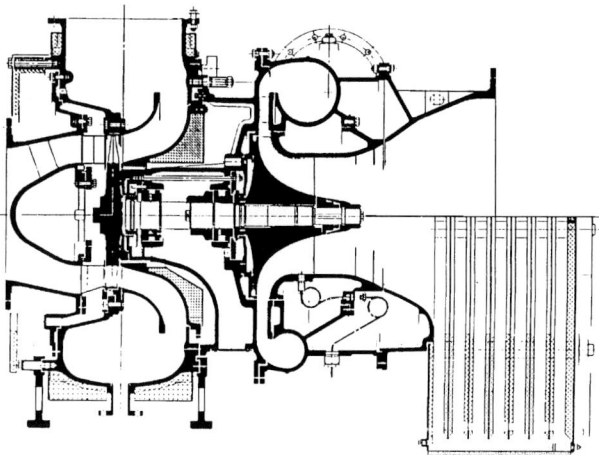

Fig. 5.39. Turbine with downstream diffuser [MAN]

for layout and tuning. This makes it possible to describe multiflow turbine housings, and their nonsymmetric admission in the case of pulse turbocharging, with sufficient accuracy, and to characterize them via actually measured values.

However, with constant-pressure turbocharging the critical high pressure ratios mentioned above cannot be reached in the turbine only. In this case the design target may be met by utilizing a diffuser downstream of the turbine which allows to increase the expansion ratio in the turbine by means of pressure recovery downstream of the turbine, which results in improved turbine efficiencies (Fig. 5.39).

In addition, **circumferentially sectional admissions** utilizing double-flow or even triple-flow volute housings nowadays are common especially for medium-speed engines. For slow-speed engines, for the most part axial turbines (diameter larger than 700 mm) are utilized as the charger drive, and sectional admissions are state of the art. Here too, the software programs mentioned are an indispensable part of the development tools.

5.4.3 Compressor design and control

Compressor selection

To enable the selection and match of a compressor, standardized compressor maps are available from the various compressor manufacturers. Figure 5.3 displays an example of such a map. In most cases, theoretical engine swallowing capacity functions for four-stroke engines are also shown in these maps, so that the compressor selection can be made on the basis of the following criteria:

– in the lower speed range of the engine, sufficient clearance to the surge limit,
– at high engine speeds, sufficient clearance to the maximum speed of the compressor, considering a reserve for operation at high altitude (Fig. 5.40).

A more precise compressor selection can be made via numeric simulations. Starting with a known compressor map and knowing the exact engine data, the ideal compressor size for a specific engine can be determined utilizing the scaling method described in Sect. 5.4.2, i.e., by changing the compressor dimensions in percentage increments. With these data, a suitable compressor can be

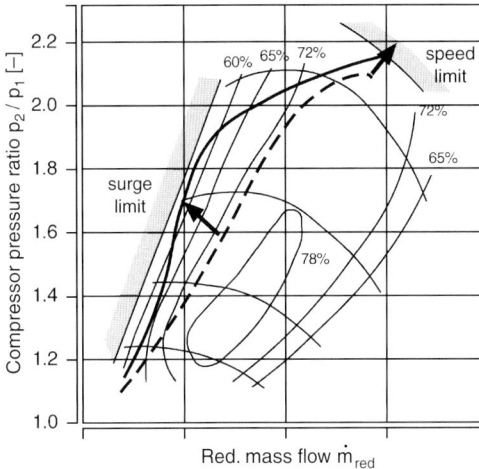

Fig. 5.40. Typical compressor map with full-load operating curves of a passenger car engine. Solid line, VTG charger; dash line, fixed-geometry charger with large turbine

selected for a specific application from basic compressor families of the different manufacturers and can be tuned by trimming.

Compressor control possibilities

In most cases, the use of a compressor without control features is sufficient for both stationary engine and automobile use. With increasingly higher charge pressure ratios and additionally increased speed ranges under load, the limits for compressors without controls have now essentially been reached. For a turbo compressor, then, basically the following possibilities for influencing the operating map exist: preswirl control, flow stabilizing measures, adjustable diffusers and adjustable compressor blades.

For **preswirl control**, the admission angle into the compressor impeller is varied with the help of – ideally, continuously adjustable – inlet guide blades, and thus the onflowing air is forced into a preswirl. Figure 5.41 shows such a device with adjustable guide blades.

Since all the inlet and outlet angles of the compressor can be designed free of pulses for only a particular flow rate, with its relevant speed and pressure ratio values, it is obvious that the admission conditions can be adjusted via a preswirl in or against the turning direction of the compressor impeller. This also reduces the danger of stalling, i.e., the surge limit is shifted, as clearly demonstrated in Fig. 5.42. This measure is especially effective at high pressure ratios.

In addition, the surge limit can be shifted "to the left" via a specially designed recirculation from "upstream of compressor impeller outlet" to the compressor inlet, which nowadays is termed a **flow-stabilizing measure**. Figure 5.43 shows such an arrangement which above all may also help to eliminate flow rate problems. Here, at low flow rates a recirculation occurs around the compressor, resulting in an effectively higher flow rate and thus improved oncoming flow to the blades of the impeller. At high flow rates, this bypass acts as an additional compressor flow area, resulting in a higher possible flow rate before reaching the choke limit.

The same applies for an **outlet diffuser** equipped with fixed or adjustable blades (Fig. 5.44). Here the volume flow of the compressor and its limits can be influenced in a wide range by the choice of the blade angle (Fig. 5.45). In general, the following rule applies: The steeper the outlet angle, the higher the flow rate through the compressor (flow area) and the smaller the pressure gain.

5.4 Matching of the turbocharger

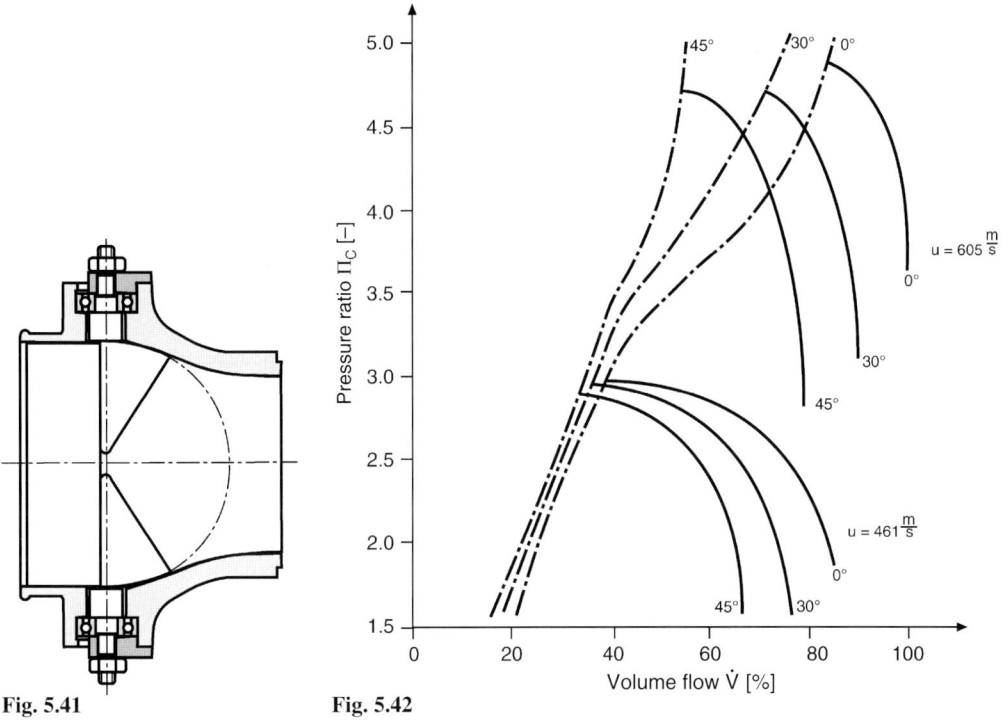

Fig. 5.41

Fig. 5.42

Fig. 5.41. Preswirl control via inlet swirl generator [KKK, now 3K-Warner]

Fig. 5.42. Shifting the surge limit via preswirl control (adjustment range of 0–45°) [KKK, now 3K-Warner]

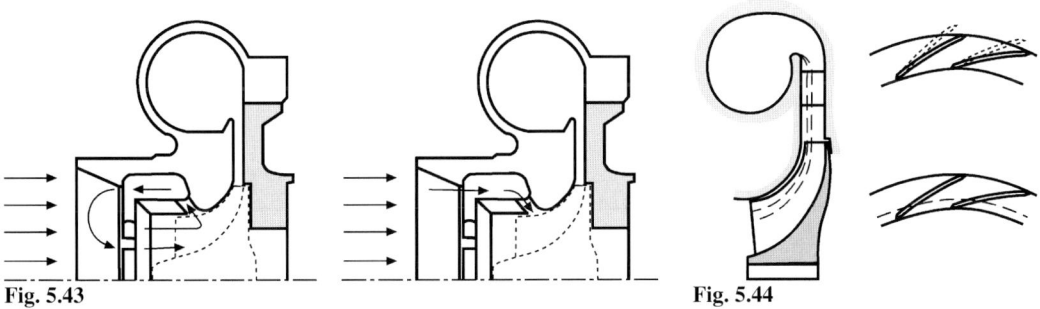

Fig. 5.43

Fig. 5.44

Fig. 5.43. Flow-stabilizing measure via recirculation around the compressor impeller [KKK, now 3K-Warner]

Fig. 5.44. Outlet diffuser with blades

Both preswirl control and diffuser blade pitch control – preferably in combination – are suited to noticeably expand the usable map of flow compressors. Thus, such compressors are suited for high degrees of supercharging in applications with a wide flow rate range, i.e., wide usable speed range of the engine.

Compressor blade pitch control is only possible for axial compressors; this is not utilized, however, even for the turbochargers of large combustion engines due to installed size and costs.

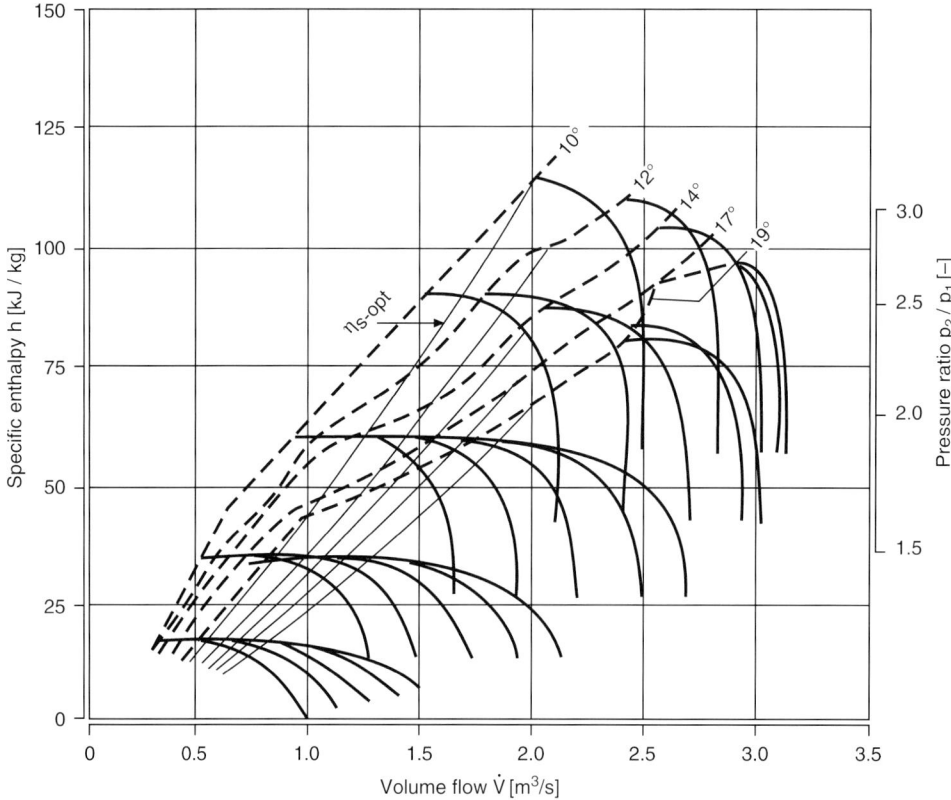

Fig. 5.45. Compressor volume flow control via specific outlet diffuser settings (adjustment range of 10–19°) [159]

5.5 Layout and optimization of the gas manifolds and the turbocharger components by means of cycle and CFD simulations

5.5.1 Layout criteria

For engines with exhaust gas turbocharging, the tasks of thermodynamic cycle simulations during the layout process can be subdivided into the following three areas (besides the engine itself, which was discussed earlier):

– intake system (manifolds, filter, charge air cooler, EGR inductor, muffler)
– exhaust system (manifolds, catalysts, particulate filter, muffler, EGR tubing)
– charge system (compressors, turbines, compound turbines, waste gate)

Besides influencing the engine itself, the manifolds of supercharged engines also influence the operating behavior of the compressor decisively. High pressure losses within the charge air system, upstream and downstream of the compressor, increase the pressure ratio necessary to achieve a desired charge pressure level. Figure 5.46a shows the consequences of such increased pressure losses and, thus, pressure ratios for the compressor operating conditions (pressure loss of case B greater than case A).

In the lower speed range, where automotive engines are operated at full-load close to the surge limit of the compressor, higher pressure losses result in a shift of the engine full-load operating curve

5.5 Layout and optimization

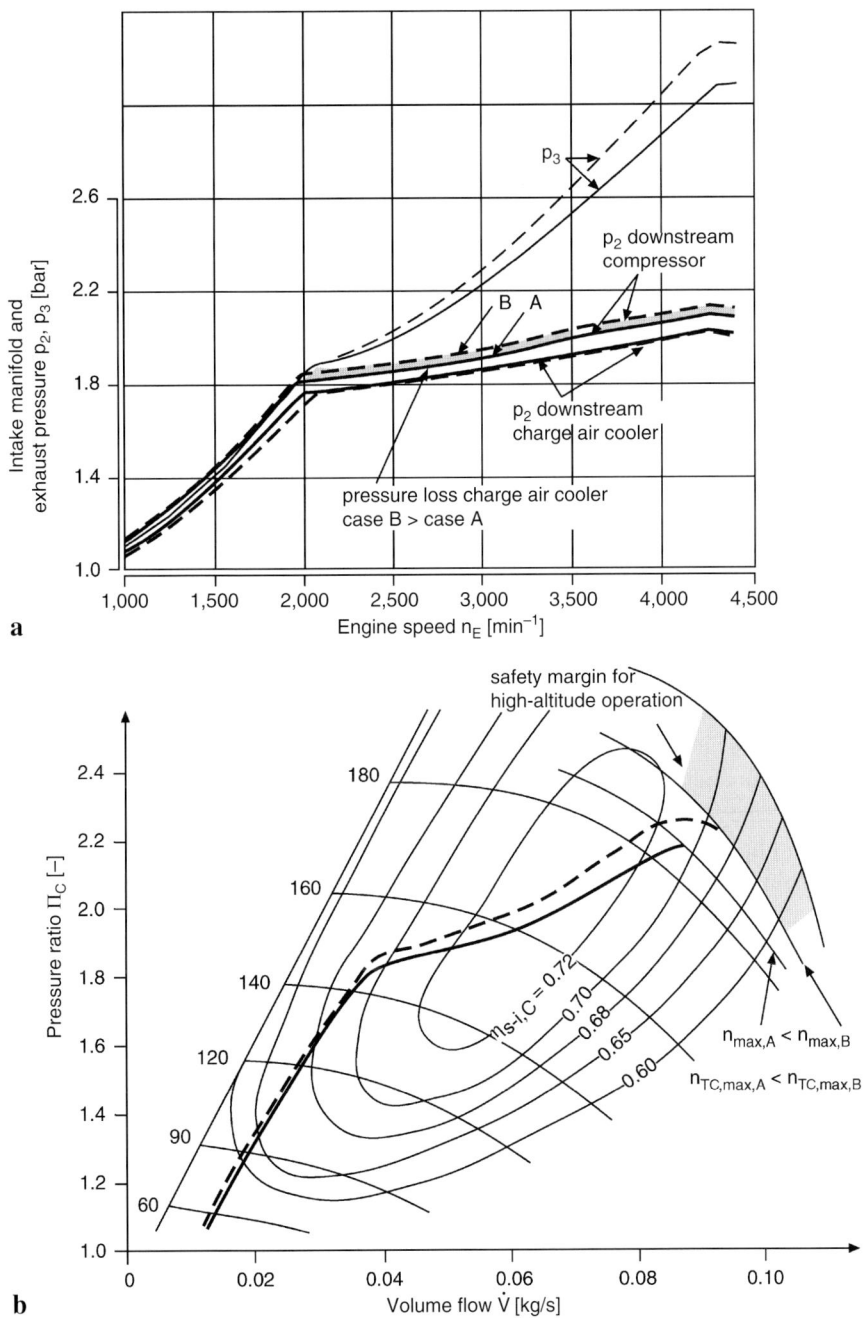

Fig. 5.46. Compressor map with engine full-load operating curves (**b**) for varying pressure losses in the intake system (**a**)

towards this limit. This can have a negative effect on the operating safety of the compressor. At high engine speeds close to the rated power, the operating range of the compressor is restricted by the choke limit. Additionally, the mechanical speed limit of the charger is approached. Furthermore, higher pressure losses may cause the maximum charger speed (e.g., at high-altitude operation) to be unacceptably approached or even exceeded and thus may limit the operating safety of the charger

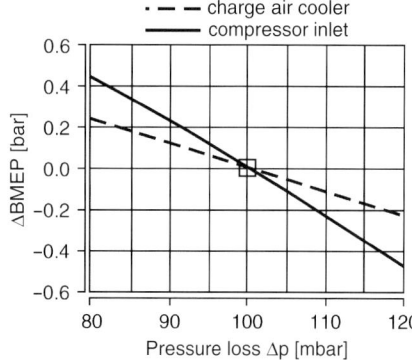

Fig. 5.47. Influence of pressure losses on the intake side on brake mean effective pressure

under these conditions. And finally, the higher compressor pressure ratios require increased drive power, which usually by itself requires higher turbine pressure ratios.

This results in increased exhaust work for the engine, reducing the mean effective pressure in the engine due to pressure losses both on the intake and on the exhaust side. This is shown in Fig. 5.47 for a turbocharged engine at rated power. To compensate for these losses, at a given air-to-fuel ratio for a specified combustion process, even more charge pressure is required. In this way, a closed feedback loop of all the negative influences of higher pressure losses on the intake side is formed, which amplifies its initial disadvantage.

For this reason, for a turbocharged engine as well, the aim is to have minimum pressure losses on the intake side, i.e., the charger cannot simply compensate for these losses. Besides minimizing pressure losses, it is also important to design the piping for optimum flow conditions, especially close to the compressor intake. At this point a speed profile as uniform as possible should be aimed at.

Additionally, unintentional or uncontrolled swirls at the compressor intake must be avoided, since they influence the velocity profiles and triangles at impeller blade entry and thus influence the compressor operating conditions or the corresponding efficiencies. This fact can be indirectly measured via the compressor outlet temperature. If it is significantly higher than that derived from intake temperature, map efficiency, and Eq. (2.14), the flow entry into the compressor has to be analyzed in detail.

For engines with charge air cooling, upstream of the compressor the charge air cooler has to be dimensioned (rating of the required cooling capacity under consideration of the temperature increase in the compressor and the simulated engine mass flow at rated power). Further, the effect of various cooler layouts – characterized, e.g., by their corresponding charge air cooler efficiencies – has to be analyzed.

On the exhaust side, special attention has to be paid to the optimum conversion of the exhaust gas energy in the turbine(s). That is, the pressure and wall heat losses between engine and turbine must be minimized. As an example, on the engine side the exhaust ports may be insulated with "port liners". The manifolds themselves are often designed double-walled, with air gap insulation. This keeps the exhaust gas temperatures upstream of the turbine at high levels, which are necessary for early catalyst light off and efficient working temperatures. Further, it minimizes the heat radiation into the engine compartment, which in most cases is narrow and poorly ventilated.

5.5 Layout and optimization

With regard to the geometric layout of the exhaust manifolds, several observations must be made. On the one hand, short manifold lengths and compact areas transfer the kinetic energy of the exhaust pulses optimally to the turbine. Depending on the engine layout and the firing order, this may result in a disadvantage for the gas exchange of the engine itself. Also, with increasing gas flows the narrow areas result in significantly increasing pressure losses. Correspondingly, for automotive engines a compromise has to be found between, on the one hand, sufficient charge pressure buildup and good response at lower speeds and under transient conditions and, on the other hand, acceptable specific fuel consumption values at high speeds.

Figures 5.48 and 5.49 show the results of such an analysis. In Fig. 5.48, the specific fuel consumption at rated power (stationary operation) is plotted in relation to the relative exhaust manifold diameter. In Fig. 5.49, the engine response is plotted for two different exhaust manifold diameters.

Similar to the intake side, downstream of the turbine minimum pressure losses have to be aimed at, since, at a given required turbine pressure ratio, otherwise the absolute exhaust gas backpressure increases, and with it exhaust pumping work and fuel consumption. The turbine itself must now be dimensioned in such a manner that sufficient power is generated in the complete engine operating range to drive the compressor such that the required degree of supercharging is achieved. In doing this, the most important layout criterion for automotive propulsion is the achievable charge pressure in the speed range below maximum torque. For chargers with fixed geometry, the design point must be fixed in this speed range. For VTG chargers, the turbine size

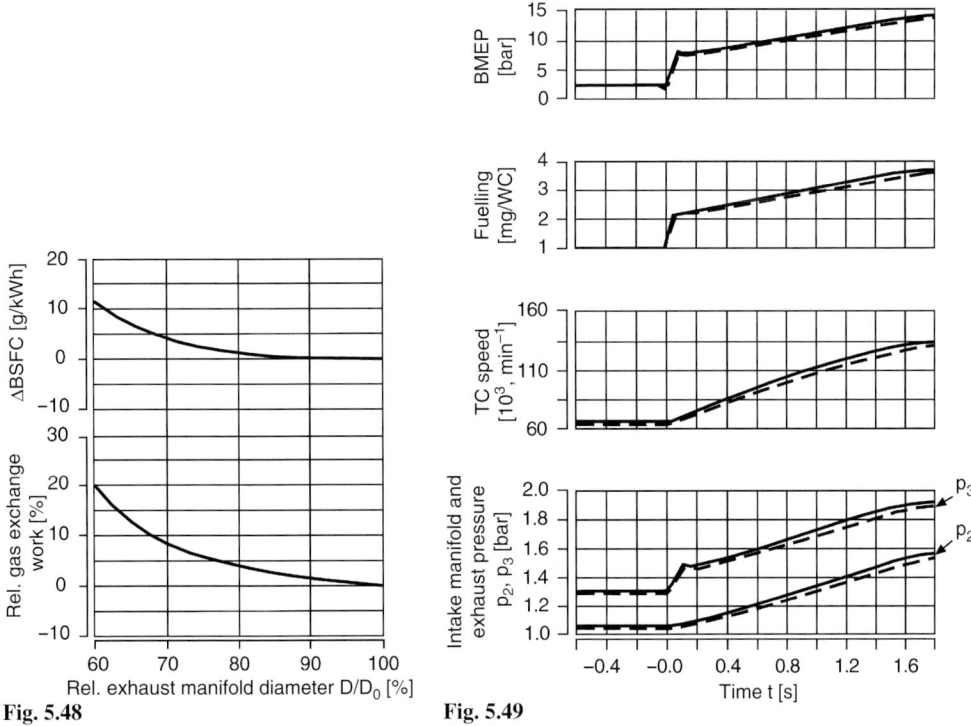

Fig. 5.48. Influence of the exhaust manifold diameter on stationary engine operation (full-load, rated speed)

Fig. 5.49. Influence of the exhaust manifold diameter (dash line, 100%; solid line, 70%) on transient engine response

has to be optimized with regard to its swallowing capacity and its turbine efficiency in this speed range.

On the other hand, for stationary applications mainly with operation near full-load (trucks, generator sets), specific fuel consumption and compliance with emission standards are the most important design criteria for the turbocharging system. Accordingly, in these cases the design has to aim at the best possible total system efficiencies under these load conditions. Thus, the turbine configuration has to be optimized in regard to efficiency, achievable charge pressure and thus combustion air-to-fuel ratio, and minimum turbine inlet pressure for best possible gas exchange work.

After engine and turbocharging components have been designed in respect to their thermodynamics, detailed engineering can start. Normally, during the actual realization of the

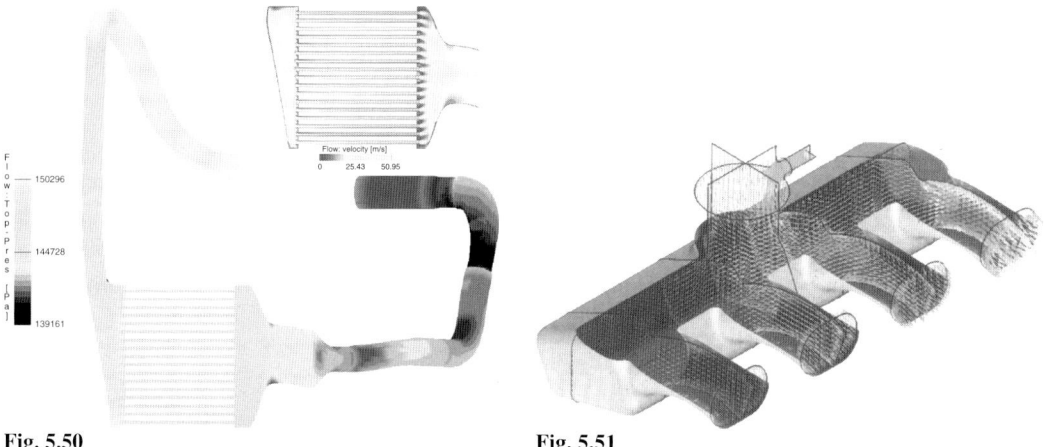

Fig. 5.50   **Fig. 5.51**

Fig. 5.50. CFD calculation results for a charge air manifold with charge air cooler

Fig. 5.51. CFD calculation results for an air plenum with EGR induction [130]

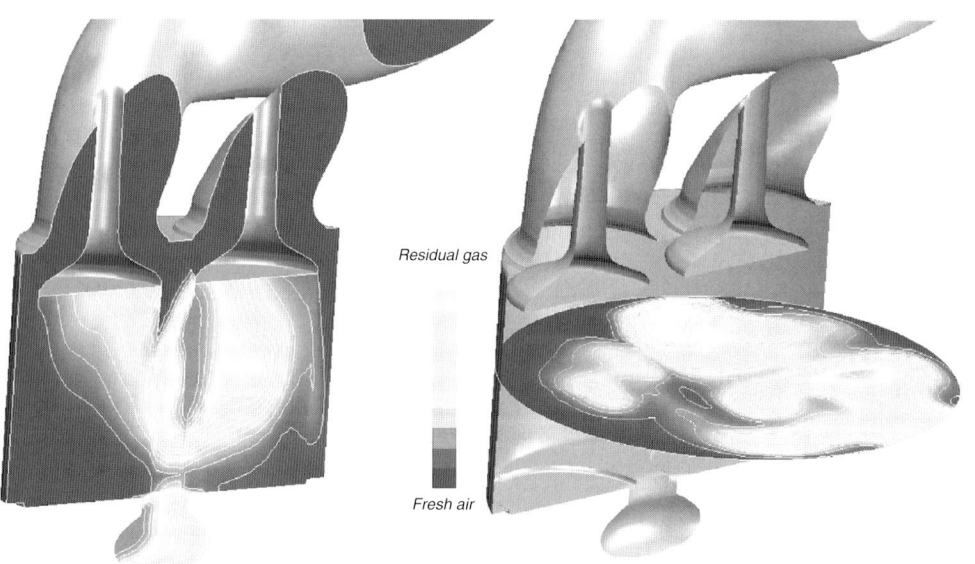

Fig. 5.52. CFD calculation results obtained by a simulation of the internal flow in the cylinder

5.5 Layout and optimization 97

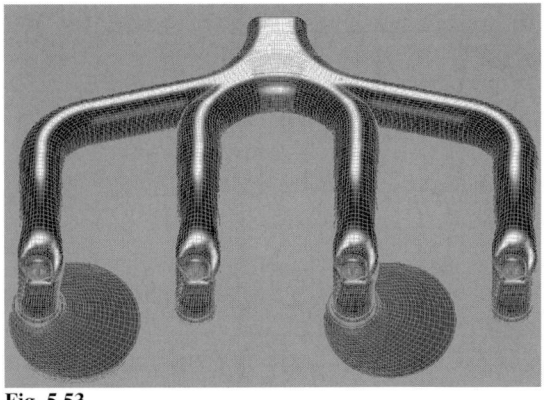 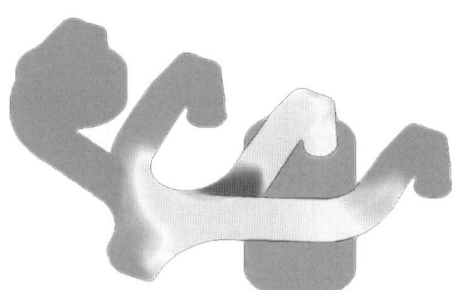

Fig. 5.53 **Fig. 5.54**

Fig. 5.53. CFD calculation grid of an exhaust manifold

Fig. 5.54. CFD calculation result of an exhaust manifold

design, detail optimizations become necessary that exceed the prediction capabilities of 1-D cycle simulations.

The 3-D CFD simulation which is then necessary covers all gas-containing components, e.g., the charge air manifold with the charge air cooler (Fig. 5.50), the intake air plenum (Fig. 5.51), the internal flow within the cylinder including variable parameters like pistons and valves (Fig. 5.52), and the exhaust manifold (Figs. 5.53 and 5.54).

5.5.2 Examples of numeric simulation of engines with exhaust gas turbocharging

4-cylinder, 4-valve gasoline engine with fixed-geometry turbocharger and waste gate (2.0 liter displacement)

The mathematical model shown in Fig. 5.55 covers the manifold between exhaust valves and turbocharger with pipe elements. The actual design of the fixed-geometry turbine of the turbocharger includes an integrated waste gate (turbine bypass for charge pressure control) in such a way that the gas paths are very similar for the partial mass flows through the turbine and through the waste gate.

In the lower speed range as well as under part-load operation, whenever the waste gate is closed, the turbine on the exhaust side can be described completely by its turbine map.

However, in operating points in which the waste gate is opened, the swallowing capacity of the turbine seems to be increased, since a part of the mass flow is channeled through the waste gate. At the same time, the virtual turbine efficiency – which is related to the complete mass flow – is reduced, since only a part of the gas flow is utilized to perform work in the turbine.

Therefore, in the mathematical model a waste gate element must be placed parallel to the turbine and between exhaust manifold and downstream exhaust pipe. As in the real component, via a differential pressure sensor, a spring, and the anticipated attenuation characteristic, this simulates the position of the control valve and thus its flow capacity (Fig. 5.56).

Experience proves that this approach is associated with significant additional measurement complexity (spring and attenuation characteristics, flow coefficient of the waste gate valve), and

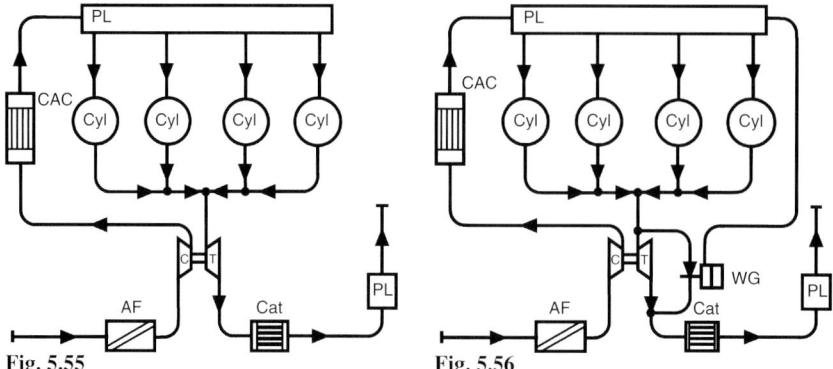

Fig. 5.55. Simulation model for a 4-cylinder, 4-valve gasoline engine with fixed-geometry turbocharger and waste gate (integrated into the turbine model). AF, air filter; C, compressor; T, turbine; CAC, charge air cooler; Cyl, cylinder; PL, plenum; Cat, catalyst

Fig. 5.56. Engine model with explicit modeling of the waste gate (WG)

the calibration of the mathematical model is very time-consuming. It is more practical to utilize the compressor's pressure ratios, either known (from measurements) or as desired, as control variables for the turbine waste gate mass flow. With these values, and the compressor mass flow actually calculated within the process simulation, the required turbine power and, thus, also the waste gate flow rate are known. In this way, even without exactly knowing the position of the control valve, an adequate accuracy of the simulation can be achieved. However, if there is a need to get specific data relating to the operating behavior of the valve (dynamics, impact due to exhaust pulsations, etc.), an exact modeling is mandatory.

On the fresh air intake side, gasoline engines normally need a compressor bypass valve as well – also called air circulation valve. It is not utilized for charge pressure control, however, but it is opened during negative load variations, in order to allow appreciable compressor mass flows when the engine's throttle is closed. It has to be considered that at a sudden closing of the throttle, due to its inertia the charger decelerates only slowly, while the compressor mass flow rapidly decreases corresponding to the new throttle position. This would cause the compressor to stall, possibly causing bearing damage. A sufficient mass flow can be assured with the air circulation valve, so that the operating points in the compressor map stay in the stable range even during negative load variations.

6-cylinder, 4-valve DI diesel engine with VTG (2.5 liter displacement)

The increasing requirements put on supercharging systems, i.e., the achievement of high charge pressures in a wide engine speed range, have led to the development of exhaust gas turbines with variable turbine inlet geometry, i.e., with variable swallowing capacity. Figure 5.57 shows the numeric model of a direct-injection diesel engine with such a turbocharger.

In the simulation model itself, this variability is represented by a set of turbine maps, each characterizing a distinct geometric position of the turbine. Typically, the maps are measured in five steps, from "fully closed" via "1/4, 1/2, and 3/4 opened" to "fully opened" (Fig. 5.58). In order to be able to calculate the required characteristic data (swallowing capacity and efficiency) from the maps, during cycle simulation additionally the actual turbine inlet guide blade position has to be defined.

5.5 Layout and optimization

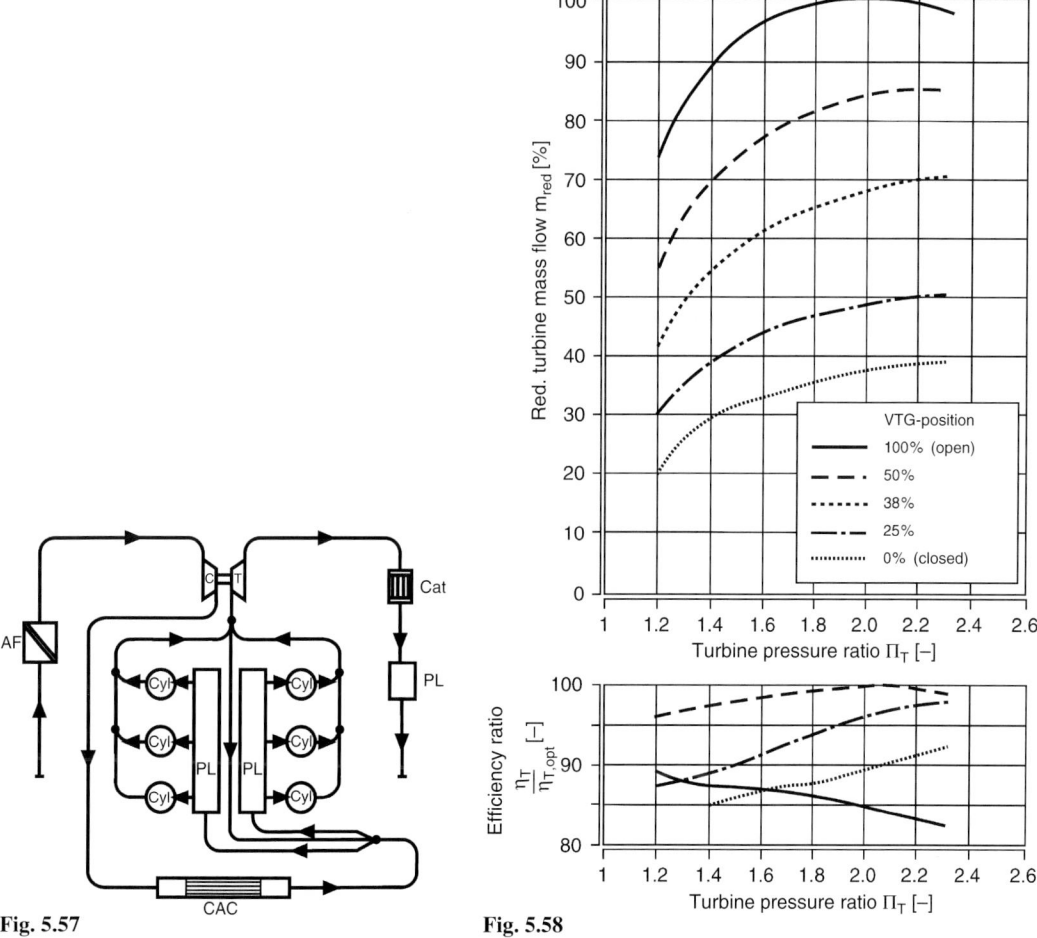

Fig. 5.57

Fig. 5.58

Fig. 5.57. DI diesel engine model with VTG

Fig. 5.58. VTG turbine maps for various opening positions

Referring to the simulation model shown in Fig. 5.57, it has to be noted that the pipe elements connecting the exhaust gas and fresh air sides represent the exhaust gas recirculation manifold. Since exhaust gas recirculation only occurs in emissions-relevant map areas, in passenger car engines this component has practically no influence on the full-load performance of the turbocharger.

6-cylinder, 4-valve DI diesel engine with twin-flow turbine (12 liter displacement)

Unlike the passenger car, under actual driving conditions the truck engine is very often operated close to full load. Thus, emissions-reducing measures as well, especially exhaust gas recirculation, must be correctly simulated under full-load conditions. Figure 5.59 shows the simulation model of a 12 liter truck engine with EGR.

Depending on the engine layout and the firing order, the pulse energy of the blow down pulses of the individual cylinders can be better utilized in the turbine if the exhaust manifolds are combined accordingly – in the case of the 6-cylinder engine, one manifold for cylinders 1, 2, and 3, as well as a separate manifold for cylinders 4, 5, and 6. The two flows in these manifolds influence each other

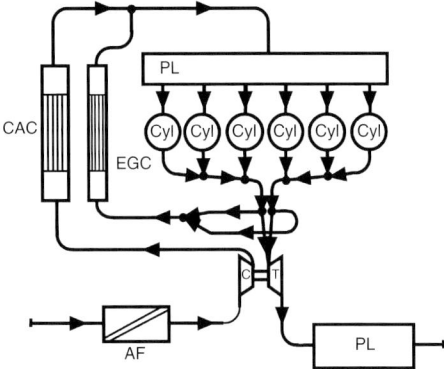

Fig. 5.59. Simulation model of a 12 liter truck engine with twin-flow turbine [46]. EGC, exhaust gas cooler

when they meet at the impeller inlet. In order to correctly reflect this behavior, the turbine map must be described in a way that simulates a partial connection. This can be done by considering the efficiencies and flow resistances as a function of the individual manifold mass flow conditions (see Sect. 5.4.2). Figure 5.60 shows efficiency and flow maps for a twin-flow turbine.

A further special feature of the numeric model shown is the exhaust gas recirculation unit. It consists of the connecting line between exhaust gas manifold and intake system, the EGR cooler, the control valve, and the EGR inductor. This inductor can be designed, e.g., as a Venturi injector, enabling EGR into the intake system even at positive static pressure differences between the intake and exhaust manifolds (Fig. 5.61). The geometrical optimization of such an injector is a typical

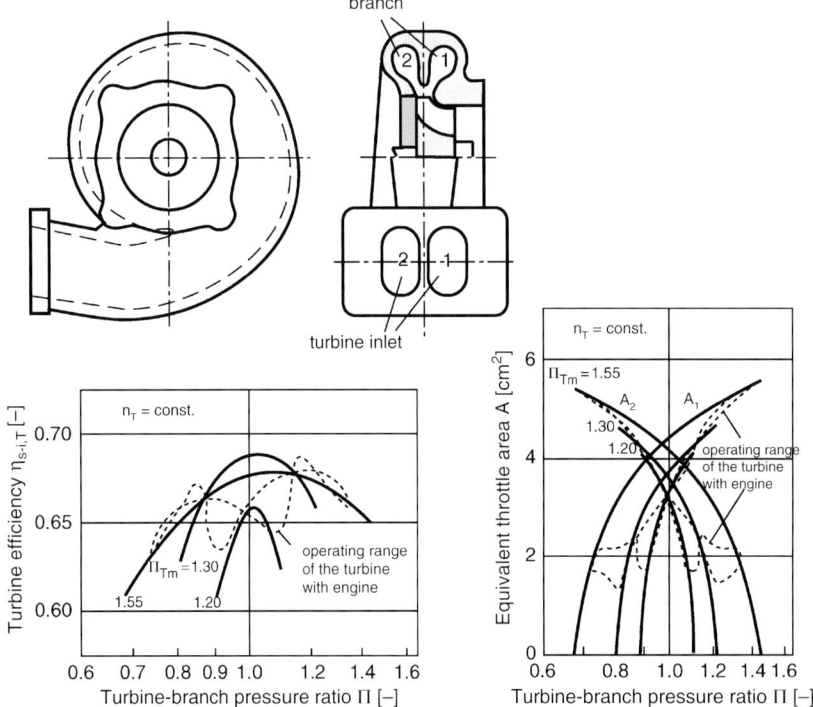

Fig. 5.60. Efficiency and flow maps for a twin-flow turbine

5.5 Layout and optimization

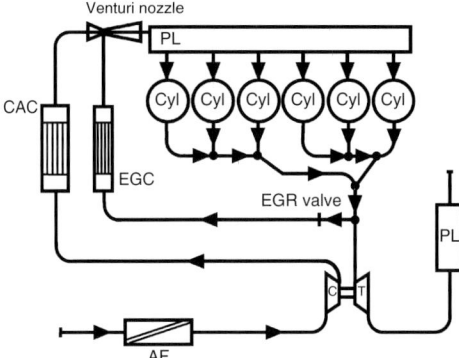

Fig. 5.61. Design of an EGR inductor in the form of a Venturi injector

task for CFD simulation, since 1-D simulation cannot resolve the details of the processes in the Venturi pipe and the interaction with the exhaust flow.

5.5.3 Verification of the simulation

The verification of the simulation models can only be performed via a comparison of engine test bench measurements with the corresponding simulation results. The standard measurements recorded on engine test benches can serve for this task: engine speed, air volume flow, medium temperatures per cycle, medium static pressures, fuel consumption, blowby mass flow, and emission data.

Especially for supercharged engines, the following additional measurement data should be collected for verification purposes:
cylinder pressure curves (high-pressure indications),
indicated pressure curves (against crank angle) in the intake and exhaust systems (low-pressure indications),
ignition timing and injection nozzle needle stroke,
turbocharger speed.

Figure 5.62 shows a compilation of the measured and the simulated steady state full-load operating data of the passenger car DI diesel engine with variable turbine geometry discussed in Sect. 5.5.2.

In general, well executed simulation models can reproduce actual engine operating data in the complete map with a maximum deviation of 2%. It is important to check the model calibration in the complete map, or at least in the total speed range at full load. Figure 5.62 shows a comparison between measured and simulated full-load operating data of this 6-cylinder, 2.5 liter DI diesel engine with VTG.

The correct measurement of the gasdynamic processes in the intake and exhaust systems of such an engine can now be made by means of the pressure indications mentioned above. The gas exchange is of special interest, since it decisively influences – via the gas exchange work (integral of $p \, dV$ during the gas exchange) – the quality and quantity of the cylinder charge and of the subsequent high-pressure process. Figure 5.63 shows comparisons between such measured and simulated pressure curves for the engine just mentioned.

As a basis for the optimization of transient processes, and to analyze the influence of individual parameters during load changes, it is additionally necessary to validate the transient behavior of a

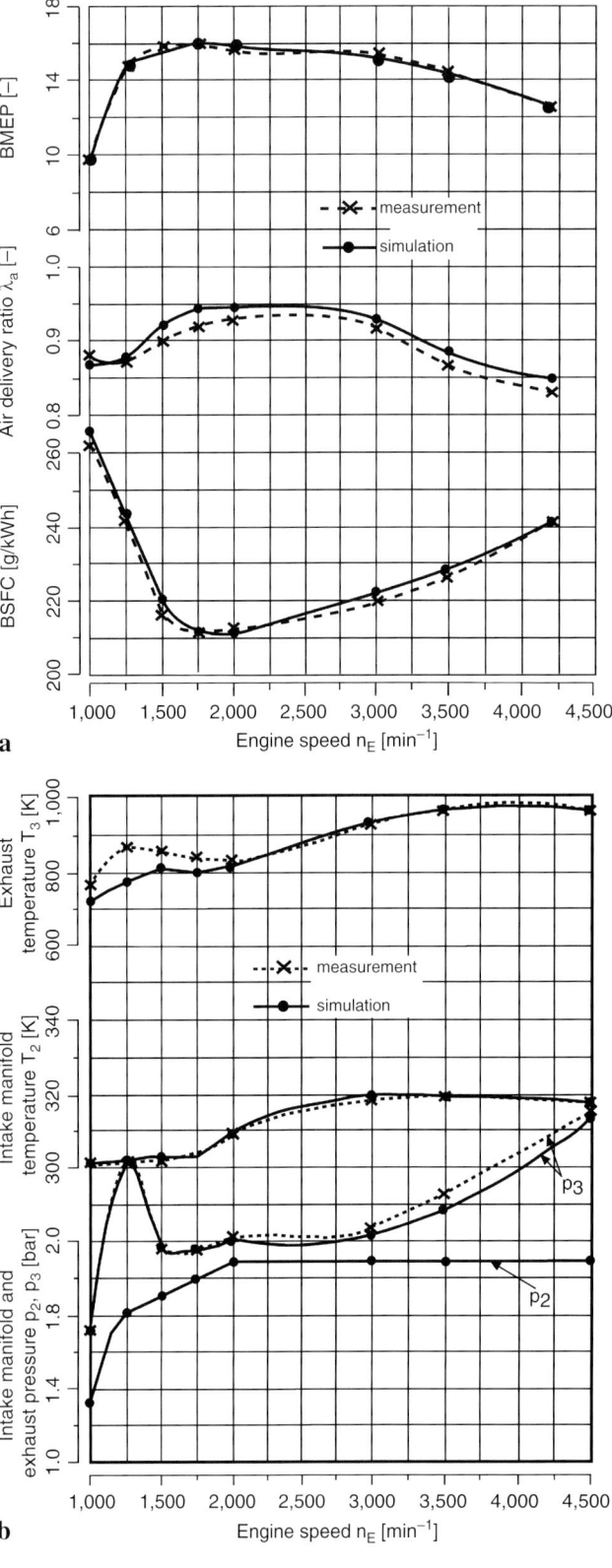

Fig. 5.62. Comparison between measured and simulated full-load data of a 6-cylinder, 2.5 liter DI diesel engine with VTG: **a** bmep, λ_a, and bsfc; **b** p_2, p_3, T_2, and T_3

5.5 Layout and optimization

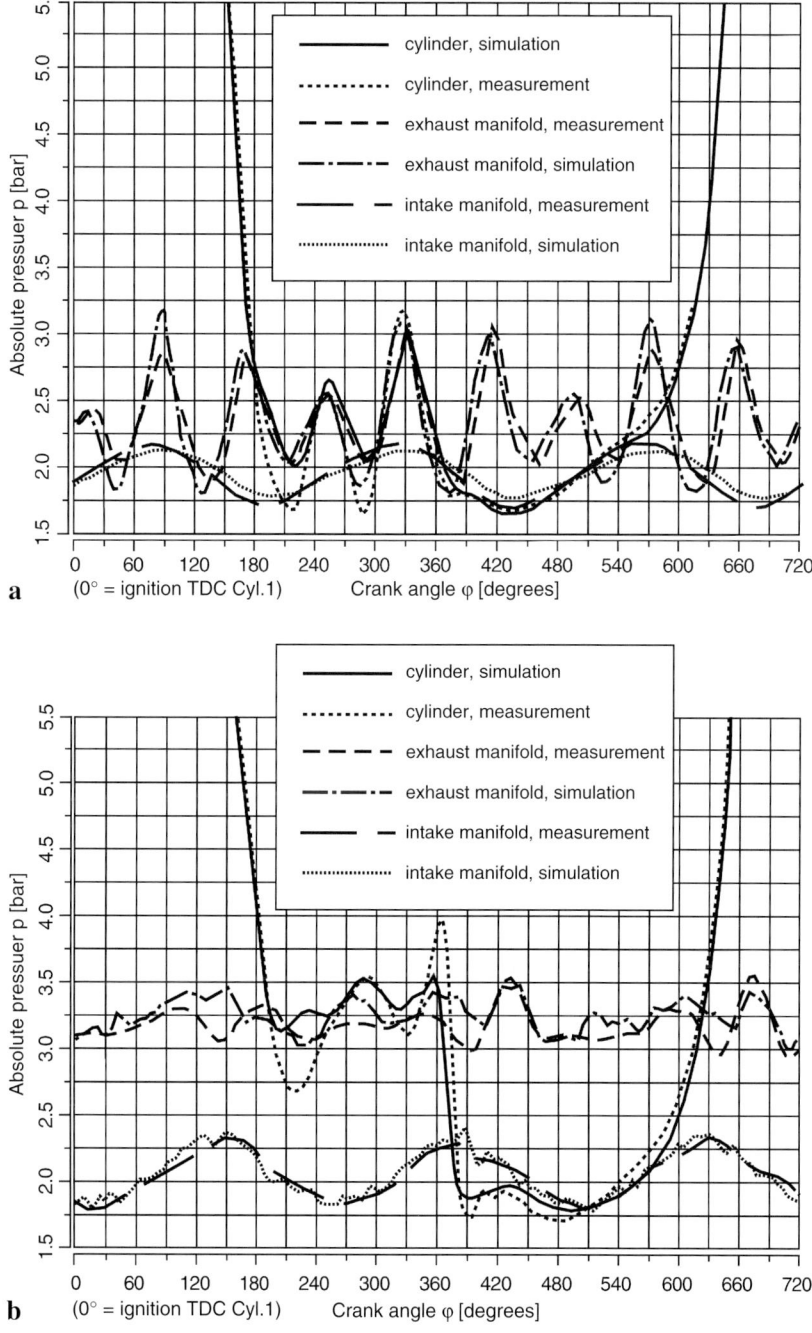

Fig. 5.63. Comparison between measured and simulated pressure data of a 6-cylinder, 2.5 liter DI diesel engine with VTG at full load and 2,000 min^{-1} (**a**) or 4,200 min^{-1} (**b**) engine speed

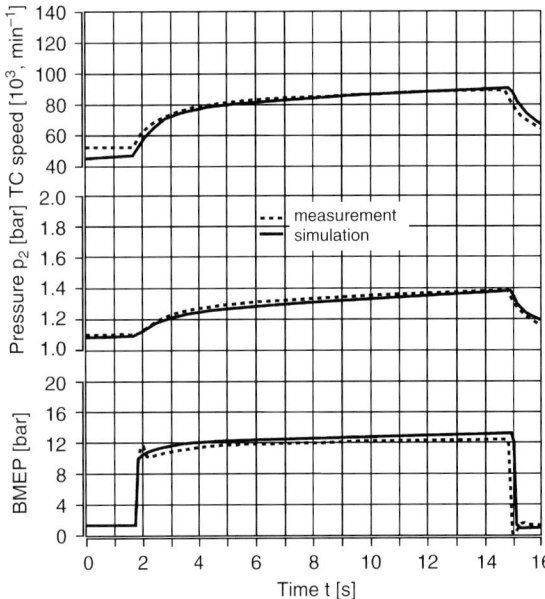

Fig. 5.64. Comparison between measured and simulated engine operating data of a 6-cylinder, 2.5 liter DI diesel engine with VTG during a load change at $1{,}500\,\text{min}^{-1}$

simulation model by means of measurement data. To make this possible, all mechanical inertias of the turbocharger's rotating assembly as well as the gasdynamic and thermal inertia effects have to be modeled correctly. Especially the thermal behavior of the exhaust gas manifolds between cylinder and turbine influences the transient response of the exhaust gas turbine.

For the same passenger car DI diesel engine, Fig. 5.64 shows a comparison between the measured operating data during a load change and corresponding simulation data. The simulations must be able to follow the changes in the combustion characteristics during the load change.

The correct acceleration characteristic of the turbocharger, and thus the response of the engine, can only be simulated with correctly modeled thermal inertia effects of the exhaust manifold. When modeling these effects, material and layout of the manifolds have to be considered (cast, steel pipe, double-wall air gap-insulated pipes, conditions for convection at the pipe surfaces).

6 Special processes with use of exhaust gas turbocharging

6.1 Two-stage turbocharging

Nowadays, most compressor impellers are made of aluminum. Their endurance strength allows circumferential speeds of about 520 m/s and thus pressure ratios of about 4.5. In exhaust gas turbochargers for slow-speed engines, pressure ratios of more than 5 are obtained through the use of titanium impellers, which allow even higher circumferential speeds. If the objective is to achieve even higher pressure ratios, and with them engine mean effective pressure values of – or even higher than – 30 bar, at least for continuous operation, multi-stage supercharging becomes necessary.

The term two-stage turbocharging describes a layout where turbochargers are connected in series, with charge air cooling between the chargers. On the other hand, a layout with two compressor and turbine stages each on a single shaft is termed a two-stage charger group [160]. Due to cost reasons, such two-stage charger groups today are no longer relevant. Nevertheless, such layouts were at one time utilized in medium-speed diesel engines. Figure 6.1 shows a compact two-stage charger group by MAN, Fig. 6.2 shows one by Hispano-Suiza.

In general, multi-stage – today mostly done as two-stage – turbocharging has the following **advantages** compared to single-stage turbocharging:
– a significantly higher boost pressure level, enabling the achievement of very high mean effective pressure values;
– an improved charging efficiency, even at unchanged charge pressure, since the efficiencies of compressor and turbine decrease with an increasing pressure ratio in a single stage. Additionally, the total efficiency can be further increased with an intercooler;

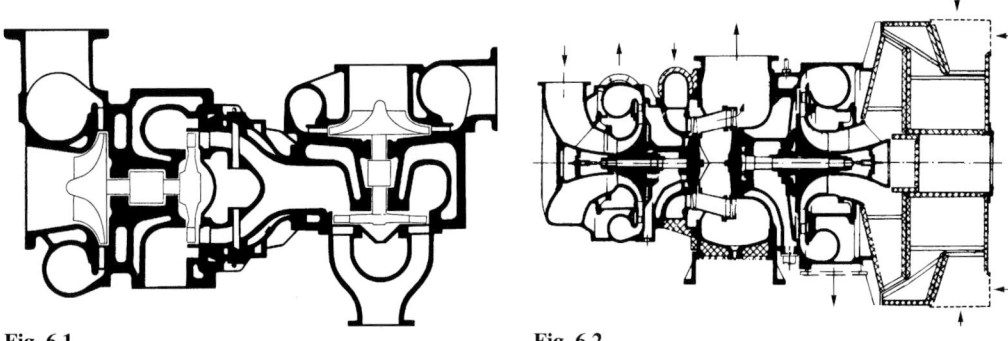

Fig. 6.1 **Fig. 6.2**

Fig. 6.1. Two-stage charger group by Hispano-Suiza

Fig. 6.2. Two-stage charger group by MAN

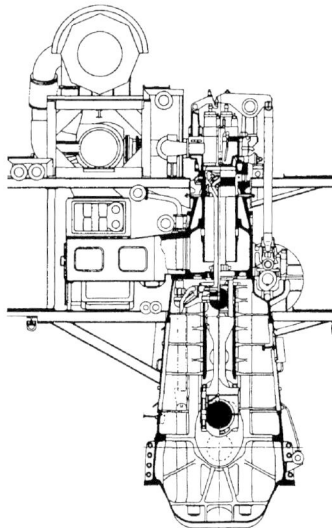

Fig. 6.3. Slow-speed two-stroke engine with two-stage turbocharging [Mitsubishi]

- wider compressor and turbine maps, and thus improved possibilities to adapt them to the desired engine operating range.

These advantages are opposed by some significant **disadvantages**:
- a much worse acceleration and load response behavior, since two rotors of each turbocharger have to be accelerated with the same exhaust gas energy;
- need for larger installation space, significant weight increase and with it higher cost;
- increased thermal inertia of the exhaust system, associated with a worse situation for exhaust gas aftertreatment (catalyst light-off behavior).

Two-stage turbocharging is more suited for slow-speed two-stroke engines than for four-stroke engines, i.e., already at a lower gain in mean effective pressure, due to the following reasons:

Weight, installed space and cost of the second charger group, including ancillaries, are less relevant in case of a very expensive large engine.

Since the swallowing characteristic of a two-stroke engine is comparable to that of a nozzle, its part-load behavior causes fewer problems.

Due to the need for an adequate scavenging gradient, the exhaust gas turbocharger efficiency strongly influences the achievable power.

Fuel consumption, decisively important in slow-speed engines, is reduced with increasing turbocharger efficiency, to a greater extend than in four-stroke engines.

For this reason, such engines are in production today. Figure 6.3 shows such a slow-speed engine by Mitsubishi.

6.2 Controlled two-stage turbocharging

The disadvantages in the response behavior of multi-stage turbocharging systems, as mentioned above, are not only avoided, but even turned into an advantage by a two-stage charger layout first introduced by KKK (now 3K-Warner). Figure 6.4 shows the principal layout. It utilizes a small high-pressure charger, whose turbine can be bypassed at high exhaust gas flows, via a streamlined and low-loss waste gate. The **waste gate flow** is also routed to the downstream low-pressure turbine of the low-pressure charger, i.e., it is utilized.

6.2 Controlled two-stage turbocharging

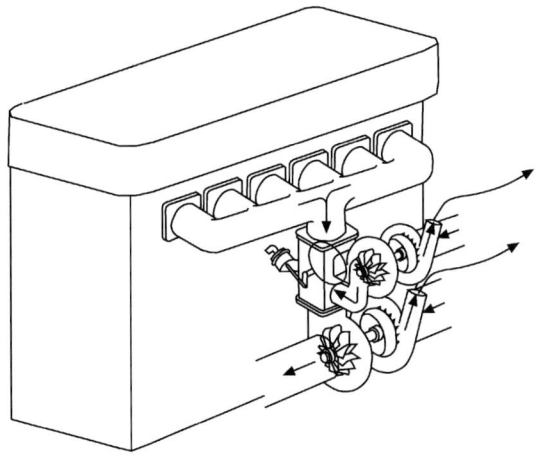

Fig. 6.4. Sketch of principle for governed two-stage turbocharging [109]

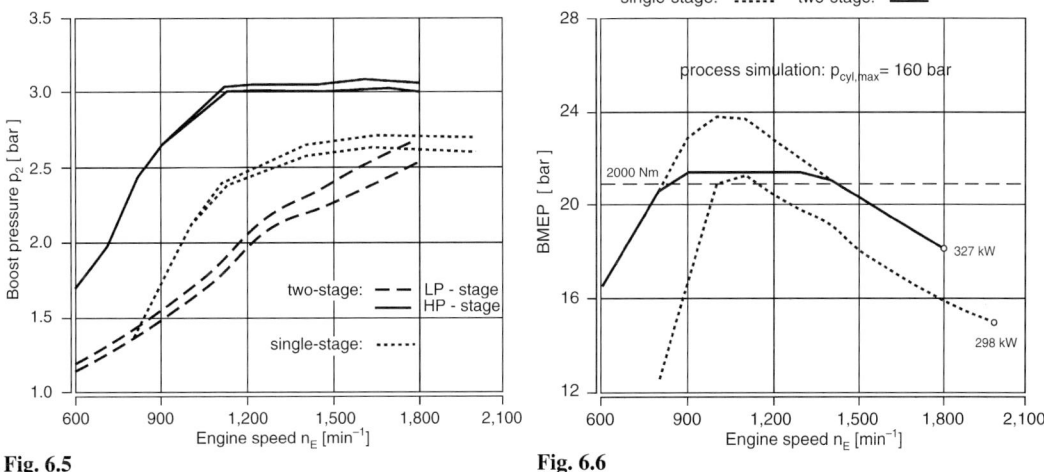

Fig. 6.5.

Fig. 6.6.

Fig. 6.5. Charge pressure curve for two-stage turbocharging of a 12 liter truck engine [109]

Fig. 6.6. Mean effective pressure curve for two-stage turbocharging of a 12 liter truck engine [109]

On the one hand, this results in a very fast response behavior of the small high-pressure charger. In addition, the complete exhaust gas energy is utilized in both turbines. At higher loads and speeds, this leads to high boost pressures at low charge air intake temperature into the engine (charge air cooling is possible downstream of the low-pressure and high-pressure compressors). Figures 6.5 and 6.6 show some striking results which were obtained with a 12 liter truck engine.

In 2005, BMW has launched mass production of a 6-cylinder diesel engine with controlled two-stage (partly register) turbocharging. The exhaust and charging system is designed such that negative effects during transients are widely eliminated, resulting in very attractive transient and steady-state boost pressure characteristics. Due to the wide speed range of this engine, a second waste gate is installed to bypass the low-pressure turbine. The details of this engine are presented in Sect. 14.2.

6.3 Register charging

For register charging, two layouts are used, namely, single-stage and two-stage register charging.

6.3.1 Single-stage register charging

For single-stage register charging, in the lower speed range of the engine one charger (or half of the chargers used) is switched off and the total exhaust flow is routed through the other charger (or the other half of the chargers). For example, in an engine with two chargers, because of the increased exhaust gas energy supply, the operating turbocharger achieves significantly higher boost pressures than would be obtained with both chargers in operation. Thus, in the lower engine speed range, higher mean effective pressures are achieved. Figures 6.7 and 6.8 show the principles

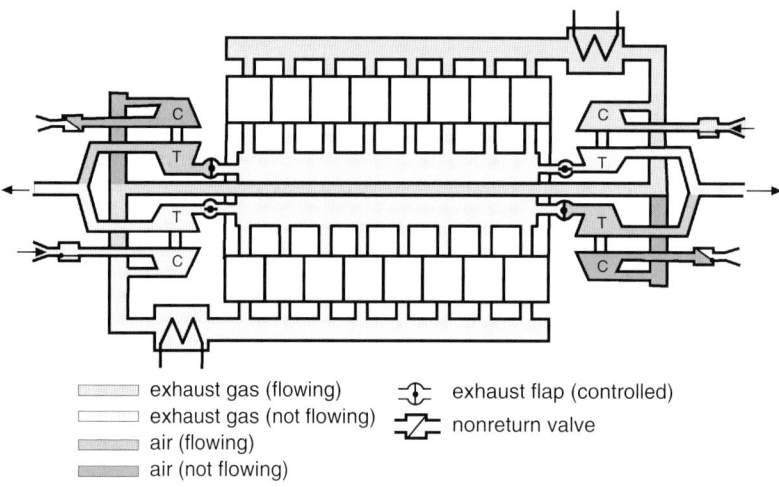

Fig. 6.7. Sketch of principle for the setup of single-stage register charging (top view)

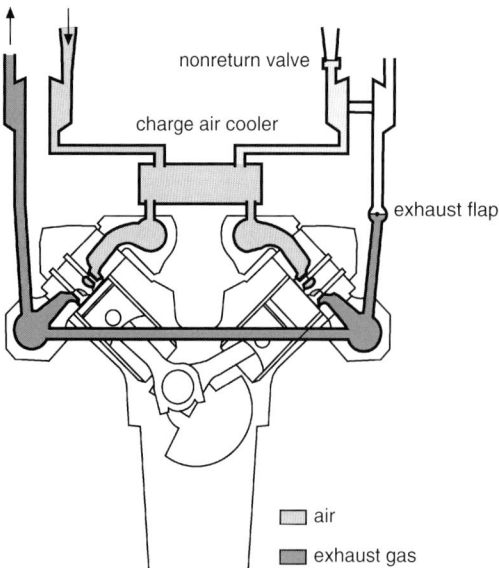

Fig. 6.8. Sketch of principle for the setup of single-stage register charging (rear view)

6.3 Register charging

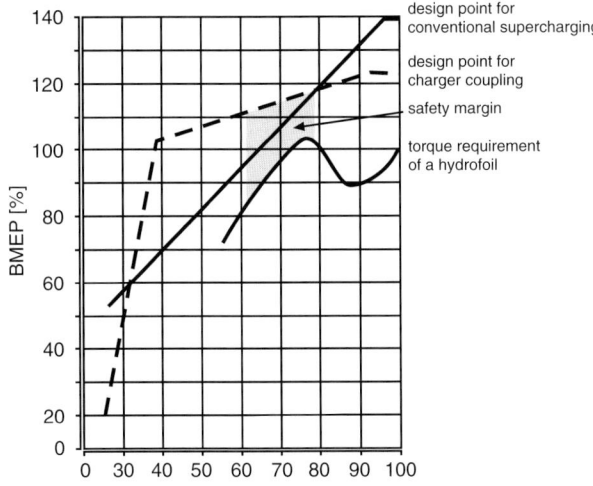

Fig. 6.9. Torque requirement and torque curves for engines with and without register charging [MTU]

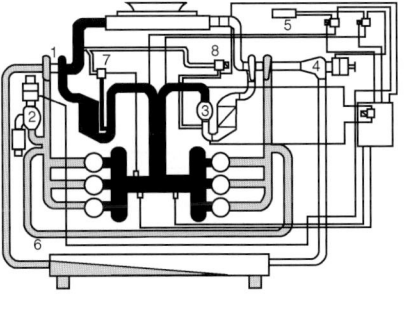

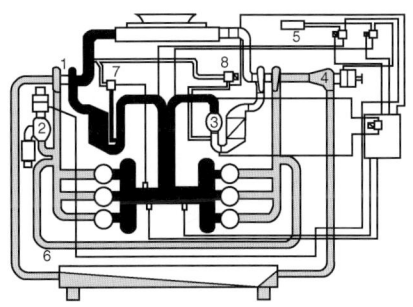

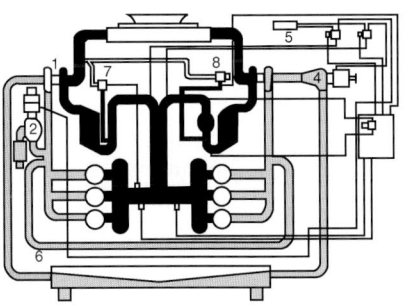

1 turbocharger
2 waste gate
3 compressor cutoff
4 turbine cutoff
5 damper element
6 connecting pipe
7 blowby valve
8 bleeder valve

■ charge air
□ exhaust gas

Fig. 6.10. Schematic of register charging [15]

of such a layout. For special applications, sufficient acceleration reserves can be obtained with this process and, correspondingly, smaller basic engines can be used (Fig. 6.9). An example are hydrofoils, which have demanding requirements regarding their emerge torque. This design is in series production at MTU for their high-speed high-power engines. Such charging systems have also been utilized repeatedly for passenger car gasoline engines. A well-known example is the 6-cylinder engine with register charging for the Porsche model 956. The principal schematics of the layout are shown in Fig. 6.10. The implementation of register charging facilitates significant improvements

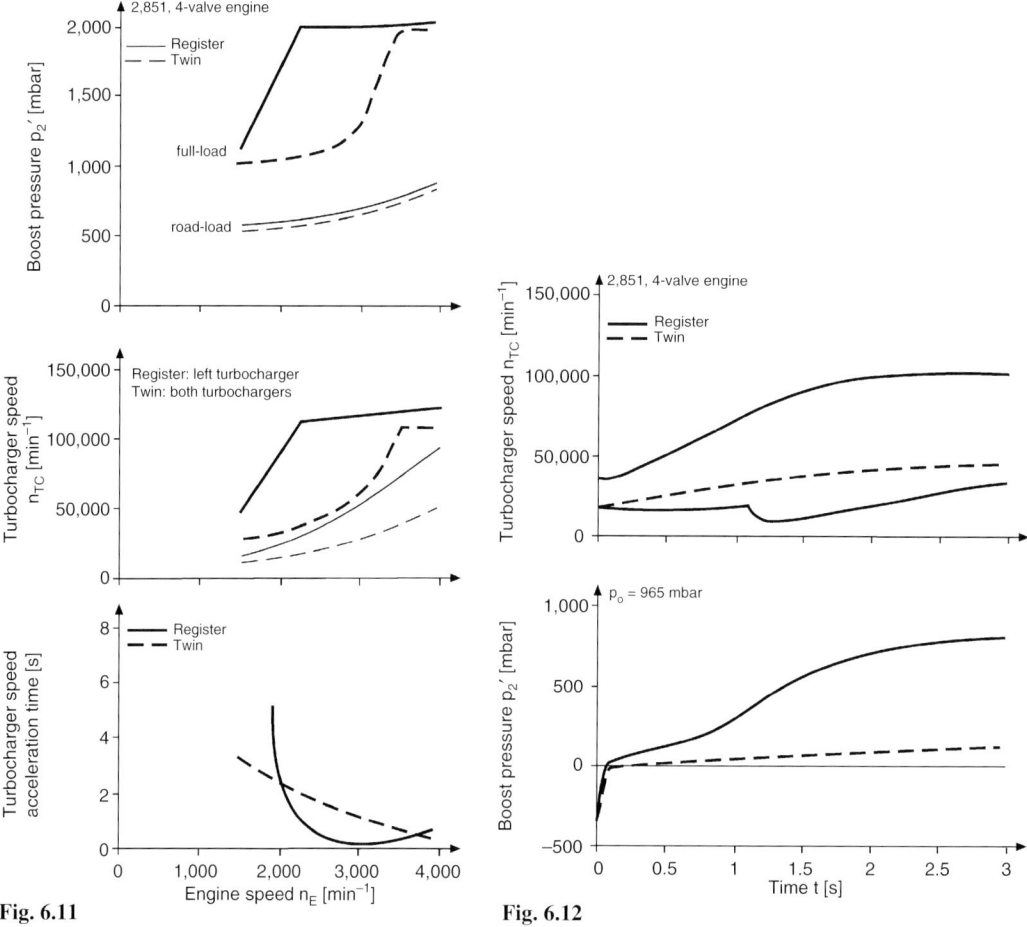

Fig. 6.11. Effects of register charging under steady-state engine operating conditions [50]

Fig. 6.12. Effects of register charging under transient engine operating conditions [50]

in the engine performance at very wide speed ranges, especially the boost pressure buildup, both under stationary (Fig. 6.11) and under transient (Fig. 6.12) engine operating conditions.

Figure 6.13 shows the results of a numeric charger layout and an analysis of the operating strategies in the entire full-load speed range of a truck engine. Starting from a desired mean effective pressure curve of the engine, the operating behavior of the first and the second charger can be tracked in the map (Fig. 6.13) with thermodynamic cycle simulations. Then the switching points can be determined, for the bleeder control of the first charger at around $680 \, \text{min}^{-1}$, for the connection of the second charger at around $900 \, \text{min}^{-1}$, and the chargers can be preselected.

6.3.2 Two-stage register charging

Typically for two-stage register charging, very compactly arranged charger groups are connected in series in order to achieve sufficiently high torque output in the complete engine speed range. Figure 6.14 shows such a compact charger arrangement with two low-pressure and two high-

6.3 Register charging

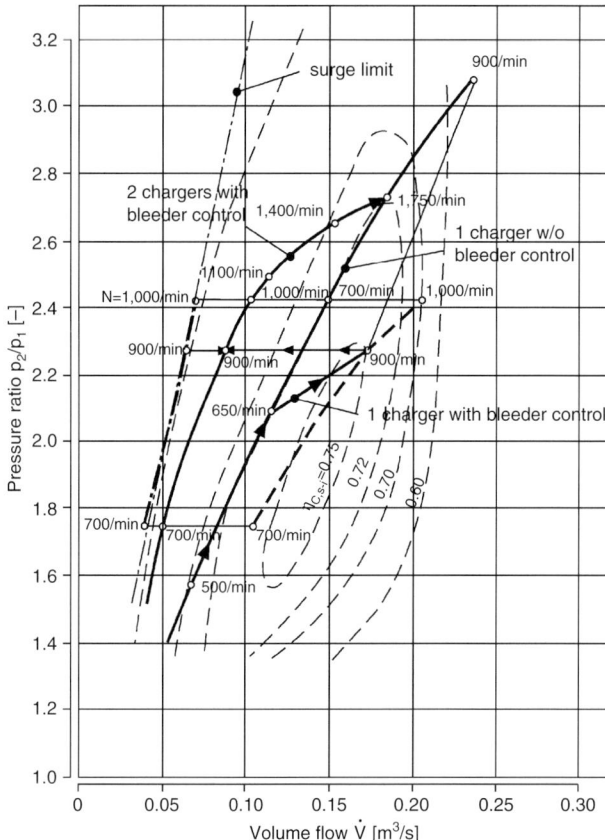

Fig. 6.13. Operating curves in compressor maps for register charging (simulation results)

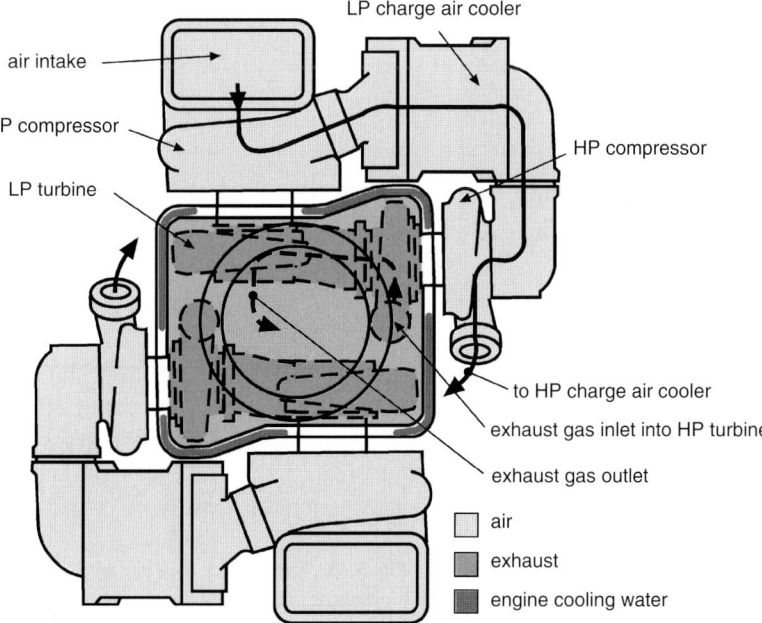

Fig. 6.14. Two-stage register charging; compact arrangement of high- and low-pressure exhaust gas turbochargers [MTU]

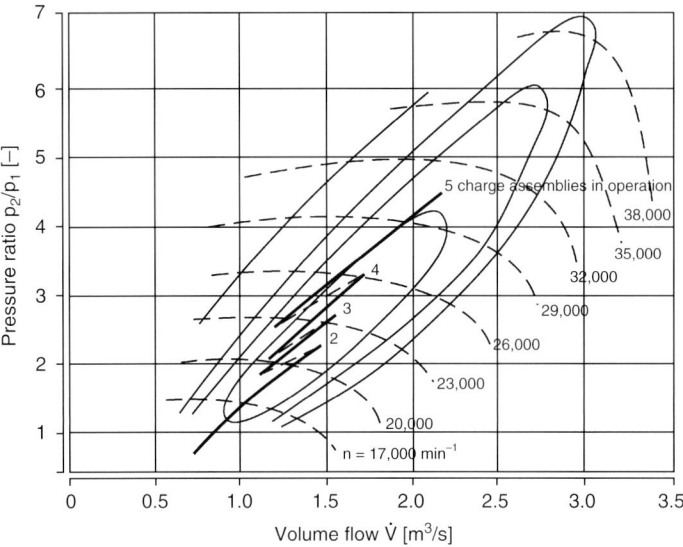

Fig. 6.15. Register switching strategy in a compressor map [MTU]

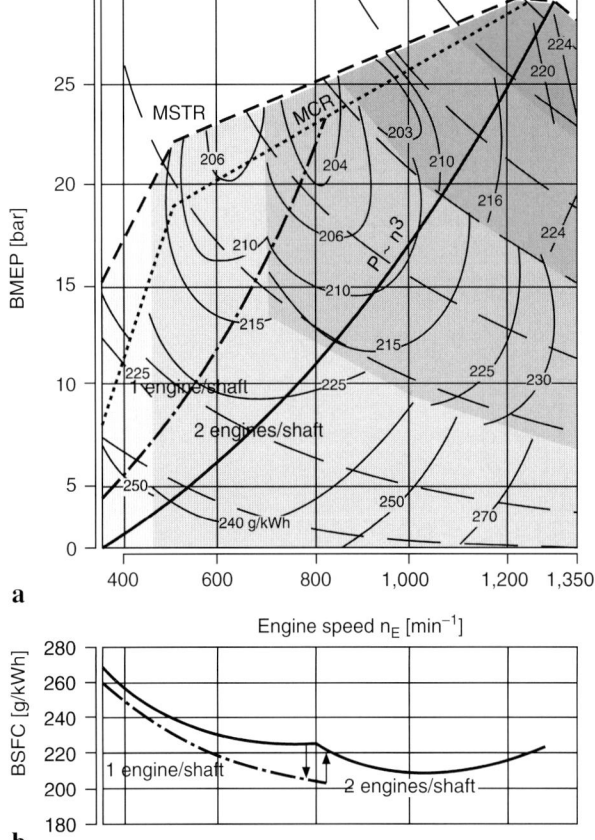

Fig. 6.16. Torque and fuel consumption of a high-power diesel engine with and without register operation [MTU]. MSTR, maximum short-time rate; MCR, maximum cruising rated power

6.4 Turbo cooling and the Miller process

pressure exhaust gas turbochargers. Figure 6.15 shows the switching strategy for the charge groups in the pressure–volume flow map. Figure 6.16a shows mean effective pressure values up to 30 bar, achieved with this arrangement, and Fig. 6.16b shows the fuel consumption values obtained with and without register operation.

It is obvious that the very high mean effective pressure and the excellent torque curve result in relatively high fuel consumption due to restricted peak firing pressures (retarded combustion) and high gas exchange losses. Such engine layouts are especially suited for high-performance ship engines. Therefore it is suited for applications where the highest power densities and engine performances are generally needed for only short periods.

6.4 Turbo cooling and the Miller process

6.4.1 Turbo cooling

As was discussed in Chap. 2, the performance of a supercharged engine can be significantly increased by means of charge air cooling. However, this is limited by the temperature of the coolant, since the charge air temperature cannot fall below it. For practical and economic reasons – e.g., intercooler size – in most cases the charge air temperature at full load is significantly higher than the coolant temperature. A method to further decrease the charge air temperature – independent of the coolant – is turbo cooling. In this case, the charge air is first compressed above the boost level required by the engine, then cooled in the charge air cooler, and finally its temperature is further decreased by expansion, e.g., in an expansion turbine. Figure 6.17 shows a layout in which the expansion turbine is located on the same shaft as the turbocharger. With this layout, patented for Daimler Benz, the charger compresses more air than the engine needs. The excess air is expanded to ambient pressure in the cooling turbine, resulting in a temperature much lower than the temperature

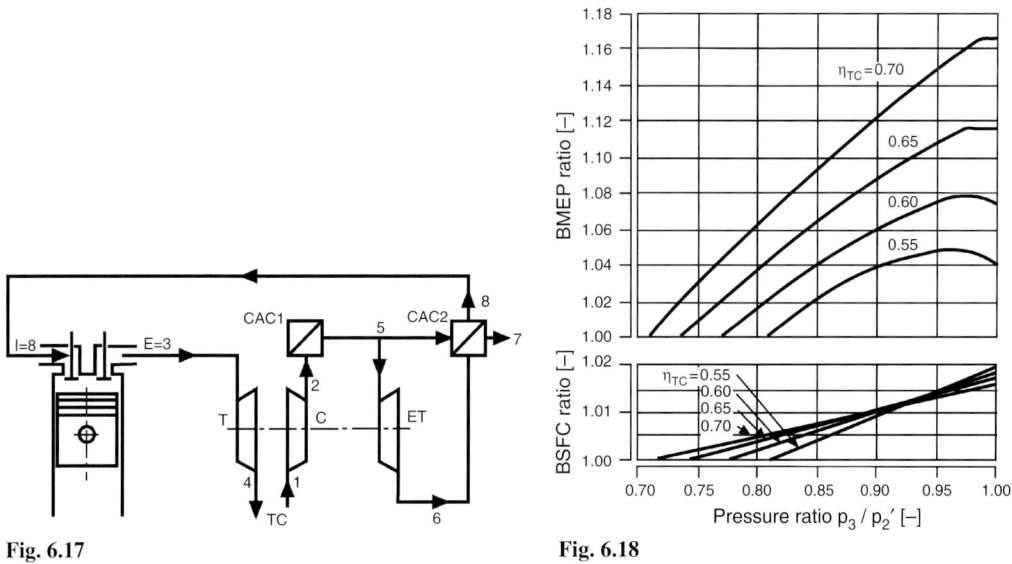

Fig. 6.17. Principal diagram of a single-shaft turbo cooling charger arrangement. CAC, charge air cooler; ET, expansion turbine

Fig. 6.18. Influence of exhaust gas turbocharger efficiency on the power increase achievable by turbo cooling [159]

downstream of the charge air cooler CAC1, and then it is utilized in the second charge air cooler CAC2 to further decrease the charge air temperature.

The more the pressure upstream of the cooling turbine exceeds the required boost pressure, the greater will be the temperature decrease achievable in the cooling turbine. Thus, the success of turbo cooling also depends on the capability – i.e., the efficiencies – of the exhaust gas turbocharger, especially its turbine.

Figure 6.18 shows the influence of the exhaust gas turbocharger's efficiency on the achievable power increase. It is obvious that significant power increases can be obtained only with very high total efficiencies of the charger, along with minimal improvements in fuel consumption.

In spite of the advantages mentioned above, turbo cooling is not worthwhile for series production diesel engines today. The complexity of the turbo cooling assembly and a second charge air cooler, in relation to the limited back-up of the boost pressure – unless two-stage compression with even higher complexity is used – is cost-prohibitive.

Turbo cooling is more promising for gas engines. In these, the achievable power usually is not limited by the possible degree of supercharging, but by knocking combustion. The knock limit depends much more on the compression end temperature – and thus, at a given compression ratio, the charge air temperature upstream of the engine – than on the degree of supercharging.

6.4.2 The Miller process

This special method of charge air cooling was originally described by Miller [103]. Contrary to common supercharging processes, it uses very early intake valve closing times which are variable depending on the load (Fig. 6.19). In this way, the cylinder is filled with fresh air only until the intake valve closes. During the remaining intake stroke the fresh air expands and its temperature decreases. Compression then starts – with a smaller charge – from a lower temperature level. Therefore, it can be termed a work process with internal cooling.

Due to the shortened effective inlet phase, the expansion stroke automatically gets longer than the intake stroke. This can also be seen as a process with elongated expansion, as described by Atkinson. This characterization also explains the major application for such an engine process, i.e., cases where process temperature limitations are advantageous, or high combustion chamber temperatures result in negative effects. For example, in slow-speed natural gas engines the temperature-dependent knock limit represents such a limitation.

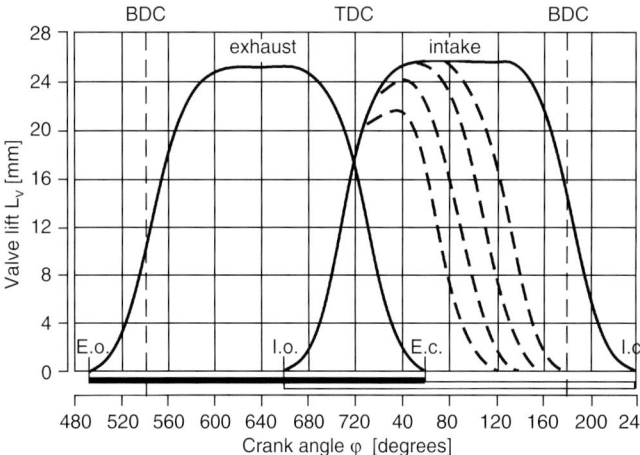

Fig. 6.19. Intake valve timing diagram for the Miller process

6.4 Turbo cooling and the Miller process

The permissible maximum pressure during combustion is one of the most important factors limiting the mean effective pressure of a supercharged engine, unless a deterioration of fuel economy is accepted. Exhaust emissions, which in part strongly depend on the process temperature – especially NO_x emissions –, represent an additional problem. Therefore, the Miller process could also be used to cool an internal or external exhaust gas recirculation.

Figure 6.20 shows pV diagrams of an ideal supercharged four-stroke diesel engine in regular configuration and of an engine using the Miller process. It can be seen that – if both are to utilize the same maximum process pressure (p_{3cyl}, p_{4cyl}) – the Miller process requires a far higher charge pressure p_2. In order to achieve the same charge air temperature at the intake valve, the Miller process requires more extensive charge air cooling. Then the temperature at start of compression (p_{1cyl}) is lower for the Miller process, and thus the temperatures remain lower throughout the process.

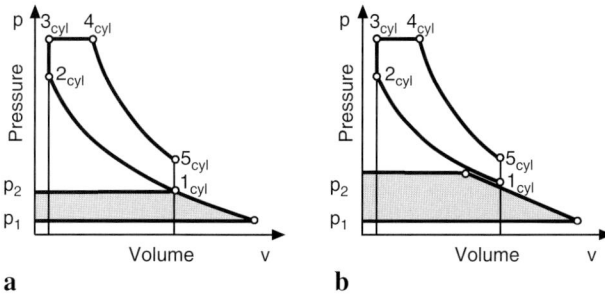

Fig. 6.20. pV diagrams of an ideal supercharged four-stroke diesel engine in regular configuration (**a**) and of an engine using the Miller process (**b**)

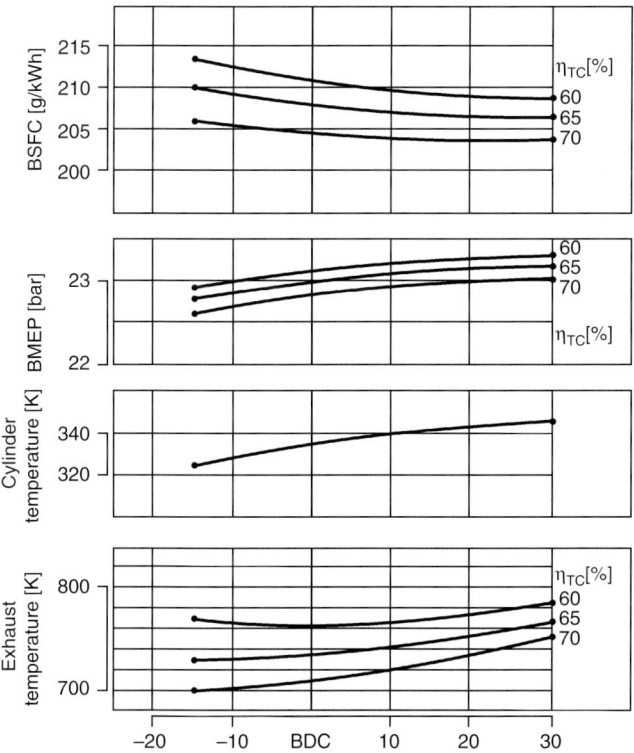

Fig. 6.21. Engine operating results of the Miller process against intake timing [159]

Figure 6.21 shows the influence of "intake valve closing" timing and turbocharger total efficiency on the operating data of a supercharged engine with a peak pressure value of about 150 bar.

For a specified power level, the Miller process therefore requires significantly higher boost pressures and turbocharger total efficiencies than the conventional engine process of a supercharged diesel or natural gas engine. Its application is thus limited to two circumstances:
- the highest permissible work pressure of the process, e.g., the mechanically permissible peak pressure given by the engine design, has been reached, but the charging system still has pressure reserves;
- process limitations, such as the knock limit or high NO_x emissions, have been reached and require a decrease in peak temperatures.

6.5 Turbocompound process

By definition, the net power of an engine using a compound process is generated not only in the cylinder but also in a downstream expansion stage. In this sense, the exhaust gas turbine or another downstream turbine provides additional power to the crankshaft in a compound engine. The purpose of the process is to utilize the exhaust gas energy to a larger extent, resulting in better fuel economy.

Calculations [154] show that, at the design point of the engine, improvements in fuel economy exceeding 5% may be reached when the compressor and turbine efficiencies are excellent. Since this is true especially at high engine load, the most prominent examples of turbocompound engines are found in maritime ships which operate at constant high load for extended periods. Therefore, reduced fuel consumption, i.e., a further increase in efficiency, is of the utmost importance.

Before the gas turbine came to dominate aeronautic propulsion, aircraft piston engines constituted another turbocompound application, because of long hours of operation at constant high load. The conditions were especially favorable at high altitude, due to the high compression and expansion ratios in the charger and the turbines at low ambient pressure.

The most powerful gasoline piston engine for commercial aircraft was the Curtiss-Wright compound aircraft engine, a double-row radial engine with 18 cylinders (Fig. 6.22). Its rated takeoff power was 2,420 kW at 2,900 min^{-1}. In this design, the charger was powered from the crankshaft at a fixed ratio, which was possible without disadvantages due to the propeller load characteristic. Three exhaust gas turbines, arranged in angular spacing of 120° were also connected to the crankshaft at a fixed ratio, supplying power to the crankshaft.

Fig. 6.22. Curtiss-Wright compound aircraft engine with 18 cylinders in double-row radial configuration

6.5 Turbocompound process

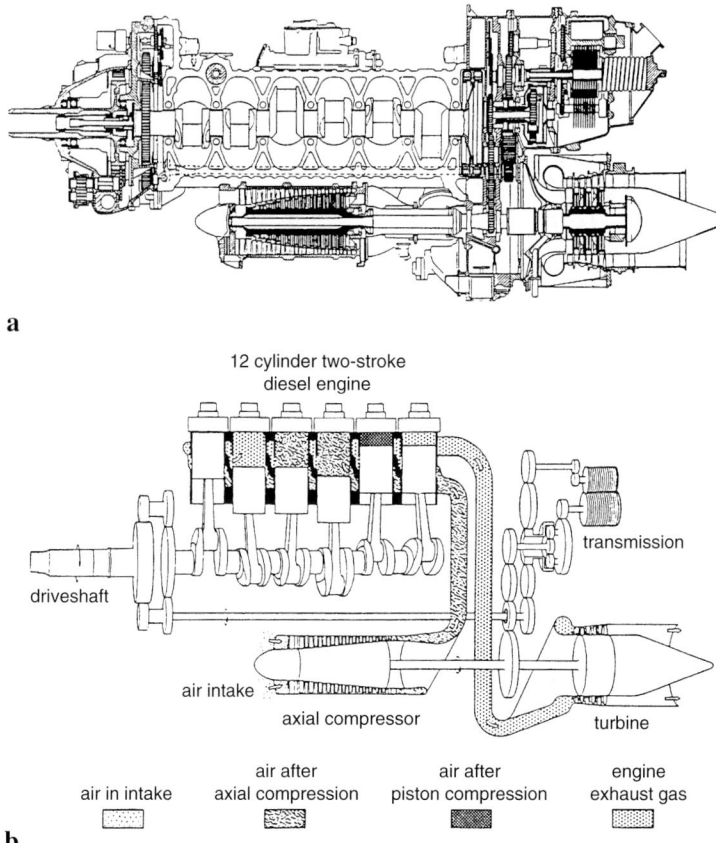

Fig. 6.23. a Napier diesel compound aircraft engine; b diagram

In addition to gasoline engines, aircraft diesel engines were also produced, e.g., the 12-cylinder two-stroke compound powertrain Nomad by Napier (Fig. 6.23).

However, finally the gas turbine, the powerful and high power-density propulsion unit also used for propeller aircraft, has relegated the piston combustion engine to use in small airplanes, where at most a simple exhaust gas turbocharger is affordable to achieve acceptable high-altitude operation.

Two methods for exhaust gas energy utilization are employed in today's compound engines, i.e., recovery by conversion into either mechanical or electric energy.

6.5.1 Mechanical energy recovery

Mechanical recovery into the engine is an application for maritime ships. A fraction of the exhaust gases is branched off upstream of the turbine of the exhaust gas turbocharger and routed to a power turbine (Fig. 6.24). At part-load, this parallel flow can be switched off via valves. The power turbine feeds its power into the engine crankshaft via a step-down gear and a hydraulic vibration-absorbing clutch. Figure 6.25 shows such a design of a slow-speed engine by ABB.

Recently, the compound diesel engine with mechanical recovery has also successfully been utilized in trucks. However, to obtain a better dynamic response during load changes, the power turbine is arranged downstream of the exhaust gas turbocharger (Fig. 6.26).

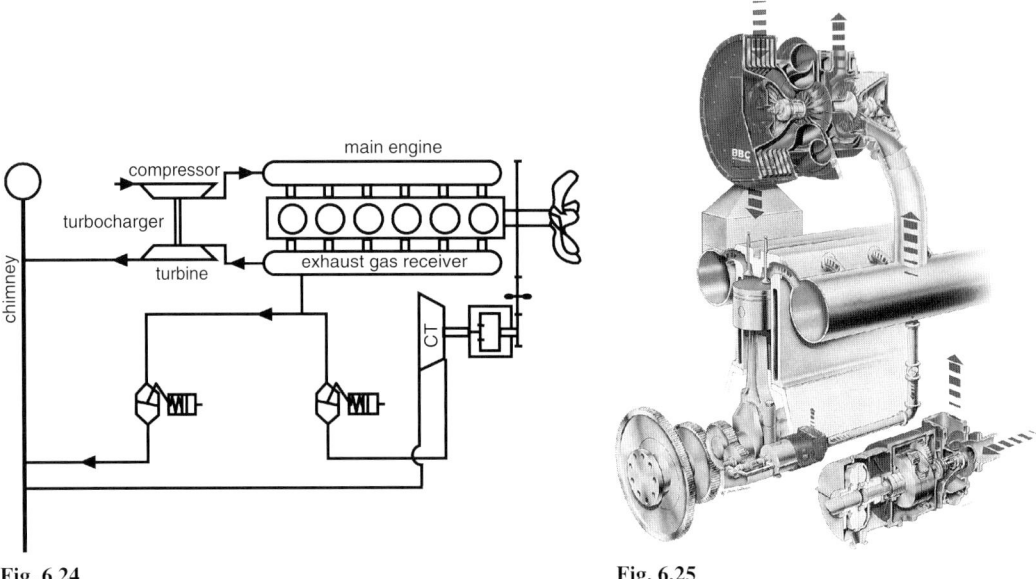

Fig. 6.24 **Fig. 6.25**

Fig. 6.24. Schematic of a layout with mechanic energy recovery; CT, power turbine

Fig. 6.25. Mechanic energy recovery for a slow-speed engine (ABB)

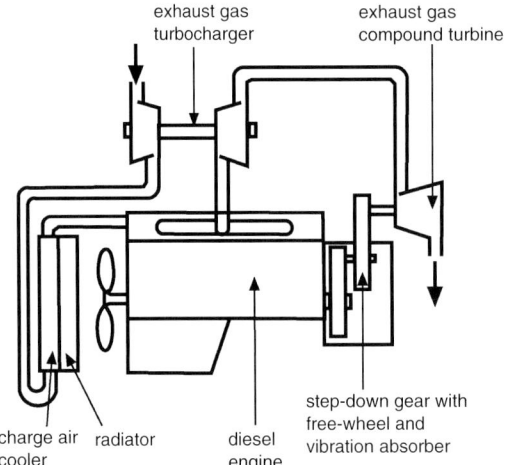

Fig. 6.26. Principal layout of mechanic recovery for truck application

Scania has equipped its 6-cylinder engine DTC 1101 (Fig. 14.53) with a compound turbine. Due to this, the fuel economy was improved under actual driving conditions by 1–3%. The general energy balance is shown in Fig. 6.27 in a truck map, expressed as a percentage gain or loss in fuel economy. As can be seen, fuel economy improvements are only possible at high loads.

The achievable improvements in efficiency essentially depend on the efficiencies of the turbomachinery used. To convert the exergy in the exhaust gas into power, the power turbine slightly increases the backpressure for the exhaust gas turbocharger turbine. This causes the gas exchange of the engine to deteriorate. The compound turbine work, which is directly dependent on

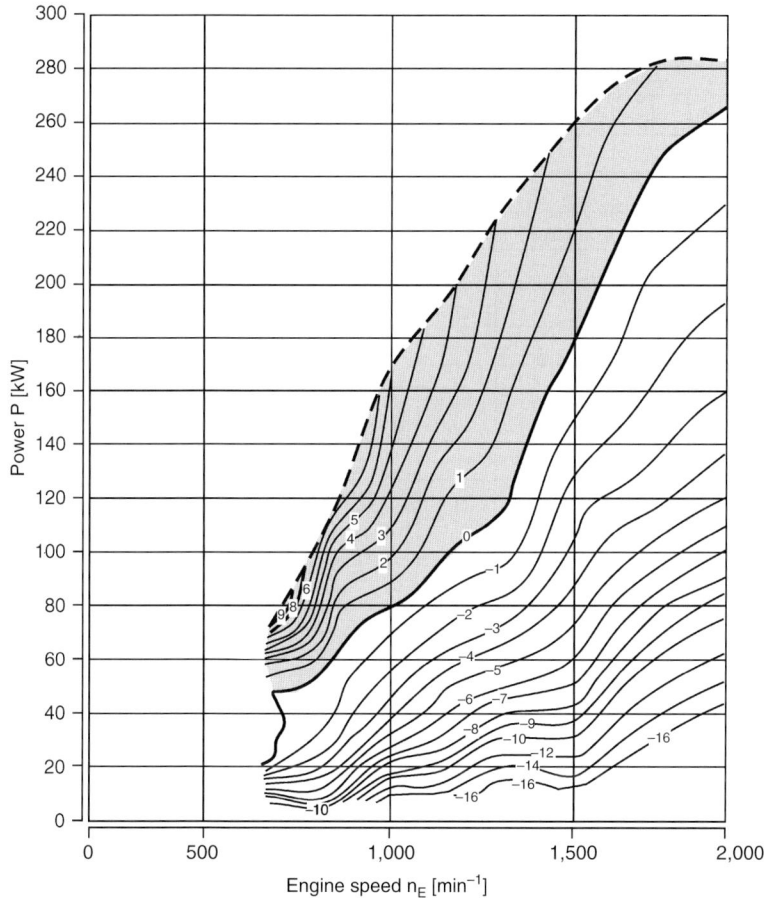

Fig. 6.27. Fuel economy gains (positive values) and losses (negative values), in percent, in the map of a truck engine with secondary power turbine

the turbine efficiency, has to overcompensate for this deterioration. Under consideration of the actual efficiencies, for each operating point an optimum relationship can be established between exhaust gas turbocharger turbine and compound turbine (Fig. 6.28). Such analyses have to be performed in all relevant load points in order to find the best layout regarding lowest fuel consumption. The analyses can also be supported by numeric simulation of part-load sections, in which the decreasing contribution of the compound turbine to the total power output can be determined. After adaptation of the turbocharger and the compound turbine, the potential for fuel economy improvements in the total load range can be determined using thermodynamic cycle simulation.

In Figs. 6.29 and 6.30, the potential for fuel economy improvements is projected, by numeric simulation, for the part-load and full-load operation of a 12 liter HSDI (high-speed direct-injection) diesel engine with and without turbocompound application.

6.5.2 Electric energy recovery

The layout favored today for slow-speed maritime engines is the electric turbocompound system, also called controlled compound turbine-generator layout (Fig. 6.31). In this case the compound

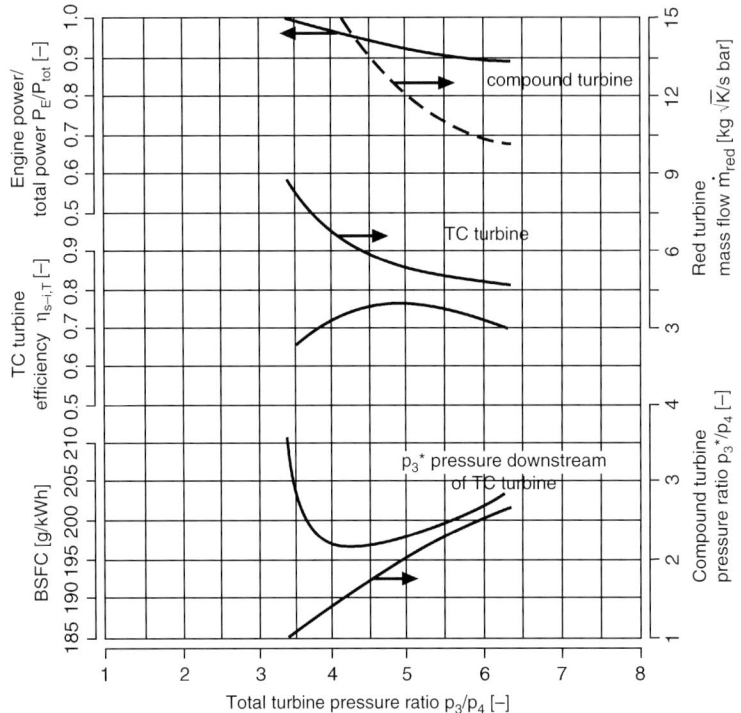

Fig. 6.28. Layout of a compound turbine by thermodynamic cycle simulation, for optimum power of the complete system at 75% of rated speed

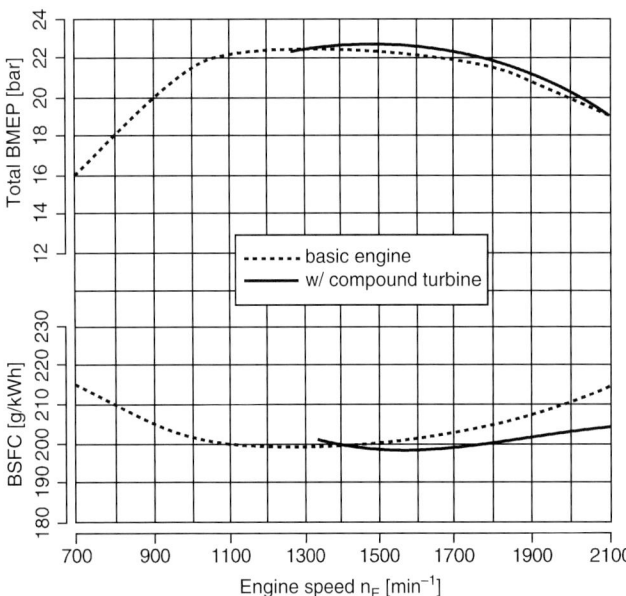

Fig. 6.29. Comparison of simulated full-load engine data, between a 12 liter HSDI diesel engine with conventional turbocharger and the same engine with turbocharger and compound turbine

turbine powers a speed-controlled electric generator. It is not mechanically linked to the main power output. It is utilized to directly generate the electric energy required by the ship, at high efficiency.

6.6 Combined charging and special charging processes

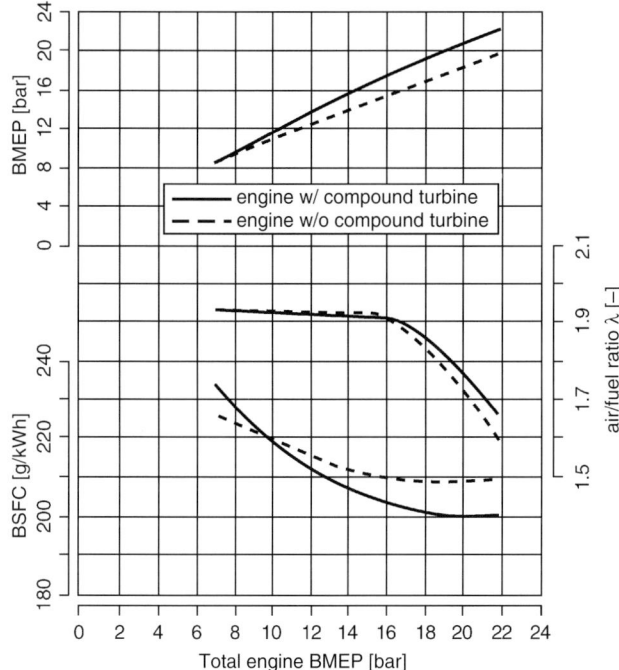

Fig. 6.30. Part-load section at 75% of rated speed, turbocharged engine with compound turbine (simulation results)

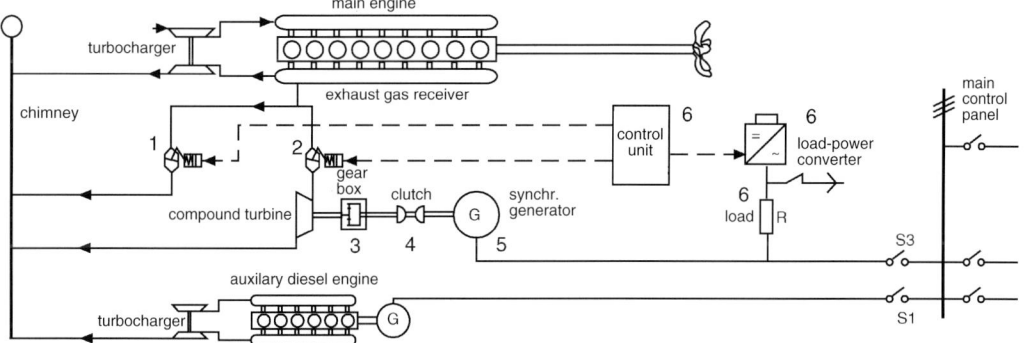

Fig. 6.31. Electric turbocompound layout for maritime engines (controlled compound turbine-generator layout) [ABB]

In addition, this solution is significantly less expensive, since the compound turbine does not have to be linked to the engine shaft, but can be operated at constant speed as dictated by the frequency of the ship's electric system. Figure 6.32 shows the main components of this system. Maritime powertrains with compound engines today achieve total efficiencies significantly above 50%.

6.6 Combined charging and special charging processes

6.6.1 Differential compound charging

Differential compound charging is only meaningful for automotive applications. The layout of the charger group includes a power split (Fig. 6.33). One power branch of the differential or planetary gearbox feeds into the vehicle's transmission, the second branch powers the compressor. The

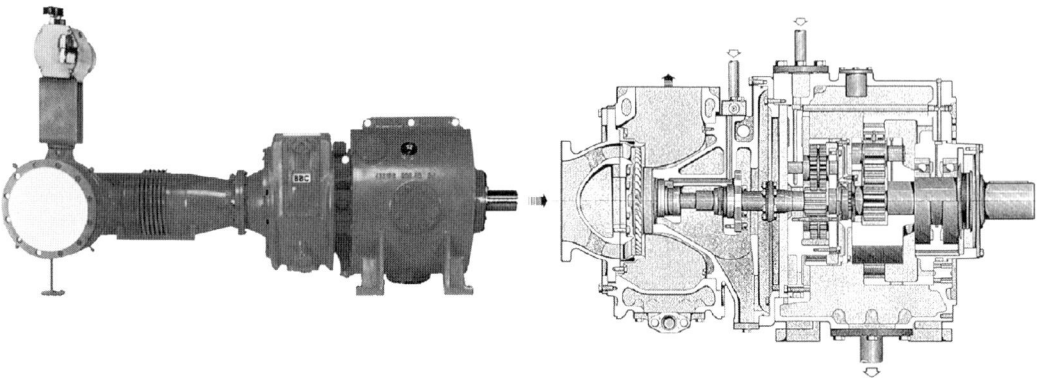

Fig. 6.32. Design example for the main components of a controlled compound turbine-generator layout [ABB]

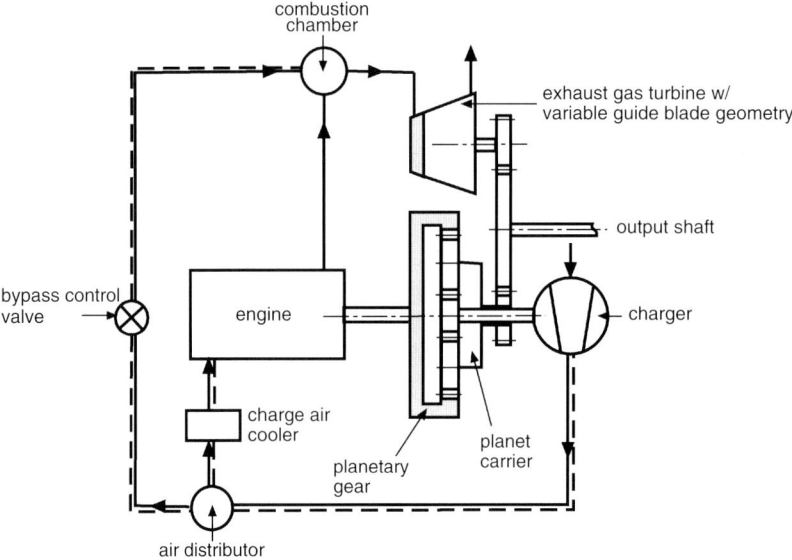

Fig. 6.33. Diagram of a differential compound charging system

torques acting on transmission and compressor are in fixed relation to each other, specified by the gear ratio. The layout is aimed at the achievement of a boost pressure curve, and thus a torque curve, which is similar to the traction hyperbola. Around 1960, Perkins Ltd. performed tests with such a system [53]. Substantial torque back-up was achieved, close to constant power against the speed range (Fig. 6.34). In the end, the possible reduction in gears was opposed by a significantly higher engine and charge system complexity. The efficiencies of the complete system were significantly below those of a comparable conventional drivetrain. Improved fuel economy, the actual target of the research, was not achieved.

6.6.2 Mechanical auxiliary supercharging

The mechanical auxiliary supercharging process utilizes an arrangement where a mechanical (displacement) charger is located upstream of an exhaust gas turbocharger. The target is to achieve

6.6 Combined charging and special charging processes

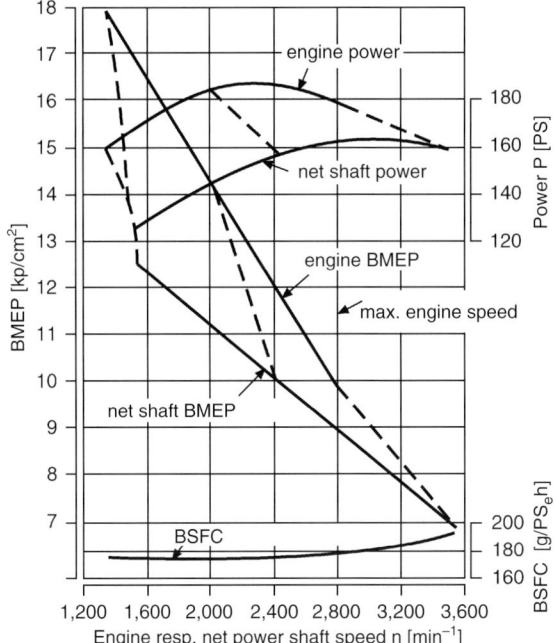

Fig. 6.34. Results obtained by Perkins Ltd. utilizing differential compound charging [53]

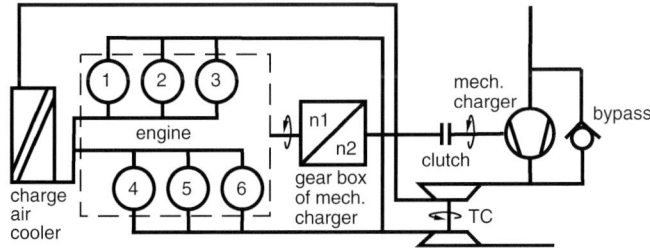

Fig. 6.35. Principal layout for mechanical auxiliary supercharging [122]

high instantaneous torque in the lower speed range of the engine, i.e., when starting, combined with low exhaust gas opacity and low fuel consumption. Figure 6.35 shows the principal layout and Fig. 6.36 the achievable improvement in the transient boost pressure and torque responses, compared to a conventionally turbocharged engine.

In addition, by using only the mechanical charger during engine braking, this layout can potentially result in very high braking power. Due to the significantly higher air flow, the engine itself has correspondingly higher braking power; furthermore, the required compressor work augments the braking power.

This is especially true if the engine is equipped with a constant throttle braking system, such as Mercedes Benz uses as standard equipment in its heavy trucks. Figure 6.37 shows the braking powers achievable with various engine braking systems. Using the system described above, braking power values as high as the maximum net horsepower of the engine are achieved [122].

Recently, VW has introduced into mass production a turbocharged gasoline engine with mechanical auxiliary supercharging for a passenger car application. This engine is presented in more detail in Chap. 14.1.

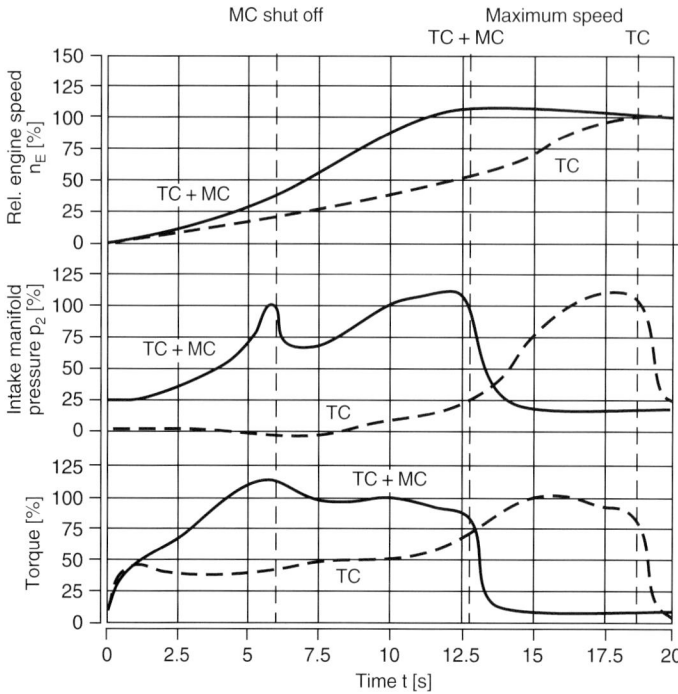

Fig. 6.36. Comparison of torque, boost pressure and engine response data; engine with mechanical auxiliary supercharging (TC + MC) vs. engine with conventional turbocharging (TC) [122]

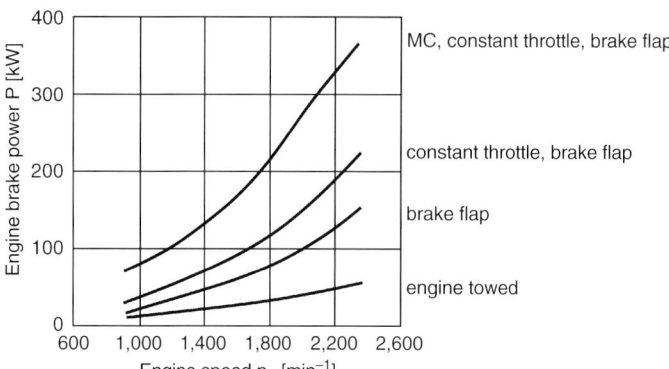

Fig. 6.37. Achievable braking power utilizing various engine braking systems

6.6.3 Supported exhaust gas turbocharging

The desire to support the exhaust gas turbocharger in critical operating ranges via additional driving power is long standing. In fact, supported exhaust gas turbocharging is the logical and further integrated advancement of mechanical auxiliary supercharging. The goals are identical. In general, supported exhaust gas turbocharging aims at additional power supply to the charger. Several drive systems, including switchable systems, are under intensive development. Two examples of systems actually produced have to be mentioned: an exhaust gas turbocharger with hydraulic support drive (Fig. 6.38 shows the hydraulic support turbine located on the rotor assembly) and systems with

6.6 Combined charging and special charging processes

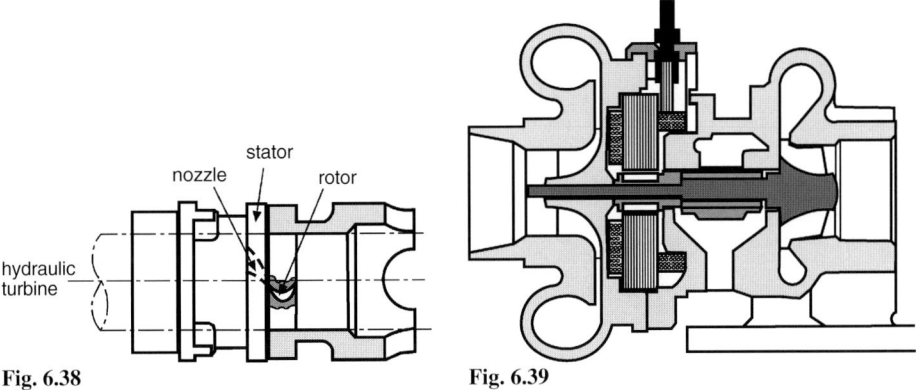

Fig. 6.38. Exhaust gas turbocharger with hydraulic support drive, by Garrett [88]

Fig. 6.39. Exhaust gas turbocharger system with electric support drive, by Garrett

electric support drive. Similar to the hydraulic support turbine, the rotor of the electric support motor is also located on the turbocharger shaft (Fig. 6.39).

With such support drives, the air supply to the engine in the lowest speed range and during transient operation can be significantly improved. Since the enthalpy flow of the exhaust gas is also increased at the turbine, performance ratios (in reference to the energy supplied to the support drive) significantly above 1 (up to 1.4) are achieved. Accordingly, with such systems both goals can be reached, i.e., significant increases of the stationary engine mean effective pressures at low speeds (Fig. 6.40), and faster transient boost pressure buildup (Fig. 6.41) [159].

6.6.4 Comprex pressure-wave charging process

The disadvantages of the exhaust gas turbocharger regarding its acceleration behavior and torque buildup have been mentioned several times. They gave ample reason to look for ways to utilize the exhaust gas energy for boost pressure generation which avoid these deficiencies. One of these is to transfer the pressure energy in the exhaust in a gasdynamic process directly to the charge air.

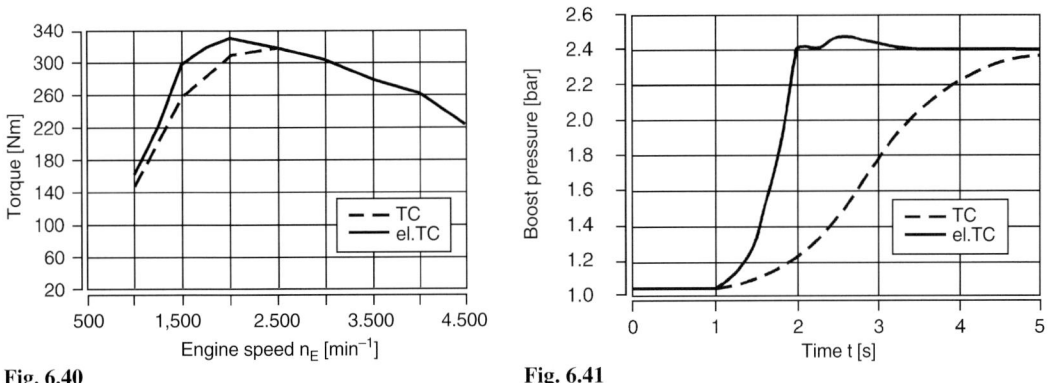

Fig. 6.40. Increase in engine mean effective pressure by utilization of an exhaust gas turbocharger with electric support drive

Fig. 6.41. Improvement of engine response by utilization of an exhaust gas turbocharger with electric support drive

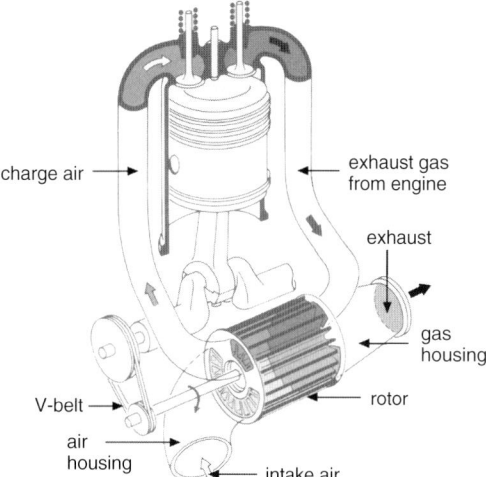

Fig. 6.42. Diagram of a pressure-wave charger [37]

Such a charger, called Comprex, was developed to series production by Brown-Boveri. The mode of action in a pressure-wave charger is based on the reflection behavior of pressure waves in a pipe. A positive or negative pressure wave running in a pipe will be changed into its opposite at an open end of a pipe, but will be amplified to double its amplitude at a closed end of a pipe.

In practice, the pressure-wave charger (Fig. 6.42) consists of a cell rotor with open channels arranged frontally at its perimeter. To control the process, the cell rotor has to be powered, but requires only as much power as needed to override bearing and ventilation losses. On the one side of the cell rotor the low-pressure and the high-pressure (or charge-pressure) fresh air ports are arranged frontally; on the opposite side the identical low-pressure and high-pressure exhaust gas ports are located. The compression energy for the charge air is extracted from the exhaust gas. The processes in the cell rotor itself can best be understood by examining the unwound cell rotor perimeter (Fig. 6.43), where the intake and outlet ports of the fixed housing are also shown.

The cycle starts at 1 in Fig. 6.43. At this point, all cells are supposed to be filled with fresh air under ambient pressure (intake state p_0). The vertical bars indicate that at this point the gas is at rest. The exhaust gases of the engine are collected in an exhaust plenum (A) and then flow towards the cell intake (HPE) at even and constant pressure. If now – due to the rotation of the cell rotor – a cell filled with air under ambient pressure is connected to the high-pressure port, the higher-pressure exhaust gas enters the cell and triggers a pressure wave in that cell which propagates at sonic speed and compresses the cell air, accelerating it towards the charge air high-pressure port (HPA).

The pressure wave must reach the opposite end of the cell rotor at that instant at which – due to the rotation of the cell rotor – the charge air port (HPA) is opened. Thus, the compressed air can flow into the charge air plenum (B) and from there to the engine. The cell reaches the closing edge of the high-pressure port at a time when it is filled with exhaust gas for about two thirds of its length, thereby preventing further entry of high-pressure exhaust gas. At this point, the cell channel is filled with a mixture of about 2/3 exhaust gas and 1/3 air, at a pressure which is lower than the exhaust gas pressure in HPE, but higher than in HPA. On its way to the low-pressure port system the exhaust gas–air mixture in the cell comes to rest (2).

As soon as the cell – in its further rotation – passes the edge of the low-pressure exhaust port, the exhaust gas–air mixture can leave the cell in the direction of the gas housing, triggering a low-pressure wave which propagates into the cell. At optimum cell rotor speed, this low-pressure –

6.6 Combined charging and special charging processes

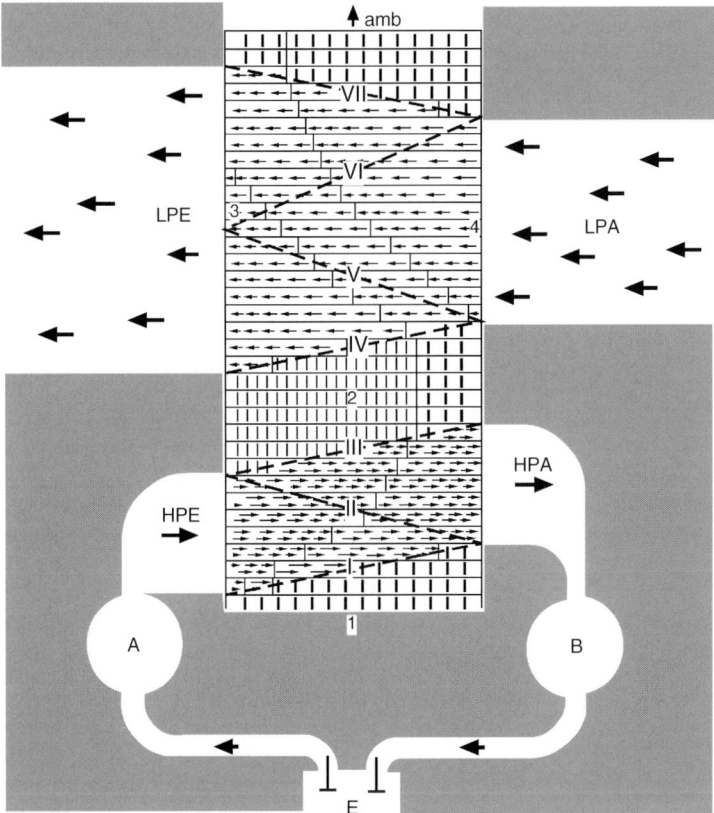

Fig. 6.43. Processes in the cell rotor of a pressure-wave charger [94]. E, engine; A, exhaust plenum; B, charge air plenum; HPE, high-pressure exhaust gas; HPA, high-pressure air; LPE, low-pressure exhaust gas; LPA, low-pressure air; amb, ambient

and thus suction – wave reaches the opposite cell end just at the instant in which the low-pressure air channel is opened. Now the cell is again filled with air from the intake system, while the exhaust gas continues to flow towards the outlet. When exhaust gas and mixed gas have left the cell, i.e., the cell is completely scavenged, the process starts again.

From this description it is obvious that the process can provide satisfactory charge pressures and efficiencies only if it is exactly controlled (the problems with the practical application of pressure-wave charging were formerly control related). Obviously, the process strongly depends on the sonic speed, and thus on the exhaust gas and air temperatures, but not on any load or speed conditions of the engine. However, since the cell rotor drive had to be linked in some form to the engine and thus to the engine speed, complicated additional gasdynamic processes were necessary to optimally adapt the process to the engine conditions. Also, the pressure-wave process is very sensitive with regard to the exhaust gas backpressure in the exhaust system.

Present research on pressure-wave chargers has eliminated the fixed ratio connection to the engine, but uses a small electric motor to power the cell rotor, since the required drive power is insignificant (Fig. 6.44). The speed of the electric motor-cell rotor group is controlled dependent on the exhaust gas temperature. Similar to the turbocharging systems with various additional drives discussed in Sect. 6.6.3, the Comprex charger achieves considerable increases in boost pressure in the lowest speed range, i.e., at small absolute mass or volume flows (Fig. 6.45).

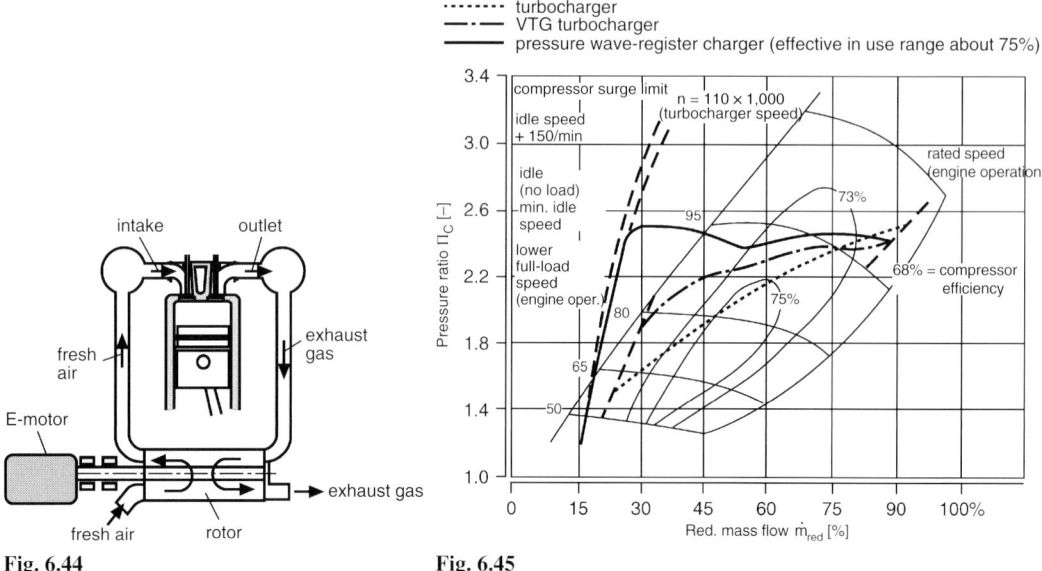

Fig. 6.44. Further development of the pressure-wave charger, now powered by an electric motor (cell rotor no longer connected to the engine via fixed gear ratio)

Fig. 6.45. Possible boost pressure characteristic of an electrically speed-controlled pressure-wave charger

Whether or not the pressure-wave charger will make a comeback, will depend on the results of this development. In the meantime, the VTG exhaust gas turbocharger is used by practically all manufacturers of supercharged diesel engines. Results from a pressure-wave charged passenger car diesel engine will be discussed in Sect. 14.2.

6.6.5 Hyperbar charging process

To achieve an exhaust gas turbocharger and thus engine response as fast as possible (to meet especially high requirements for torque, i.e., boost pressure buildup), a combustion chamber can be added upstream of the turbine. With the injection of the fuel, the enthalpy supply to the turbine can be significantly increased instantaneously. Such a layout (Fig. 6.46) is termed the

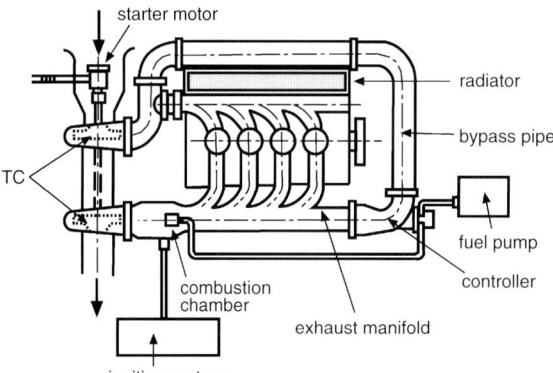

Fig. 6.46. Hyperbar charging process [100]

6.6 Combined charging and special charging processes

Hyperbar charging process [100]. However, the faster response of the charging system has to be paid off by a deterioration in the total efficiency of the system due to the afterburner process. The fuel consumption is increased so dramatically that the process was never introduced into series production.

6.6.6 Design of combined supercharging processes via thermodynamic cycle simulations

Two-stroke engine with combined supercharging

In contrast to four-stroke engines, two-stroke engines cannot force the gas exchange via piston movement, since the exhaust and intake strokes of the piston are absent. Accordingly, in a two-stroke engine the exhaust gas has to be expelled and the fresh gas inducted with the help of an external **scavenging pump**.

In order to achieve a desired torque curve in a two-stroke engine it is important to match the gas exchange and the charging system of the engine, especially in the lowest full-load speed range. There, the low gas cycle scavenging efficiency demands relatively high air delivery ratios and thus mass flows. Therefore, for a two-stroke automotive engine the design point of a mechanical scavenging or supercharging compressor must be in that speed range.

In addition, for the two-stroke engine, turbocharging is an indispensable method to increase its power density. If the efficiencies of the turbocharger are high enough such that a positive pressure gradient is maintained between intake and exhaust manifolds, a scavenging pump becomes unnecessary. For this, the total efficiencies of the charger have to surpass approximately 55%, which are common values for slow-speed two-stroke maritime diesel engines.

Besides the requirements of the scavenging process and the supercharging system, also sufficient cooling of the combustion chamber – especially the exhaust valve(s) – must be considered particularly for medium- and slow-speed two-stroke engines. This can be achieved by late closing of the valve. A significant fraction of the induced fresh gas mass is scavenged through the engine in this way (scavenging efficiency of only 0.6–0.7) and the components are very effectively cooled by the passing charge air.

Additionally – with regard to the scavenging and supercharging systems for slow-speed engines – due to the high turbocharger efficiencies, mechanically driven compressors are usually only utilized at part-load (or during engine startup), unless electrically driven turbochargers are available. Here, charge air cooling is also indispensable to obtain the best efficiencies and to meet maximum durability requirements.

On the other hand, the efficiencies of turbochargers for small passenger car engines are much lower. Therefore, the small two-stroke engine always needs a scavenging pump in addition to the turbocharger. The pump assures the necessary scavenging pressure gradient between the fresh gas and exhaust gas sides. In general, such a layout is called a combined charging and scavenging system.

Various layouts are possible for such combined scavenging and charging systems:

1. turbocharger **upstream** of scavenging pump without charge air cooling
2. turbocharger **downstream** of scavenging pump without charge air cooling
3. options 1 and 2 with charge air cooling directly **upstream** of the engine
4. options 1 and 2 with charge air cooling **between** the two compressors
5. options 1 and 2 with charge air cooling **between** the two compressors and **directly upstream** of the engine.

Options 1 and 2, without charge air cooling, have minimum design complexity, but – as mentioned above – also minimum power density. They will hardly be used in the future. If charge air cooling is used directly upstream of the engine (option 3), the specific power and the torque of the engine can be increased, both by increasing the charge air density (up to 40%) and by reducing the thermal load. Additionally, NO_x emissions are reduced, making charge air cooling practically indispensable for achieving low engine-out emission levels.

The advantage of option 4 is the lower required drive power for the second compressor. The charge air cooler located downstream of the first compressor lowers the inlet temperature into the second compressor by up to 100 K (depending on the pressure ratio). This reduces the drive power for the second compressor by up to 25%. Consequently, the specific fuel economy of the engine is also improved. Option 5 combines the advantages of the other options, but adds the disadvantages of high design complexity and very high system costs.

The design of the scavenging and charging systems must be carried out under consideration of the layout of gas exchange timing. Usually, this is done with the help of thermodynamic cycle simulations. Design criteria are the scavenging pressure and, especially, the exhaust backpressure, i.e., a sufficient scavenging pressure gradient. Low scavenging gradients require relatively elongated opening periods for the gas exchange control devices (ports or valves). These, however, result in a deterioration of the fresh gas scavenging efficiency at low speeds. Therefore, as a first step the influence of port or valve timing has to be analyzed, and then the timing has to be selected according to the desired torque characteristic (Fig. 6.47).

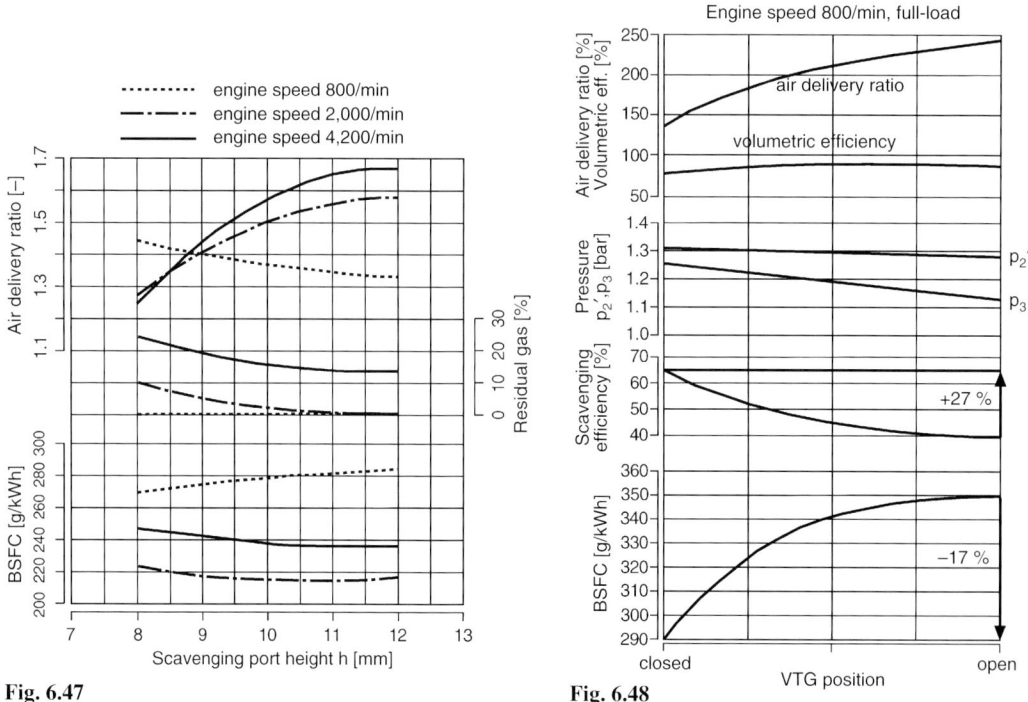

Fig. 6.47

Fig. 6.48

Fig. 6.47. Analysis of the influence of scavenging port height on the engine operating behavior of a supercharged passenger car two-stroke engine by means of thermodynamic cycle simulations

Fig. 6.48. Influence of variable turbine geometry on full-load low-speed operation of a passenger car two-stroke engine (simulation results)

6.6 Combined charging and special charging processes

Once the timing is selected, the scavenging compressor and the turbocharger can be dimensioned. Both fixed-geometry chargers and VTG chargers may be used. By closing the inlet guide blades of a VTG, it is possible to significantly increase the flow resistance of the turbine especially in the lower speed range. This reduces the scavenging losses and significantly improves

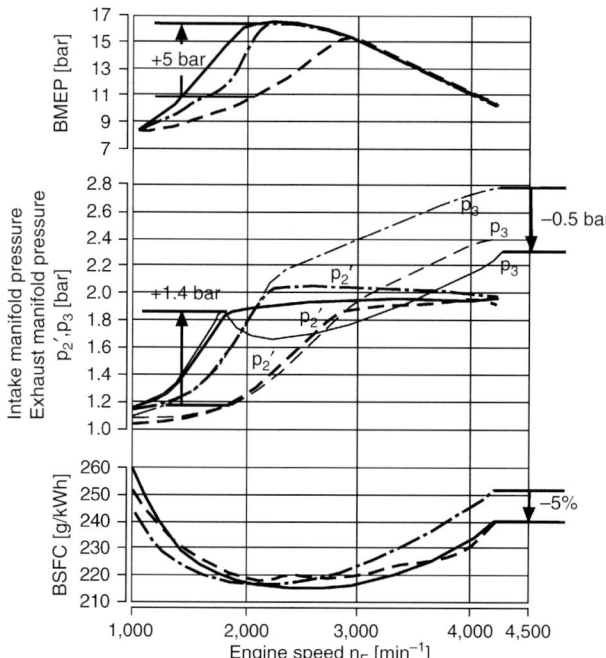

Fig. 6.49. Comparison of the full-load engine characteristic of a passenger car four-stroke diesel engine for various turbocharger layouts: solid line, VTG charger; dash dot line, fixed-geometry charger with small turbine; dash line, fixed-geometry charger with large turbine

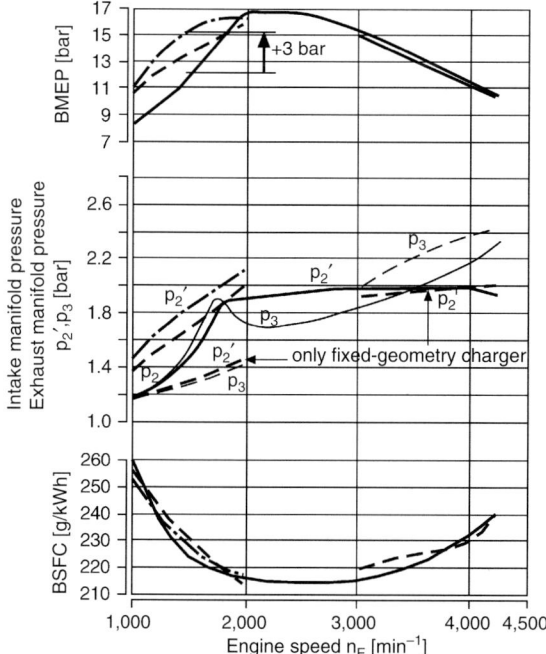

Fig. 6.50. Comparison of full-load torque curve for normal VTG charger (solid line) and fixed-geometry turbocharger plus mechanically driven auxiliary supercharger at transmission ratio of 2.2:1 (dot dash line) and 2.4:1 (dash line)

both the achievable mean effective pressure and the specific fuel consumption. The advantage of the VTG is shown in Fig. 6.48 for a full-load low-speed operating point of a 3-cylinder passenger car two-stroke diesel engine, as obtained by numeric cycle simulations (AVL-BOOST).

Four-stroke engine with combined supercharging

For four-stroke engines as well, low end torque requirements are a very important design criterion for the supercharging system. The minimum turbine swallowing capacity is the criterion for the dimensioning of turbines with fixed as well as with variable geometry. When comparing various chargers, the boost pressure and thus the mean effective pressure buildup in the lowest speed range, and the engine full-load fuel consumption at high speeds, can be simulated and evaluated (Fig. 6.49).

Combined supercharging of four-stroke engines offers additional possibilities for increasing the mean effective pressures at low engine speeds. In this case it is advantageous for the turbocharger compressor to be supported by the mechanical charger, thus less turbine power is necessary. In this way, larger turbines can be applied, which significantly improve, on the one hand, the engine's fuel economy, especially at high speeds, and on the other hand, the speed range of the engine.

The layout of such complex supercharging systems can always be supported by numeric simulations. These are also used to determine the gear ratios for mechanical chargers (Fig. 6.50).

7 Performance characteristics of supercharged engines

7.1 Load response and acceleration behavior

The air flow through the engine is nearly independent of the load and strictly determined by the engine speed in a naturally aspirated diesel engine. With this characteristic it exhibits the best possible load response behavior. Its power and torque increase only depend on the rate of increase in the amount of fuel injected – if injection timing and combustion efficiency are considered nearly constant and the speed change is small in comparison to the change in the amount of fuel injected.

The exhaust gas turbocharged gasoline engine represents the opposite, i.e., the worst case. In gasoline engines with external mixture formation, load is controlled by mixture quantity, which is controlled by throttling the amount of mixture aspirated. Therefore, at low load, low pressures (down to 0.5 bar) occur in the complete manifold system downstream of the throttle, and the speed of the turbocharger drops significantly due to the low amount of exhaust gas supplied.

The entire process used to achieve full load, i.e., full torque, as fast as possible in this engine is very complex and especially time-consuming. By opening the throttle, the pressure in the intake system downstream of it has to be raised to ambient pressure, and in parallel the amount of fuel injected has to be increased.

In the course of this process, the amount of exhaust gas and its temperature also rise, increasing the turbine power of the exhaust gas turbocharger. The charger's rotating assembly is accelerated and the boost pressure increased. After that, the process accelerates progressively, since the turbine power increases faster then the required compressor power. As can be seen, the complete process cannot occur in a very short period of time.

Figure 7.1 shows an example. From an initial vehicle velocity of 40 km/h, i.e., low part load, full load is applied. First, the intake manifold is filled to ambient pressure. This process takes about 0.2 s – the same would be the case in a naturally aspirated gasoline engine. In the case of the exhaust gas turbocharged engine, subsequently the exponential boost pressure buildup just described occurs, in this extreme case taking another 5 s.

For the same vehicle, Fig. 7.2 shows a transient engine acceleration from an initial operating point which approximately corresponds to ambient pressure in the intake manifold. It can be seen that, without first having to fill the intake system and with a higher baseline exhaust gas energy, the boost pressure buildup occurs much faster, i.e., in about 0.6 s.

Engines with mechanically powered and fixed coupled chargers react like naturally aspirated diesel engines; if a clutch between charger and engine is included in the layout, the pressure buildup also depends on the clutch characteristic.

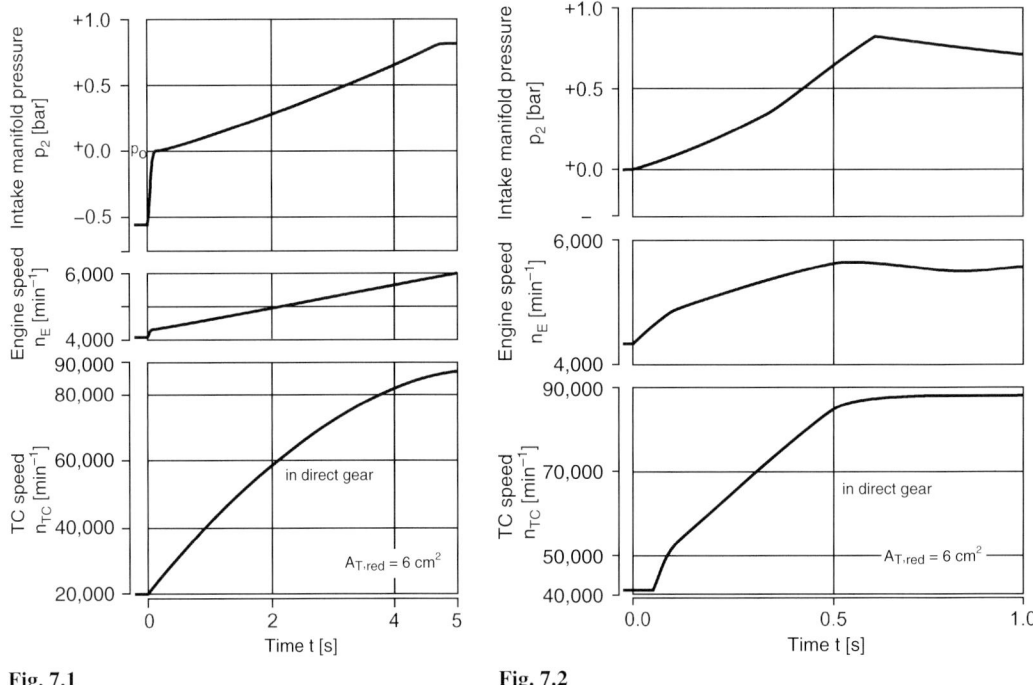

Fig. 7.1. Full-load application from low part-load for an exhaust gas turbocharged gasoline engine [60]

Fig. 7.2. Full-load application from ambient intake manifold pressure for an exhaust gas turbocharged gasoline engine [60]

7.2 Torque behavior and torque curve

For any turbocharged engine, the achievable torque depends on the state of the charge air, i.e., its pressure and temperature, and on the excess air required for the actual combustion process. Against speed, engine torque is therefore determined by the charger characteristic. In an exhaust gas turbocharged engine, an equilibrium is reached between turbine power supply and compressor power need, increased by the charger's power loss.

These relationships were described in Sect. 2.6 for stationary processes. According to this, mechanically supercharged engines have maps as, e.g., shown in Fig. 7.3. For exhaust gas turbocharged engines the pressure–volume flow maps are valid as summarized in Fig. 7.4, depending on the mode of operation. Figure 7.4a shows the map of an uncontrolled exhaust gas turbocharger of a truck diesel engine with a limited speed range. The map of an exhaust gas turbocharged passenger car diesel engine is plotted in Fig. 7.4b, followed by Fig. 7.4c with the wide map of an exhaust gas turbocharged gasoline engine, the last two both with waste gate. Due to the thermodynamic coupling to each individual engine type, a pressure equilibrium is reached, which results in far better boost pressure curves against engine speed than a turbo compressor could achieve with fixed speed coupling.

Thus, the torque curve of exhaust gas turbocharged and mechanically supercharged diesel engines is primarily defined by the given fuel injection quantities against speed. For the gasoline engine, with its approximately constant air-to-fuel ratio against load and speed, the torque curve is determined by the boost pressure curve (see Sect. 8.6.2).

7.3 High-altitude behavior of supercharged engines

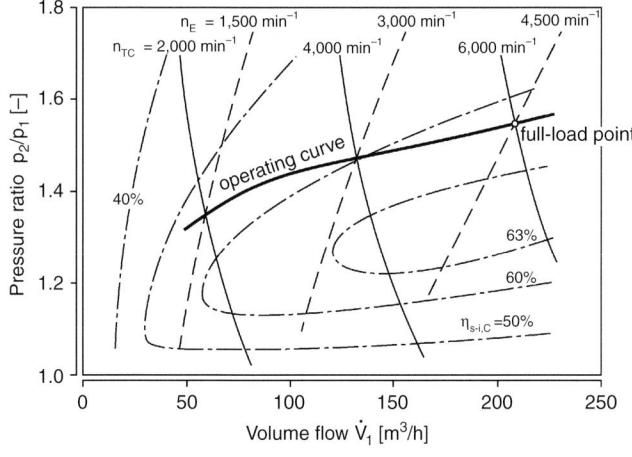

Fig. 7.3. Pressure–volume flow map of a mechanically supercharged engine

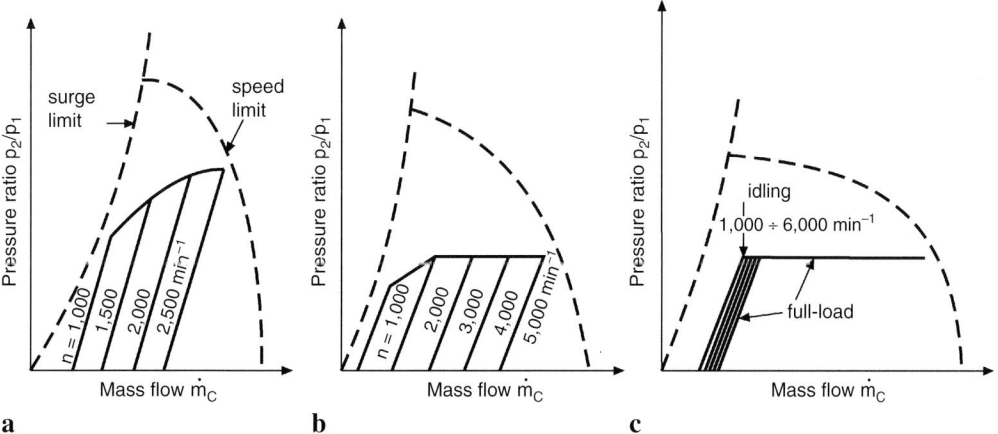

Fig. 7.4. Principal pressure–mass flow maps of exhaust gas turbocharged engine categories: **a** truck diesel engine with uncontrolled exhaust gas turbocharger, **b** passenger car diesel engine with exhaust gas turbocharger and waste gate, **c** passenger car gasoline engine with exhaust gas turbocharger and waste gate

7.3 High-altitude behavior of supercharged engines

Looking at their behavior at high altitude, distinctions have to be made between mechanically supercharged and turbocharged engines. As described in detail in Chap. 4, both displacement and turbo compressors can be utilized for mechanically driven chargers with fixed speed ratio.

As is generally known, turbo compressors are characterized by their nearly constant pressure ratios (depending on speed) between surge limit and choke limit. At high-altitude operation, and at unchanged speed, such a charger will maintain the pressure ratio and deliver the volume flow which can be aspirated by the engine. However, due to the reduced ambient pressure at high altitude, the boost pressure decreases (at unchanged charger speed) and the engine power is reduced by the same degree as it would be reduced in a naturally aspirated engine.

Contrary to the turbo compressor and in accordance with their speed characteristic curves, displacement compressors supply nearly any pressure ratio at a given speed. But, due to the lower ambient density of the intake air at high altitude, the mass pumped per volume unit is reduced and thus also the achievable engine power.

In summary:

– The displacement compressor supplies a constant volume flow through the engine, and therefore can also generate higher pressure ratios. At high altitude, however, it would have to aspirate more volume, i.e., to be operated at a higher speed, in order to maintain the engine power independent of altitude.
– The turbo compressor can supply nearly any required volume flow – independent of speed – but for the higher pressure ratios desired at high altitude its speed would have to be increased correspondingly.

For the exhaust gas turbocharger, which is coupled to the engine only thermodynamically, the situation is different.

At first, with increasing altitude and thus decreasing ambient pressure the utilizable turbine expansion ratio Π_T increases, due to a decrease in p_4. All other data assumed constant, this leads to an increase in turbine power and thus automatically to an increase in compressor power which, naturally, can only be transformed into an increased compressor air flow at increased pressure ratio. This results in a nearly full compensation for the reduced air density at the intake of the charge air into the engine.

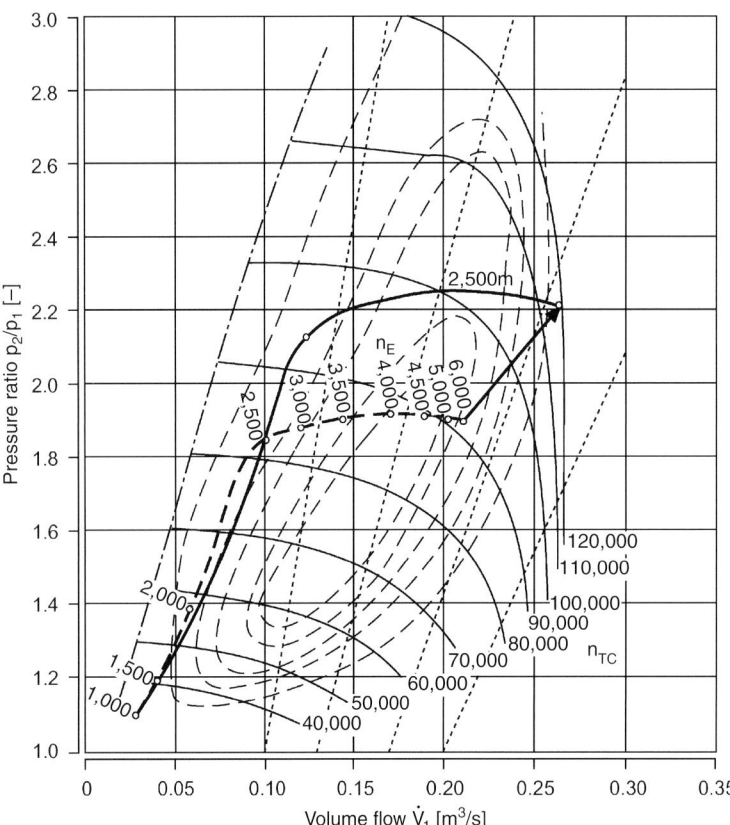

Fig. 7.5. Pressure–volume flow map of an exhaust gas turbocharged gasoline engine with low-altitude (dash line) and high-altitude (solid line) full-load curves

7.4 Stationary and slow-speed engines

Table 7.1. Typical effects of changed ambient conditions on engine parameters of exhaust gas turbocharged diesel engines [160]

Engine parameter	Rel. change per 1,000 m altitude increase	Rel. change per 10 K increase in ambient temperature
bmep	−1 to −2%	−0.5 to −1.0%
bsfc	+1 to +2%	+0.5 to +1.0%
n_{TC}	+6 to +8%	small increase w/o CAC[a] small decrease with CAC
Ignition pressure	−3 to −4%	−1.5 to −2%
λ	−6 to −7%	−3 to −4%
T upstream turbine	increase by 30 K w/o CAC increase by 15 K with CAC	increase by 20 K w/o CAC increase by 5 K with CAC
Thermal load	larger increase w/o CAC small increase with CAC	larger increase w/o CAC small increase with CAC

[a] CAC, charge air cooling

Therefore, the power of an exhaust gas turbocharged engine is more or less independent of the altitude – at least as long as the compressor and turbine efficiencies stay about constant and the compressor and turbine maps show sufficient flow reserves.

For an exhaust gas turbocharged gasoline engine, Fig. 7.5 shows the changes which occur in the charger map at an altitude of 2,500 m. The volume capacity increases by about the same amount as the density decreases, and the same increase is seen for the resulting pressure ratio. It can also be seen that at this altitude the speed limits of the charger are reached.

Nowadays, for truck and passenger car diesel engines, high-altitude compensation is implemented up to about 2,500 m. At higher altitudes, torque and power are reduced – by injecting less fuel in such a way that the exhaust temperature remains approximately constant – to avoid excessive speeds of the charger. In the exhaust gas turbocharged gasoline engine, in which the boost pressure has to be controlled in any case, specified absolute boost pressure levels are maintained through the control of the required fuel quantity. Here too, the exhaust gas temperature is the decisive parameter.

While for naturally aspirated and, as discussed earlier, mechanically supercharged engines, according to the decrease of the air density, which reaches about 9% per 1,000 m, the engine output decreases at the same percentage, the exhaust gas turbocharged engine is able to compensate this change in ambient conditions.

For these changes in ambient conditions, Table 7.1 provides some guiding lines regarding the change of important operating parameters in exhaust gas turbocharged four-stroke diesel engines (at constant fuel injection quantity).

7.4 Stationary and slow-speed engines

The transient behavior of the exhaust gas turbocharging system is also of importance for stationary and slow-speed engines. Here, sudden load changes also occur, by load addition or changes in operating status. As mentioned, each change in the boost pressure of the exhaust gas turbocharger is necessarily linked to a change in speed. Thus, in these engines significant accelerations occur very frequently as well, from low idle (lowest permissible engine speed) to peak torque at medium speed or even to rated power, i.e., highest torque at rated speed. To achieve this, the exhaust gas turbocharger has to be accelerated to higher speeds via an increased exhaust gas quantity as well as a correspondingly increasing exhaust gas temperature. Here, air delivery lags behind the power requirement.

The acceleration behavior of various exhaust gas turbocharger designs is the first parameter to be analyzed. This can be done by either applying a reference value developed by Zinner [160, 161], or by numeric simulation of the transient behavior (see Sect. 7.5).

At a sudden increase in power demand, the exhaust gas turbocharger lag will be the larger the higher the degree of supercharging is chosen. For example, if the mean effective pressure of a turbocharged four-stroke diesel engine is 20 bar, and assuming that such an engine in naturally aspirated configuration could reach about 9 bar, the power deficit at a sudden load increase is 11 bar. In this case, and this applies to all quality-controlled engines, it is not possible to immediately release the full-load injection quantity, but it has to be adjusted to the instantaneously available boost pressure.

This used to be done with a so-called charge-pressure-dependent full-load stop. Today, with electronically controlled injection pumps, the boost pressure status of the engine is continuously monitored and the injection amount is adjusted to the actual boost pressure. At the same time, for improved charger acceleration, the air-to-fuel ratio λ is electronically controlled down to a tolerable minimum. In any case, depending on engine size and design, it takes some time until full load is obtained.

7.4.1 Generator operation

For slow-speed engines, load increase in generator operation represents a critical situation. Figure 7.6 shows how the electric and engine data react if a load is suddenly applied to the generator. Although, due to the inertia of the system, the current requirement I can be met immediately, the full recovery of engine and charger speed takes about 6 s. Improvements can be achieved especially by a reduction of the charger inertia – i.e., possibly a layout with several chargers with smaller rotors instead of one large charger – as well as increases in the compressor and turbine efficiencies.

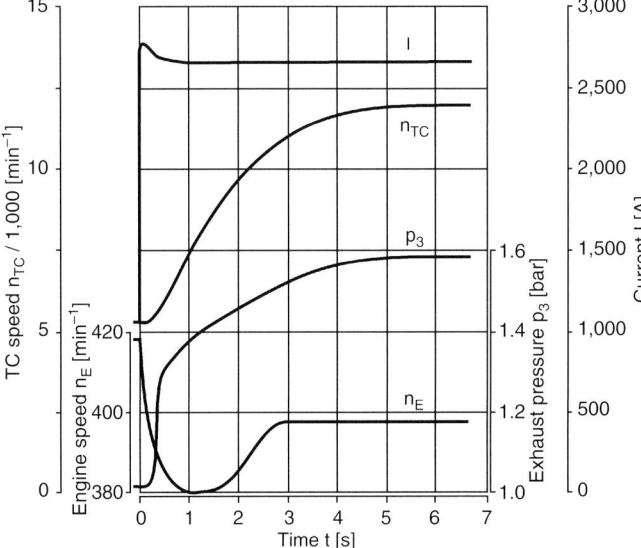

Fig. 7.6. Reaction of the electric and engine data when a sudden load is applied to a generator [159]

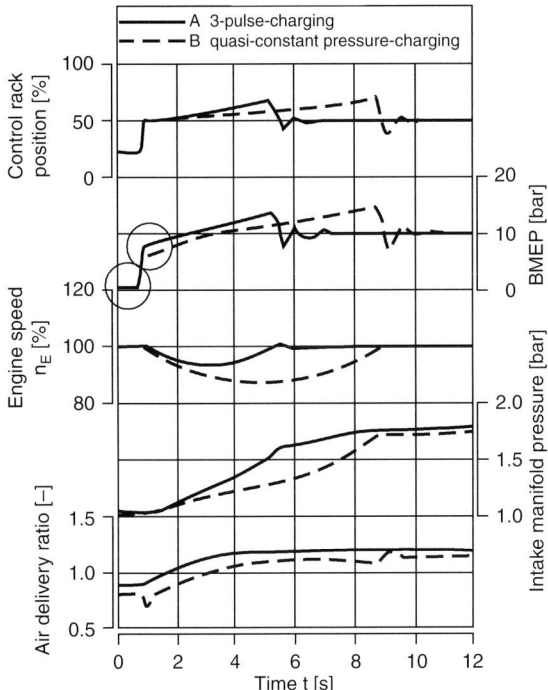

Fig. 7.7. Comparison of load response for pulse and constant-pressure turbocharging [159]

As Fig. 7.7 shows, for an engine with quasi-constant-pressure turbocharging and a 3-pulse turbocharging system, the layout of the charger group itself can significantly contribute to an improved load response. The **speed undershoot** is significantly minimized at pulse turbocharging – duration of about 5 s and speed decrease of about 8% – compared with that at constant-pressure turbocharging – duration of about 9 s and speed decrease of 15%.

7.4.2 Operation in propeller mode

The acceleration capability of the powertrain is also very important for propeller operation. Here the load is predetermined by the power-consumption capability of the ship or aircraft propeller. In addition, during transient operation of the system, specific standards for both noise and pollutant emissions must also be met.

For a medium-speed, highly supercharged four-stroke diesel engine with a power of about 4,000 kW, Fig. 7.8 shows an acceleration process, with a comparison of pulse turbocharging and constant-pressure turbocharging. Starting from idle at 135 $\min^{-1}$, the time was measured to reach full-load mean effective pressure of 16 bar at 400 $\min^{-1}$ by releasing the fuel injection quantity permissible at any given time. The advantage of utilizing gasdynamic processes in exhaust gas turbocharging is apparent.

As a second example, an acceleration process is shown in Fig. 7.9 for a slow-speed two-stroke engine that is directly connected to a ship propeller. The engine is equipped with constant-pressure turbocharging, and its piston bottoms are utilized as additional mechanical air pumps. At low loads, the additional air is directly routed to the charger via injectors. At higher loads, it is routed parallel to the exhaust gas turbocharger, directly into the charge air manifold. With constant-pressure turbocharging, at part load the turbine can generate only very little energy. Therefore, without the

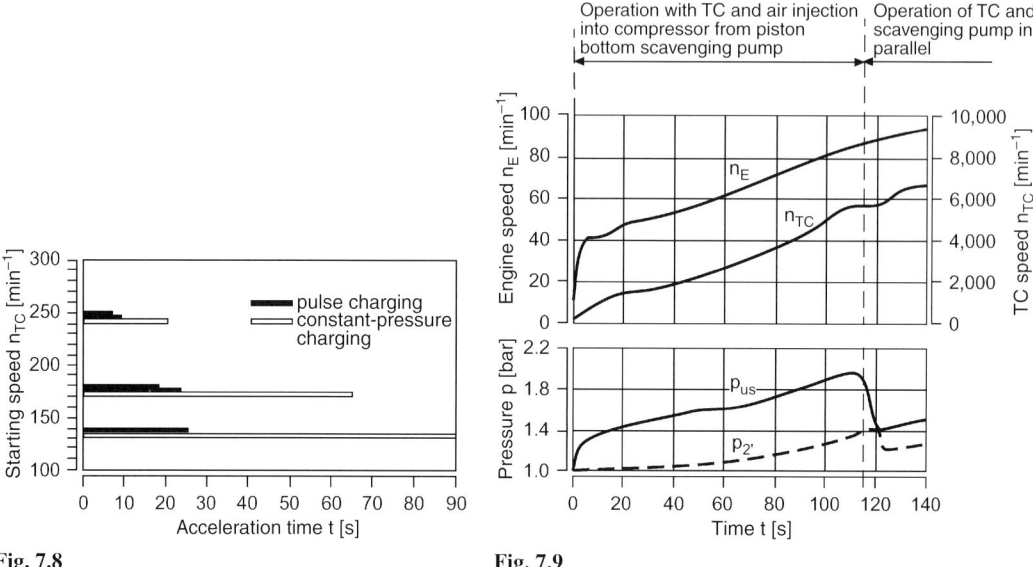

Fig. 7.8. Acceleration process of a slow-speed four-stroke engine with pulse and constant-pressure turbocharging [159]

Fig. 7.9. Acceleration process of a slow-speed two-stroke engine directly connected to a ship propeller [159]. p_{us}, pressure in the compressor injection nozzle

additional piston bottom pumps, the positive scavenging gradient necessary for scavenging could not be maintained. The injection of the additional air into the charger is also necessary to prevent compressor surging.

It can be concluded that the acceleration process is relatively slow, on the one hand, due to the large masses of the exhaust gas turbochargers, with compressor diameter of 760 mm, on the other, due to the very low turbine excess power of constant-pressure turbocharging systems (also see Sect. 7.4.3).

7.4.3 Acceleration supports

The acceleration support methods discussed here are especially suited for slow-speed engines. They are only meant to support acceleration and for adjustments when sudden load changes are required. They can be classified as follows:

– acceleration support necessary only at start and under sudden load increases from idle (rare utilization);
– acceleration support frequently needed which justifies certain additional design complexity (Sect. 7.4.4).

For the first group, preferably additional compressors, either externally powered or driven by the engine, or alternatively the addition of pressurized air from a tank can be considered. However, the additional air quantity acts differently for two-stroke and for four-stroke engines.

The pressure–volume flow map of a two-stroke engine is shown in Fig. 7.10. Here, at low loads an additional air mass, b_2, causes a push back of the air mass a_1, supplied by the exhaust gas turbocharger, into the surge area of the compressor. At high loads, this additional air mass is

7.4 Stationary and slow-speed engines

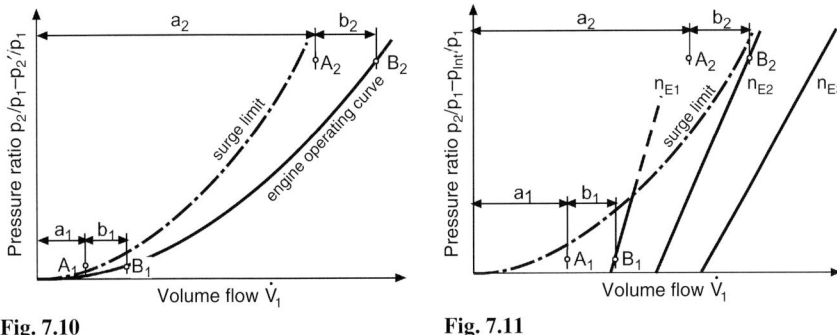

Fig. 7.10. Schematic charger surge limit and engine operating curve for a two-stroke engine with additional air injection

Fig. 7.11. Schematic charger surge limit and operating curves for a four-stroke engine with additional air injection

tolerated. A_1 is left of the surge limit. Therefore, in a two-stroke engine the induction of additional air – for the optimum acceleration of the exhaust gas turbocharger at the low end – has to be done via injection into the compressor (Fig. 7.9). Only at high loads, direct injection upstream of the engine is feasible.

For a four-stroke engine, the conditions are exactly opposite (Fig. 7.11). In the idle range, additional air induction is accepted by the turbocharger compressor. At high loads, its induction would force the compressor into the surge range (A_2 is left of the surge limit). Therefore, in a four-stroke engine the additional air must be added directly into the charge air manifold, only in the lower speed and load range.

7.4.4 Special problems of turbocharging two-stroke engines

For two-stroke engines it is especially important that acceleration support systems, with their increased design complexity, be constantly available. Depending on the actual layout, their exhaust gas turbocharger may not provide the necessary positive scavenging gradient (the boost pressure must be higher than the pressure upstream of the turbine) for all operating conditions. In order to solve this problem, a first possible method is the addition of a mechanically powered compressor in series with the compressor of the turbocharger. At low exhaust energy, i.e., insufficient scavenge pressure by the turbocharger, the mechanical compressor takes over. With increasing engine power, the exhaust gas turbocharger's fraction of the compression work increases, while that of the mechanical compressor decreases. An example of this setup in an automotive engine is described in Sect. 6.6.6.

An especially elegant but complex solution is powering the exhaust gas turbocharger from the engine crankshaft via a transmission and a freewheel clutch (Fig. 7.12). General Motors used this solution in their two-stroke locomotive engines EMD 567 and 645. At low engine power, when the exhaust gas energy is not sufficient, the shaft of the exhaust gas turbocharger is powered by the engine. With increasing engine power the exhaust gas turbocharger covers an ever increasing fraction of the compressor power. Once the compressor power is completely supplied by the exhaust gas turbocharger, the mechanical connection is disengaged by the freewheel clutch.

In slow-speed two-stroke engines, e.g., MAN models KSZ and KEZ (Fig. 7B), an electrically powered auxiliary supercharger is utilized which automatically engages or disengages at a specific load point. With this layout, the use of the piston bottoms as auxiliary pumps can be avoided. As an

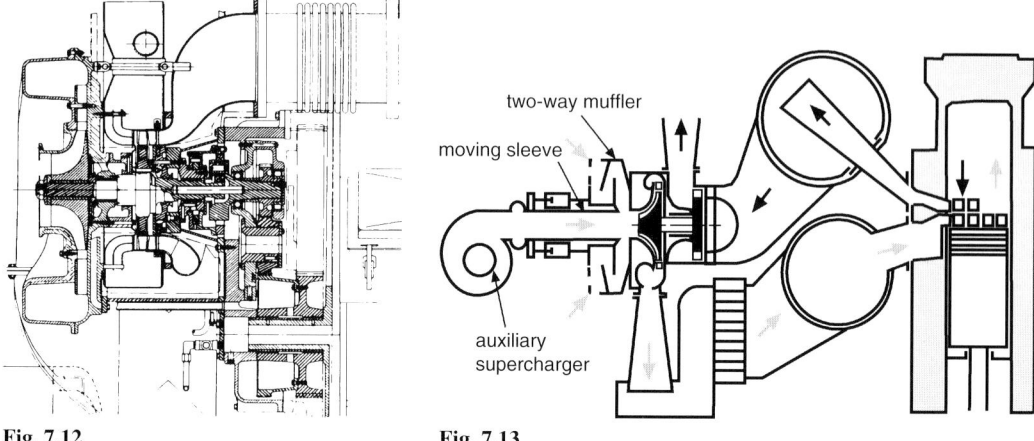

Fig. 7.12. Additional exhaust gas turbocharger drive, from the crankshaft via transmission and freewheel clutch [GMC]

Fig. 7.13. Electrically powered auxiliary supercharger for the largest two-stroke engines [MAN]

alternative, Siemens has developed an electric motor directly coupled to an exhaust gas turbocharger shaft. Governed by power electronics, it is engaged at startup and part-load operation. Under all other conditions it runs, without using power, without being disconnected.

The described systems offer possibilities for transient performance improvements to be utilized in future, modern engine designs with the most sophisticated charger technology. This especially since new emission standards in marine port areas may very soon require additional steps to reduce exhaust emissions at low loads. More information on this follows in Chap. 13.

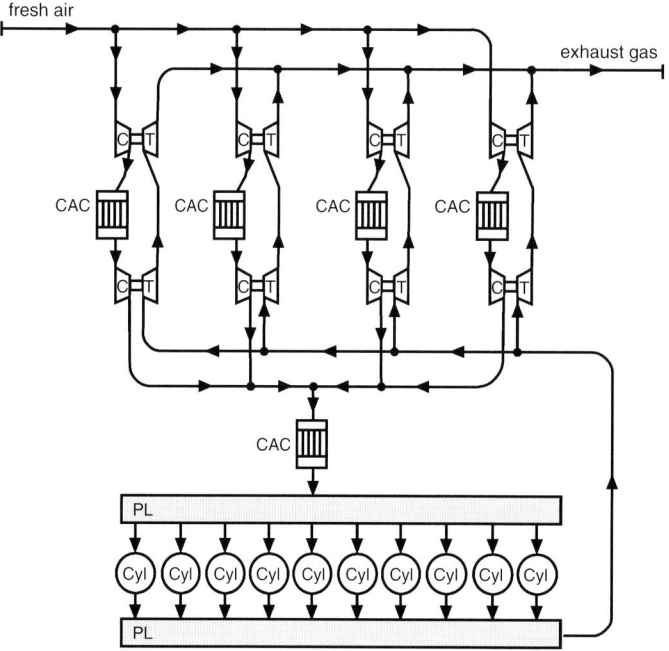

Fig. 7.14. Ship propulsion system with register-charged turbo engines [55]

7.5 Transient operation of a four-stroke ship engine with register charging

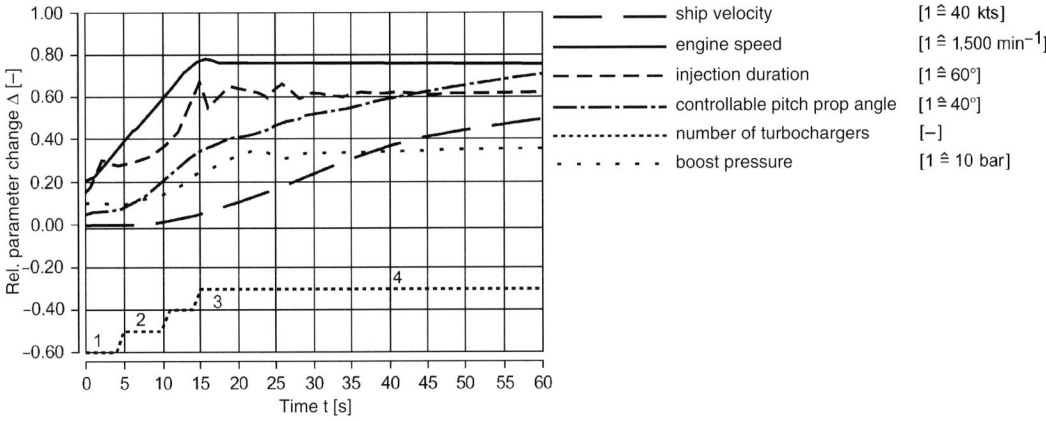

Fig. 7.15. Control strategy for optimum acceleration of a ship engine with register charging [55]

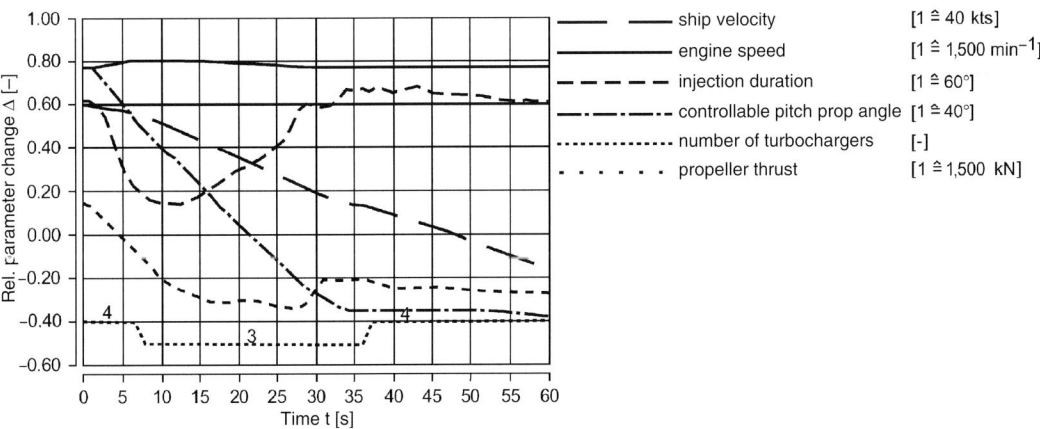

Fig. 7.16. Control strategy for optimum load cutoff of a ship engine with register charging [55]

7.5 Transient operation of a four-stroke ship engine with register charging

Not only for on-road vehicles but also for other applications the transient operating behavior of supercharged engines is of substantial importance in meeting operational requirements. A high-power ship powertrain and its dynamic operating behavior are discussed here as an example [55]. The complete system (Fig. 7.14) consists of the combustion engine and four two-stage register charging groups. Additionally, water-cooled charge air coolers are arranged between the compressor stages of each group. For transient operating processes, the individual register groups can now be connected with the exhaust gas manifold via corresponding control elements.

Figures 7.15 and 7.16 summarize the control strategy for the engagement and disengagement of the various register groups, the resulting response of the engine, and the acceleration and deceleration of the ship. It has to be mentioned that such a rapid acceleration and deceleration of the engine is only possible in interaction with a variable-pitch propeller.

8 Operating behavior of supercharged engines in automotive applications

The term "automotive" covers not only on-road vehicles but also off-road vehicles, tractors, and, e.g., locomotives. Here we first will only differentiate between passenger car and truck requirements. Off-road and tractor concerns can be treated as a subgroup of truck requirements. And requirements for locomotives largely correspond to those of stationary engine operating conditions. The special requirements of the latter vehicle categories will be discussed in Sect. 8.3.

For the application of a supercharged reciprocating piston combustion engine in a vehicle, it is totally insufficient to simply increase its specific power and torque. Far more criteria have to be met, as Table 8.1 tries to characterize even if it is certainly not exhaustive.

Therefore, we have to analyze what requirements today's engine designs pose on a supercharging system and to what extent the supercharging systems described here can meet these requirements.

In this context, the term "engine" is actually too limited. What's more, the criteria listed have to be met in various vehicle categories, e.g., passenger cars or trucks. Depending on the use, the criteria have to be weighted quite differently or, if need be, must be augmented. Then they can be summarized in a list of specifications for new engine concepts and can be checked against their anticipated use. In relation to the supercharging system, such requirement specifications are explained in the following sections.

8.1 Requirements for use in passenger vehicles

Already now, but even more so in the future, passenger cars require quite high driving performances. These are linked to equally high requirements regarding their driving and operating comfort, i.e., harmonic power curve and buildup, wide drivable speed range, high engine elasticity, and excellent smoothness of operation.

Table 8.1. Optimization criteria for reciprocating piston combustion engines

Economy	Environmental friendliness	Operating behavior
fuel economy	exhaust gas emissions	torque and power characteristic
durability	oil consumption	load response behavior
ease of maintenance	noise emissions	starting and drive-off
acquisition cost	recycling-compatible production	engine braking power
lifecycle cost	cold-start behavior	smoothness and balance

In addition, high economy must be achieved at low acquisition and lifecycle costs and at extremely low fuel consumption, which is one of the main reasons for supercharged passenger car engines.

Achievement of exhaust emission standards, with some safety margins, is naturally a prerequisite. These standards will certainly become even more stringent in the future. On the basis of the scenarios briefly discussed above, the – again strongly abbreviated – requirements for a supercharging system can be summarized as follows:

- a **wide capacity range** of the compressor, due to the wide speed range of passenger car engines;
- an only **limited torque increase** from the **lowest full-load speed**, to achieve a controllable starting and drive-off behavior;
- a relatively **flat torque curve**, to achieve acceptable drivability via the gas pedal, independent of the engine speed;
- **low cost** of the supercharging system.

8.2 Requirements for use in trucks

Trucks are used for quite more purposes than passenger cars – from delivery trucks to long-haul 18-wheelers. Here we will concentrate on the long-haul scenario. In this application, the maximum speeds permissible have to be maintained wherever possible, which requires high excess engine power and high braking power. In addition, the lowest possible fuel consumption values must be realized. This is economically of the utmost importance – for a long-haul truck, fuel cost represents up to 30% of the operating costs. Furthermore, it must be possible to start – fully loaded – on an incline, which poses high demands on the torque curve and torque response. Further indispensable requirements are low initial cost, durability, long maintenance intervals, and high reliability.

Thus, different requirements apply regarding the application of supercharging systems in trucks:

- A **narrow capacity range** of the charger, due to the narrow power speed range of truck engines;
- **high to very high attainable pressure ratios**, from 3.5 to 4.5, which approach the strength limit of today's aluminum cast radial compressors, due to the high degree of supercharging required and the significant torque back-up desired, up to 30%;
- a **durability** of the supercharging system that is compatible with the durability of the engine, nowadays at least 0.6 to 1 million km.

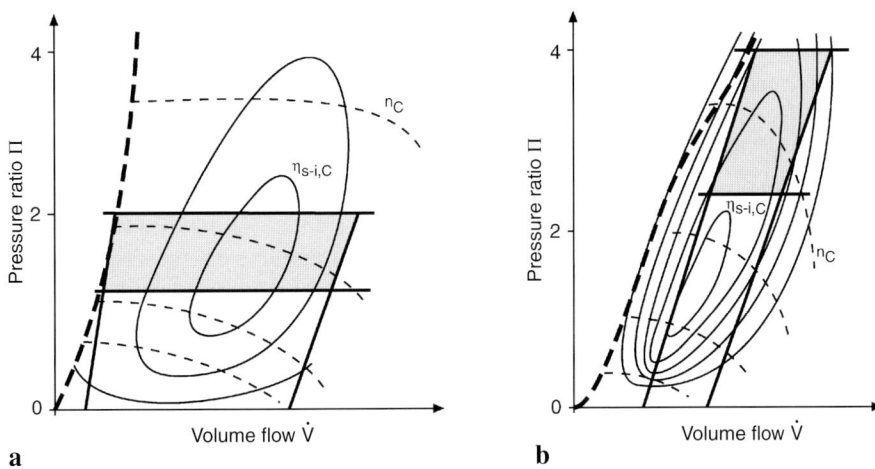

Fig. 8.1. Map characteristics for passenger car (**a**) and truck (**b**) applications

Plotting these requirements in a pressure–volume flow map, which is the relevant diagram to characterize superchargers, for passenger car applications a wide and flat map and for truck applications a narrow and high map are obtained (Fig. 8.1).

8.3 Other automotive applications

In addition to the requirements listed above for truck engines, engines for **agricultural applications** must meet the following requirements with regard to their supercharging system: Large torque increase at falling engine speeds all the way down to the lowest full-load speed, in order to avoid engine stall at load peaks (e.g., when plowing).

If this is not feasible, engines with higher rated power have to be utilized, whose normal load response can deal with such load peaks.

For **locomotive applications**, a distinction has to be made between diesel-hydraulic and diesel-electric powertrains.

For the diesel-hydraulic powertrain, largely the same requirements as for truck engines apply, especially the need for a torque increase at lower speeds. Here this is necessary to approach the traction hyperbola in the individual transmission gears as closely as possible.

In contrast, the diesel-electric powertrain is mostly operated along the load curve, and therefore the criteria established for stationary engines apply.

Depending on the design of a ship, to some extent very diverse requirements apply for **maritime applications**. For example, under normal operating conditions, at least the additional acceleration power for engine, powertrain, and propeller must be generated, while for hydrofoils additionally the emerge resistance has to be accommodated.

8.4 Transient response of the exhaust gas turbocharged engine

What do the requirements mentioned above mean for an exhaust gas turbocharging system? For turbocharged automotive engines, the problems with load response, described in Sect. 7.4.1 and more closely specified above for automotive applications, are especially inconvenient, because they can interfere with the operator's desire to control the power and torque of his vehicle by means of the gas pedal. This is especially true in critical starting situations.

But also under other (perhaps even critical) driving situations, the driver of a road vehicle depends on the predictability of his powertrain. An example is a passing maneuver. For such a situation, naturally aspirated or mechanically supercharged engines are best suited. This is one reason why passenger cars with gasoline engines today are predominantly equipped with naturally aspirated engines.

Further, in some engine designs, e.g., the supercharged gasoline engine, the boost pressure has to be limited to avoid knocking combustion.

As a consequence of the above statements, for supercharged automotive engines, or any other supercharged engines with predominant transient operation, the **boost pressure buildup** and the **control of the boost pressure** are of major importance. This applies especially for **the time needed to increase the boost pressure when the load is increased**.

To take a closer look at this problem, comparative load response tests were performed utilizing a very fast-reacting Comprex pressure-wave charger (Sect. 6.5.4) and a fixed-geometry exhaust gas turbocharger equipped with waste gate. Figure 8.2 shows the typical differences in boost pressure buildup when the load is increased, once from low part-load (exhaust gas temperature of 150 °C),

8.4 Transient response of the exhaust gas turbocharged engine

and once from medium part-load (exhaust gas temperature of 450 °C). As can be seen, with buildup times of about 0.5 s the pressure-wave charger shows excellent results.

Following up on these results, load change response tests were performed for typical truck and passenger car application patterns. Both bench tests and road tests with the corresponding vehicles were performed. The resulting driving impressions were characterized and discussed on the basis of the boost pressure buildup characteristics.

8.4.1 Passenger car application

The power and torque curves of a passenger car diesel engine, equipped once with a pressure-wave charger and once with an exhaust gas turbocharger, are shown in Fig. 8.3. The engines have the same power output, and – at least under steady-state conditions – the pressure-wave charger does not have a significantly different torque curve.

Figure 8.4 shows the results of transient acceleration in first and fourth gear on a test bench. Here significant differences in load response can be observed. While it takes the exhaust gas turbocharger about 3 s for full boost pressure buildup, the fast pressure-wave charger can perform this in about 1 s.

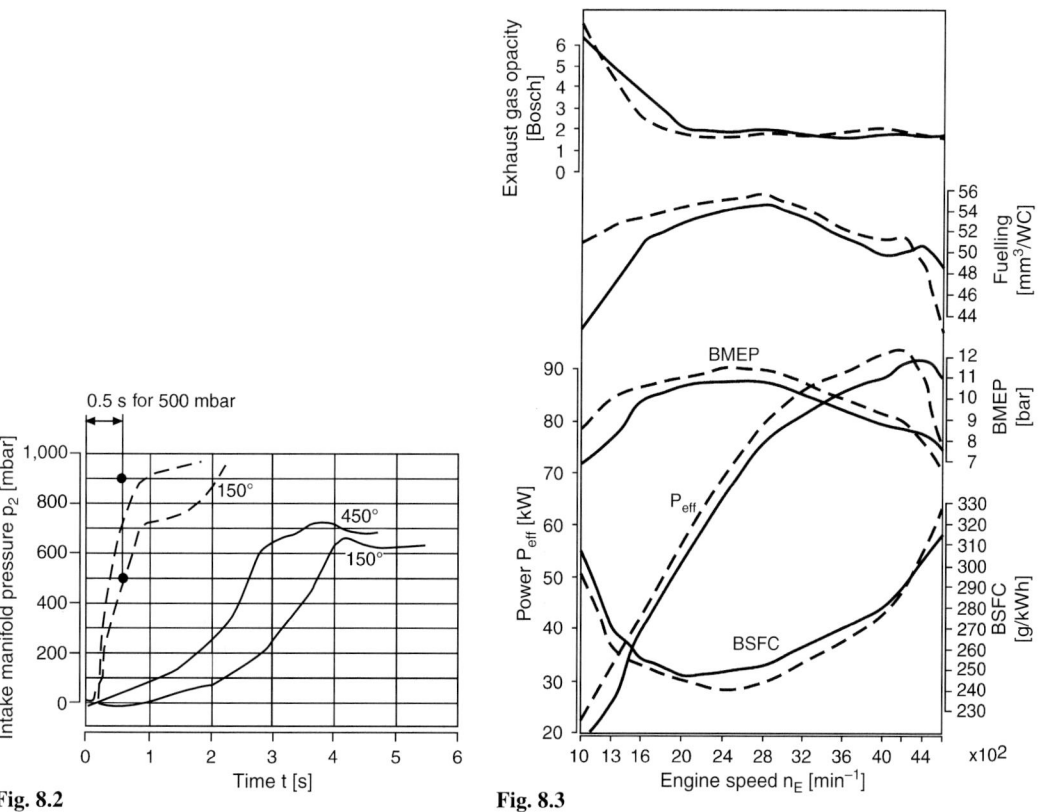

Fig. 8.2. Response behavior of a pressure-wave charger (dash line) and an exhaust gas turbocharger (solid line), load change from low load and from medium load [157]

Fig. 8.3. Power and torque curves of a passenger car diesel engine equipped with pressure-wave charger (dash line) and with exhaust gas turbocharger (solid line)

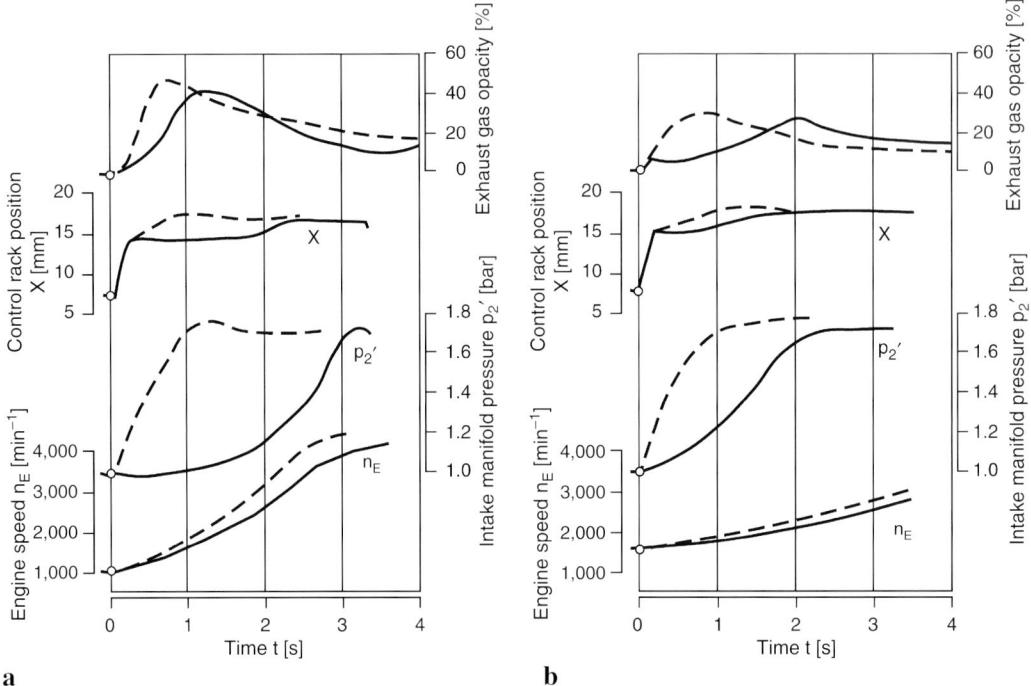

Fig. 8.4. Transient acceleration bench test of a passenger car diesel engine with pressure-wave charger (dash line) and turbocharger (solid line), in first (**a**) and fourth (**b**) gear

The impact of this on the driving performance of a passenger car is very significant as plotted in Fig. 8.5. It shows the starting and acceleration behavior of a vehicle, equipped with both engine variants and a 4-speed automatic transmission. Again, in the case of the pressure-wave charger the boost pressure buildup occurs within 1 s. This results in a harmonious power increase and accordingly harmonious speed increase with approximately constant acceleration. Since this is what the driver expects, it is rated as **good driving behavior**.

When the vehicle is equipped with the exhaust gas turbocharged engine, the full boost pressure buildup does not only take about 3 s but also is **nonlinear**. This results in a significant dip in speed increase and acceleration which the driver neither expects nor accepts. It is obvious that the pressure buildup characteristics of the supercharging system play a decisive role in an application in a passenger car. In short, for passenger car applications of supercharging systems the following rule applies:

In passenger cars, supercharging systems with **boost pressure buildup times significantly larger than 1 s result in a nearly unacceptable load buildup behavior**.

It is therefore imperative to reduce the boost pressure buildup times of the exhaust gas turbocharger by suitable measures.

8.4.2 Truck application

Figure 8.6 shows the power, torque, and fuel consumption curves of two inline 6-cylinder truck engines rated at 224 kW each. The pressure-wave charger enables a significantly advantageous torque curve in comparison with the exhaust gas turbocharger. The transient load response bench

8.4 Transient response of the exhaust gas turbocharged engine

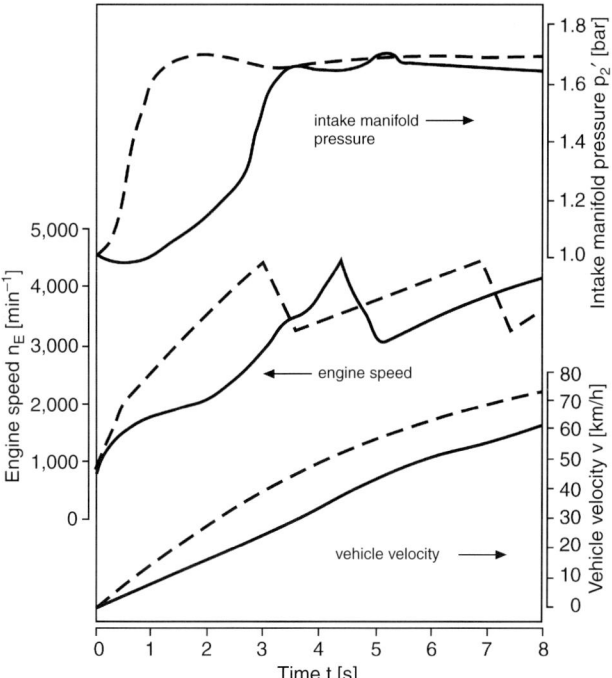

Fig. 8.5. Driving performance of a passenger car with pressure-wave charger (dash line) and turbocharger (solid line), and automatic transmission

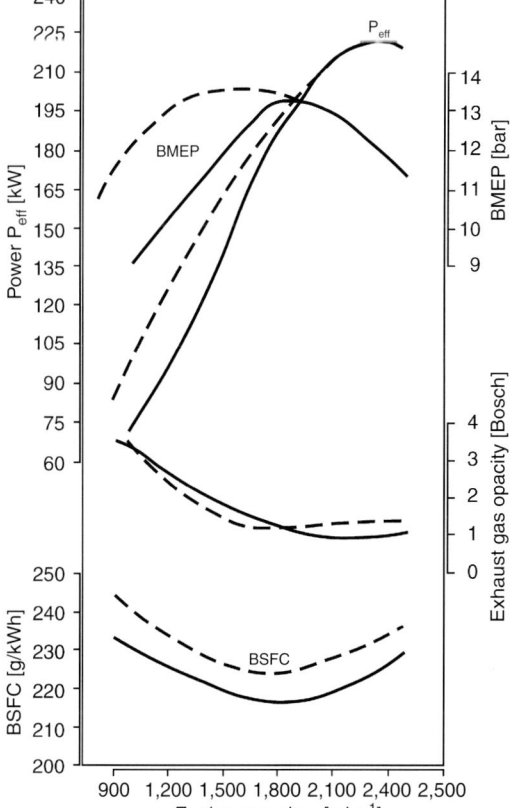

Fig. 8.6. Power, torque and fuel consumption curves of two truck engines of 224 kW rated power, with pressure-wave charger (dash line) and with turbocharger (solid line)

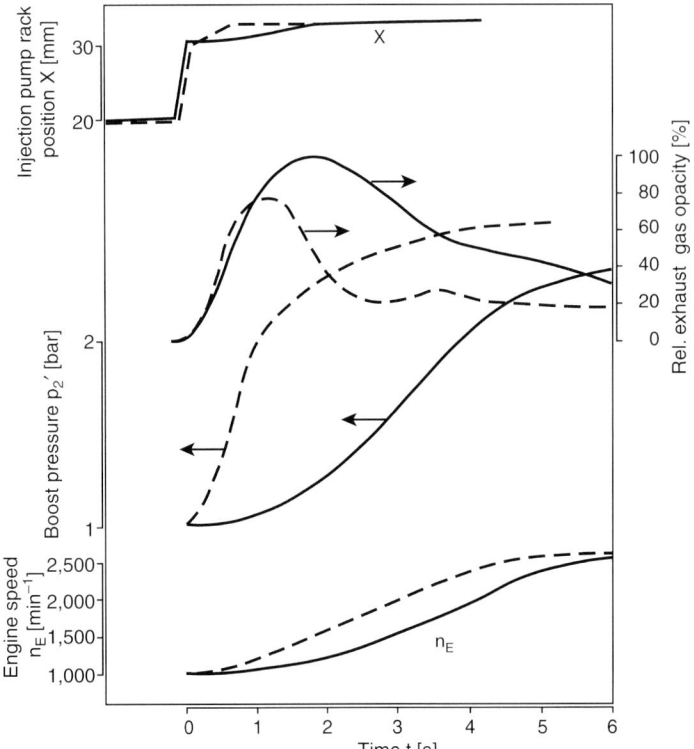

Fig. 8.7. Load response bench test of the engines shown in Fig. 8.6, with pressure-wave charger (dash line); with exhaust gas turbocharger (solid line)

test is summarized in Fig. 8.7. While the pressure-wave charger can quickly build up its full boost pressure – i.e., the boost pressure resulting at a given engine speed under steady-state conditions – in about 1 s, the exhaust gas turbocharger takes about 4 s for this task. This is an immense difference.

Figure 8.8 shows a starting process, at an incline and with full payload, with both engine configurations in identical vehicles. As can be seen, due to the large mass of the truck the seemingly sluggish boost pressure buildup of the exhaust gas turbocharged engine is practically unnoticeable in the two lower gears. In contrast, the bad driveability in the third gear becomes very obvious – caused by the exhaust gas turbocharged engine's lack of torque at low speeds.

With this, the following conclusions can be drawn with regard to the exhaust gas turbocharged truck engine:
– high torque at low speeds is very important to assure power continuity on upshifts;
– in comparison to a good torque curve, the load response time of the exhaust gas turbocharger is not very critical for driving performance;
– for the starting process itself, fast boost pressure increase is only necessary if the basic torque of the engine, i.e., its naturally aspirated torque, is not sufficient for starting under difficult conditions.

Of course, for a complete engine assessment, these summarized conclusions have to be augmented by additional criteria, e.g., pollutant emissions, lowest possible fuel consumption, noticeable power dip of turbocharged engines while shifting, and engine braking behavior.

8.5 Exhaust gas turbocharger layout

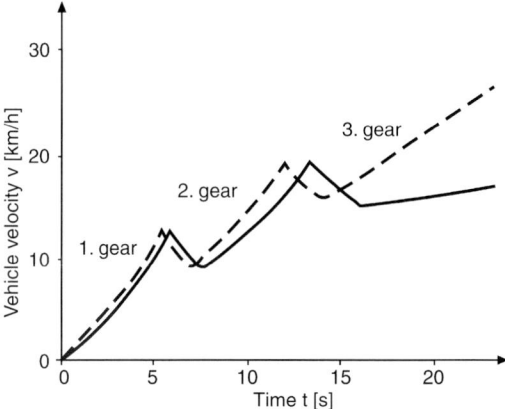

Fig. 8.8. Starting process at incline, full payload, two identical trucks, engines from Fig. 8.6 with pressure-wave charger (dash line), turbocharger (solid line)

8.5 Exhaust gas turbocharger layout for automotive application

8.5.1 Steady-state layout

Engine exhaust gas temperatures change with changing engine speed. This represents a layout problem for the automotive application of turbochargers since it means that the turbine has to process larger and highly fluctuating volume flows in comparison to the compressor. Turbines with fixed geometry are limited in their capability to do this (Fig. 8.9), i.e., to provide for volume flow ratios of 3:1 and also for similar full-load speed ratios. This can easily be recognized by the fact that the turbine pressure ratio exceeds the compressor pressure ratio. Figure 8.10 shows an actual exhaust gas turbocharger compressor map of a fixed-geometry charger, into which – corrected for the different volume flow – the correspondingly occurring turbine pressure and turbine pressure ratio curves are also plotted.

For this reason, so-called relief (blowoff) or waste gate control is often utilized for engines with a wider speed range, i.e., passenger car and small truck engines.

Waste gate control describes an arrangement in which a valve is located in the exhaust manifold upstream of the turbine. Whenever required, a certain exhaust gas quantity can be routed by the

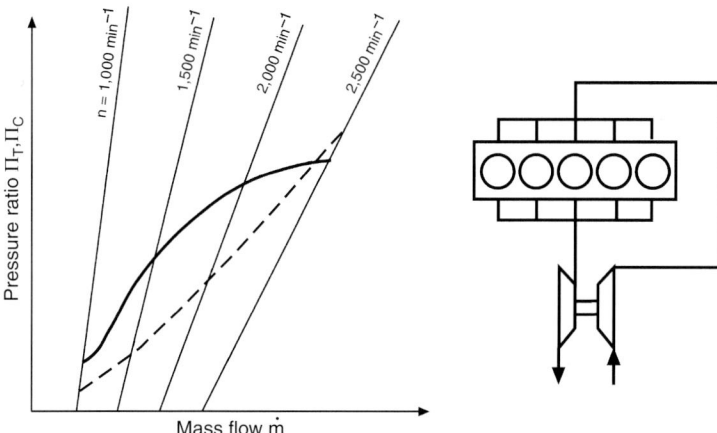

Fig. 8.9. Comparison of pressure ratio–mass flow performance characteristics of a compressor and a turbine with fixed inlet geometry (solid line, Π_T; dash line, Π_C)

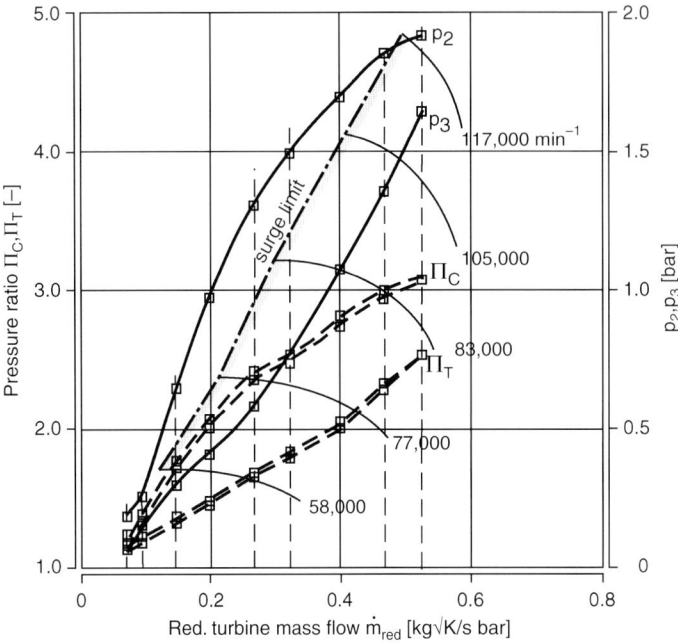

Fig. 8.10. Actual turbine and compressor map of a fixed geometry truck turbocharger

valve directly into the low-pressure exhaust system, bypassing the turbine. This leads to a reduction in turbine volume flow. However, compared to a turbine designed for rated power, for the remaining exhaust gas this has the disadvantage of a higher turbine inlet pressure, with which the turbine still has to generate the power required to drive the compressor. A principle diagram of a compressor and turbine map with possible layout strategies for the turbine is sketched in Fig. 8.11.

Figure 8.12 shows a sketch of the principle and Fig. 8.13 an actual compressor and turbine map of an exhaust gas turbocharged gasoline engine equipped with a waste gate turbocharger. It is obvious that the boost pressure has to be reduced with increasing volume flow, i.e., power, to avoid negative pressure gradients between charge air and exhaust gas flow.

Considering the facts discussed up to now, it can be concluded that typically a fixed-geometry turbine can be effectively matched to an engine – provided it is for minor volume flow ranges of about 3:1, i.e., for truck applications. For volume flow ratios above 4:1, even with waste gate control, either high boost pressure cannot be realized in the low-speed range or the turbine cannot handle the flow rates at high engine speeds (i.e., due to volume flow and turbine inlet pressure reasons, p_3 becomes intolerably high). This is even more so if the boost pressure has to be limited – which for turbocharged passenger car engines is always the case due to powertrain stress or combustion problems (knocking combustion in gasoline engines).

It was therefore necessary to look for an effective solution which on the one hand increases the exhaust gas turbine power at low exhaust gas flows and low exhaust gas temperatures. On the other hand it should generate the required turbine power at high engine speeds, preferably utilizing the total exhaust gas mass and with the lowest possible exhaust gas turbine inlet pressure. A solution already in series production for some time is a variable turbine geometry (VTG).

In an **exhaust gas turbocharger with variable turbine geometry** the fixed-geometry turbine housing is replaced by one with adjustable blades. With this, the conditions for flow entry into the

8.5 Exhaust gas turbocharger layout

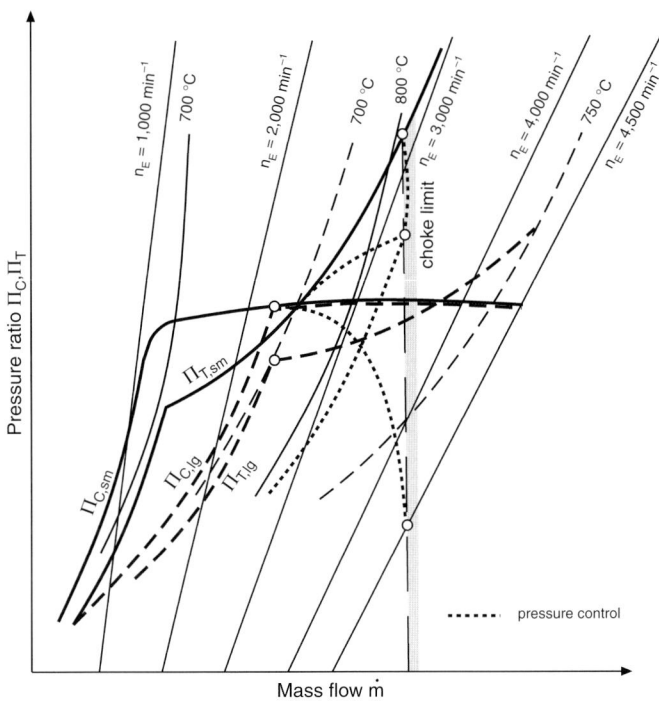

Fig. 8.11. Turbine layout strategies using waste gate control. Indices sm and lg indicate small and large turbine, respectively

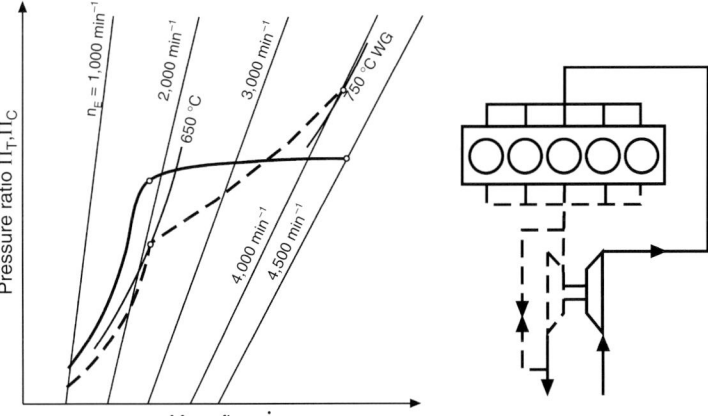

Fig. 8.12. Sketch of principle of a waste gate charger (solid line, Π_C; dash line, Π_T)

turbine can be varied to a large extent. As a result, such an exhaust gas turbocharger can already generate high boost pressures at low full-load speeds. Further, at high volume flows the required turbine power can be achieved with comparatively good efficiency, i.e., low exhaust gas turbine inlet pressures. This is shown in Fig. 8.14 in comparison with a waste gate charger.

Nowadays, nearly all passenger car diesel engines are using a VTG turbocharger in series production. The first truck engines with VTG are also in series production and more will soon follow. Figure 8.15 shows an actual compressor and turbine map of a small truck engine. Boost pressure and exhaust gas turbine inlet pressure are tuned for minimum fuel consumption at full load.

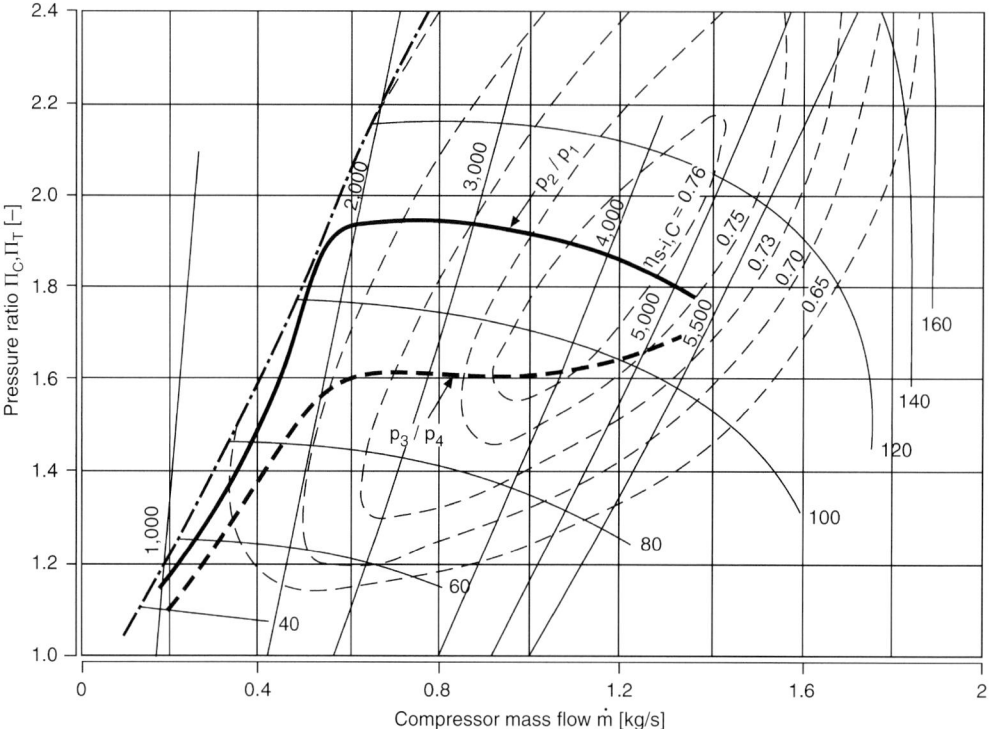

Fig. 8.13. Compressor–turbine map of an exhaust gas turbocharged gasoline engine with waste gate charger

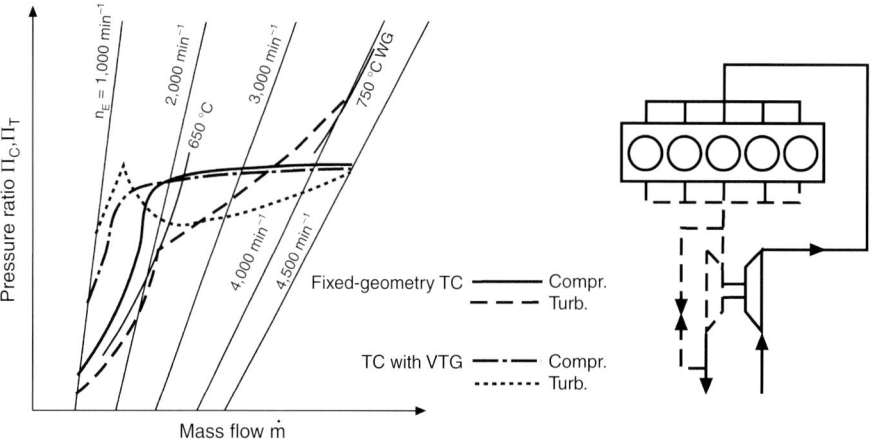

Fig. 8.14. Comparison of principle map characteristics for a fixed-geometry and VTG exhaust gas turbocharger

8.5.2 Transient layout

Up to now, only the general relationship between engine and supercharging system under steady-state operating conditions has been examined. However, for many applications mostly transient operating conditions dominate the load histories.

8.5 Exhaust gas turbocharger layout

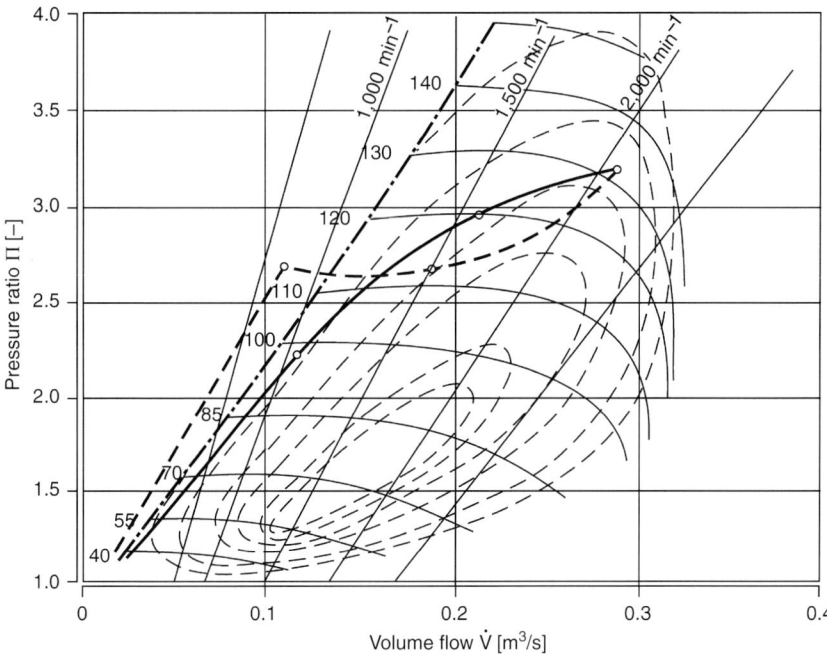

Fig. 8.15. Combination of actual compressor and turbine maps, truck engine with VTG charger (solid line, Π_C; dash line, Π_T)

As was discussed as an empirical test result in Sect. 8.4, besides generating the rated boost pressure, this also addresses the time needed to build up the pressure when sudden load increases occur. In addition, the effects on the driving behavior which have been established in tests must be analyzed. To do this, let us examine the diagrams in Fig. 8.16.

Figure 8.16a shows the compressor power (thick line) required for the desired boost pressure curve, plotted against the mass flow, which itself is approximately proportional to the engine speed.

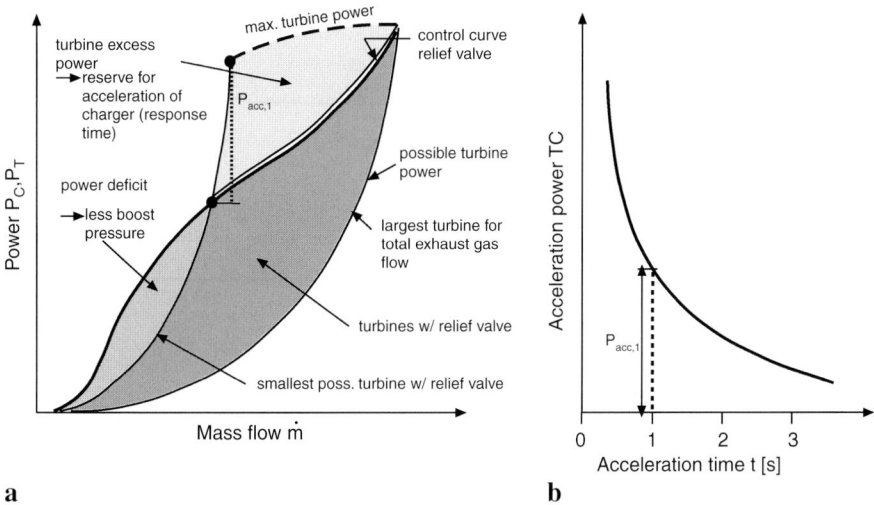

Fig. 8.16. a Compressor power requirement for optimum boost pressure, and possible turbine power delivery depending on turbine layout. **b** Required charger acceleration power against acceleration time

Additionally, the power curves for two extreme turbine layouts are shown.

The right-hand limiting curve describes a turbine layout which can achieve the required compressor power only at maximum air flow, i.e., rated engine power (see its intersection with the power curve of the compressor). The left-hand limiting curve represents a turbine layout which, at maximum blowoff quantity along the choke limit of the turbine, would be barely able to generate the maximum compressor power.

As can be seen, even this smallest turbine with waste gate cannot generate the compressor drive power required for the desired boost pressure at lowest engine speeds, i.e., the engine can be operated only with less torque than desired. The power deficit is shown in light grey. On the other hand, it can be seen that sufficient excess power at the turbine, necessary for acceleration of the charger, is achieved only at relatively high mass flows. In addition, the diagram shows the turbine excess power requirements, necessary for specific turbocharger speed acceleration values and thus pressure buildup time spans.

The value $P_{acc,1}$, plotted in Fig. 8.16a, represents the acceleration power required for boost pressure buildup in 1 s. Transferred to Fig. 8.16b, this results in the flow range in which this rapid boost pressure change can be realized. In our example, this would be possible from about half of total air flow, i.e., from about half of rated engine speed.

Examining the situation for an exhaust gas turbocharger with variable turbine geometry in the same diagram setup (Fig. 8.17), it becomes obvious that the conditions are significantly improved, both for the generation of sufficient boost pressure at low engine speed and for the excess power required for rapid boost pressure buildup.

Figure 8.18 shows acceleration bench tests comparing a fixed-geometry exhaust gas turbocharger, a pressure-wave charger (known for its rapid response and here taken as benchmark), and a VTG exhaust gas turbocharger, utilizing the same passenger car diesel engine and for accelerations in first and fourth gear. As can be seen, at pressure buildup time spans of about 1 s in both gears, the VTG charger can reduce the corresponding values of the fixed-geometry turbocharger by about 2/3.

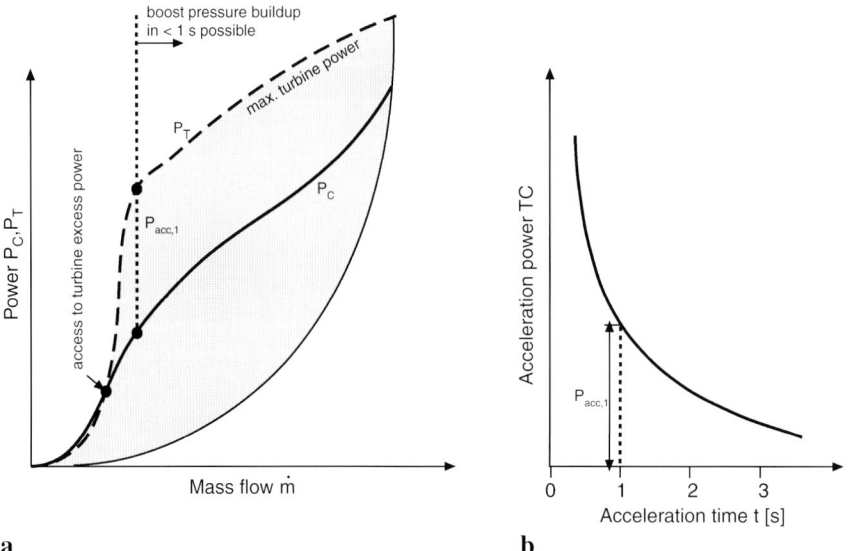

Fig. 8.17. **a** Compressor power requirement for optimum boost pressure, and possible turbine power for a VTG charger. **b** Required charger acceleration power

8.5 Exhaust gas turbocharger layout

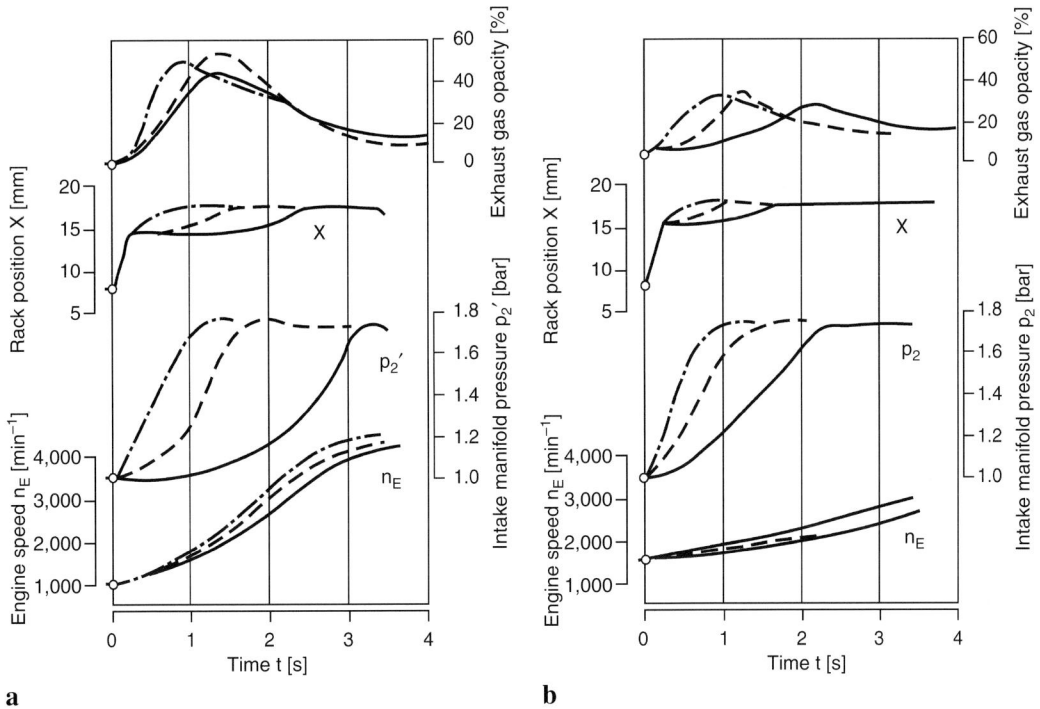

Fig. 8.18. Acceleration bench tests comparing a fixed geometry exhaust gas turbocharger with waste gate (solid line), a pressure-wave charger (dash line), and a VTG exhaust gas turbocharger (dot dash line), utilizing the same passenger car diesel engine and for accelerations in first (**a**) and fourth (**b**) gear

Extended driving tests lead to the conclusion that pressure buildup times of about 1 s are considered pleasant by normal drivers. Therefore, a torque buildup time of 1 s must be aimed at. With the exception of the starting phase, this target value can be reached by using a VTG charger.

Thus, the transient layout criteria can be summarized as follows:

- If mostly transient conditions are of major relevance, both for truck and passenger car applications the layout of an exhaust gas turbocharger leads to certain problems. However, these problems differ in trucks and cars:
  In truck applications, due to the high degree of supercharging utilized today, which result in pressure ratios of $\Pi_C = 3.5$–4.5, the turbine and compressor map is narrow.
  In passenger car applications, due to the wide speed range of modern engines, the full utilization of the turbine and compressor maps becomes problematical.
- In addition, the turbine layout, with or without waste gate, must accommodate these requirements:
  at maximum speed a sufficient distance has to be maintained from the choke limit of the turbine;
  an adequate speed reserve for high-altitude operation has to be provided;
  at low engine speeds a boost pressure as high as possible has to be achieved.
- As a final basic requirement, the turbine must be able to provide sufficient acceleration power for short pressure buildup times.

Naturally, waste gate and VTG chargers are most successful in meeting all these requirements.

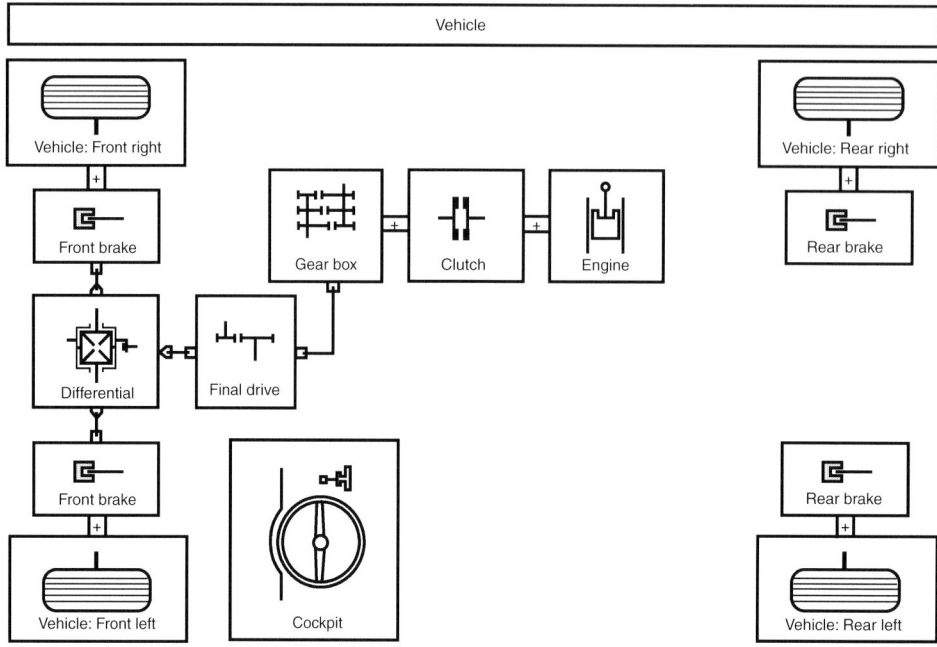

Fig. 8.19. Simulation model of a passenger car powertrain with diesel engine for drive cycle simulation (AVL-CRUISE)

Table 8.2. Comparison of measured and simulated exhaust emissions of a passenger car equipped with a 2.5 liter HSDI diesel engine in the MVEG test

Parameter	Measurement (vehicle dynamometer)	Simulation (AVL-CRUISE)
Fuel consumption [liter/100 km]	6.75	6.80
Particulate emissions [g/km]	0.030	0.035
NO_x emissions [g/km]	0.385	0.360
Test length [km]	11.13	11.01

8.5.3 Numerical simulation of the operating behavior of the engine in interaction with the total vehicle system

The support of the preliminary layout of supercharged engines by numerical simulations is not limited to the charger layout and control itself, but – by using suitable methods – can be expanded to include the complete powertrain. Appropriate program systems enable the modeling of the complete vehicle (GPA [125], CRUISE [35]), or the required simulation models can be created via methods such as MATLAB-SIMULINK.

An example for complete system modeling with the simulation tool AVL-CRUISE is shown in Fig. 8.19. Here, a passenger car powertrain with a 2.5 liter DI-diesel engine is modeled. Such complete system models have to be verified by comparison with measurement data. Since these methods are preferentially used for the assessment of complete concepts (e.g., regarding their fuel consumption and emissions values in a driving cycle), the comparison between simulation and measurement data has to include not only the thermodynamic conditions but also the bench test emission values. Table 8.2 shows such a comparison for the mentioned vehicle.

8.6 Special problems of supercharged gasoline and natural gas engines

8.6.1 Knocking combustion

In a gasoline engine, knocking combustion may occur according to the following process. At full load, after normal ignition at the spark plug, the combustion propagates at a high pressure level and even increases further. Thus, the unburned mixture located ahead of the flame front is compressed isentropically. Its temperature is increased and it either self-ignites or ignition is caused by a "hot spot" in the combustion chamber. The mixture fraction which has until then remained unburned now combusts extremely rapidly, similar to a detonation. This causes pressure vibrations in the combustion chamber which are in some cases very violent. Such detonation waves are characterized by very high heat transmission coefficients. This leads to **mechanical and thermal damage** of the engine. Therefore, knocking combustion has to be avoided by all means. It thus limits the supercharging capability of gasoline engines.

In an examination of knocking combustion, the following main parameters have to be considered:

mixture condition (λ, cylinder charge)
compression ratio
ignition timing
combustion chamber geometry

For supercharged engines, the first three parameters are of special importance since they strongly influence the process via pressure and temperature upstream of the turbine. Their impact was analyzed utilizing an exhaust gas turbocharged 2.8 liter engine at an especially knock-critical speed.

The **mixture condition** in the cylinder of a spark-ignited engine is defined by the parameters compression end temperature $T_{2\text{cyl}}$, compression end pressure $p_{2\text{cyl}}$, the relative air-to-fuel ratio λ, the amount of residual gas in the cylinder, and the fuel octane number. These parameters, systematically varied, lead to the relationships shown in Fig. 8.20.

Here, the boost pressure p_2' possible at the knock limit is plotted against the mixture temperature T_2' upstream of the intake valve, for the two air-to-fuel ratios of $\lambda = 0.9$ and $\lambda = 1.1$ and for two fuel octane numbers, RON = 91 (regular gasoline) and RON = 100 (premium gasoline). The strong influences of both octane number and charge air temperature can be clearly seen.

We would normally anticipate that the **compression ratio** ε would significantly influence the tendency towards knocking due to its effect on the compression end pressure and temperature. However, in tests where ε was reduced from 8 : 1 to 6 : 1 the knock tendency did not decrease as much as expected. A rough estimation of the compression end temperature according to the equation

$$T_{2\text{cyl}} = T_{1\text{cyl}}\varepsilon^{\kappa-1} \cong T_2\varepsilon^{\kappa-1} \tag{8.1}$$

shows that a significant decrease in the compression end temperature and, thus, the knock tendency could theoretically be expected.

It becomes obvious that when the amount of residual gas is taken into account, the compression end temperature decreases much less with decreasing compression ratio ε than when residual gas is not taken into account. This is caused by the relative increase in the amount of residual gas at reduced compression ratio (increase in combustion chamber volume), which results in higher charge temperatures.

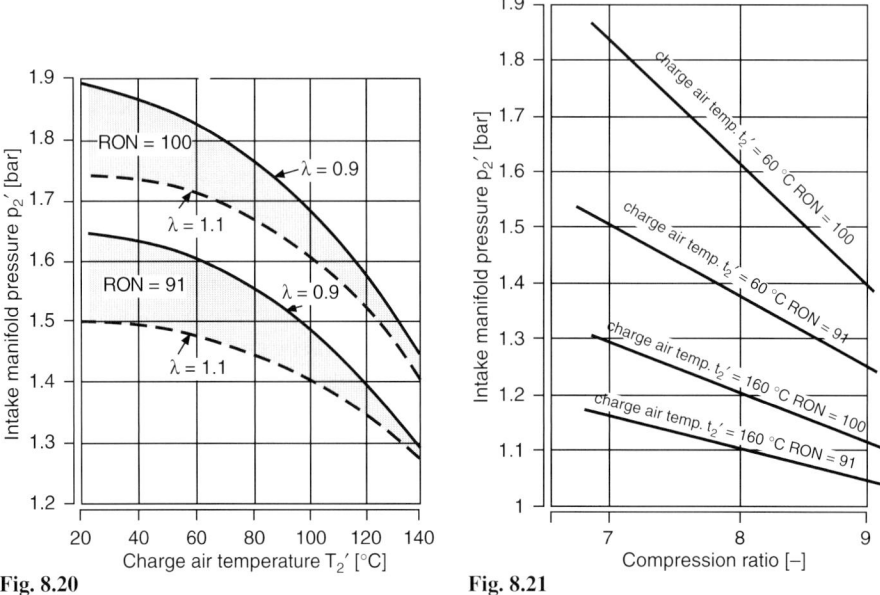

Fig. 8.20. Maximum possible boost pressure depending on p'_2, octane number and air-to-fuel ratio λ, against charge air temperature (for an exhaust gas turbocharged gasoline engine)

Fig. 8.21. Boost pressures achievable at knock limit, depending on octane number and charge air intake temperature, against compression ratio ε (for an exhaust gas turbocharged gasoline engine)

For the same test engine, the boost pressures achievable at the knock limit, depending on octane number and charge air intake temperature are shown in Fig. 8.21 as a function of the compression ratio ε. As can be seen, due to the relationships discussed above, a reduction in the compression ratio only allows small boost pressure increases if the engine is not equipped with charge air cooling. With the application of an efficient charge air cooler, however, the influence of the compression ratio on the achievable boost pressure significantly increases. This can be seen in the upper curve in Fig. 8.21.

Regarding its influence both on combustion knock and the efficiency of the high-pressure cycle, **ignition timing** is a significant factor. Figure 8.22 shows the influence of ignition timing on the achievable mean effective pressure, depending on the boost pressure and both with and without charge air cooling.

Even at a very low compression ratio of 8:1, the best ignition timing can be selected only up to a boost pressure of 1.4 bar. Due to the knock limit, higher boost pressures can only be realized with retarded ignition timing. Here too the dominant influence of charge air cooling can be seen.

Together with the intake port and valve geometry, the shape of the **combustion chamber** of a modern gasoline engine has to assure that sufficient mixture motion is achieved. To avoid knocking combustion, intensive mixture motion is especially effective. On the one hand, it mixes and thus cools down the residual gas mixture fractions, on the other hand, it increases the flame velocity. Nowadays, by designing the intake ports accordingly, a "tumble" is created in the combustion chamber.

Besides selecting the optimum combustion chamber shape, it is also necessary to assure sufficient and targeted cooling of the valves and walls of the combustion chamber. This will avoid any hot spots in the chamber where the mixture could prematurely ignite or at least could heat up significantly.

8.6 Gasoline and natural gas engines

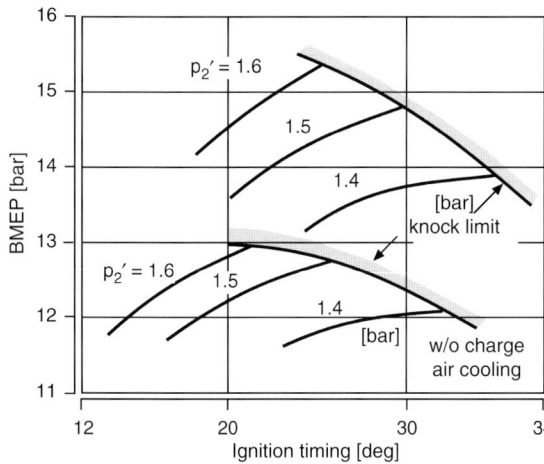

Fig. 8.22. Influence of ignition timing on achievable mean effective pressure, depending on boost pressure; with and without charge air cooling (exhaust gas turbocharged gasoline engine)

The four-valve technology of today's gasoline engines offers the best boundary conditions for supercharging since it enables a significantly better cooling of the hot exhaust valves due to increased valve seat and valve shaft areas, especially if the given potential for cooling has been fully exhausted and is augmented by piston cooling. In this context, further advanced research and development is necessary.

8.6.2 Problems of quantity control

As is generally known, the gasoline engine can only be operated within a relatively narrow range of the air-to-fuel ratio λ; with the 3-way catalyst – which is nowadays indispensable – only at $\lambda = 1$.

Therefore, load control cannot be attained in the same way as in the diesel engine, i.e., by only changing the amount of fuel at constant air flow (quality control), but it has to be done via changing the amount of mixture at constant $\lambda \approx 1$ (quantity control).

Accordingly, a mapping of swallowing curves against engine speed, as is characteristic for diesel engines, is not applicable for supercharged gasoline engines. As Fig. 7.4 shows in comparison to diesel engines with and without waste gate, the boost pressure in the intake manifold downstream of the throttle is kept at the desired pressure level, as long as possible, by the boost pressure control device. The reduction in the aspirated mixture mass – due to a load reduction – is achieved by closing the throttle located between the turbocharger and cylinder.

Only when the boost pressure cannot be maintained any longer due to lacking turbine power, the characteristic curve in the compressor map drops down. As can be seen in the map, for different engine speeds this occurs at different times. This leads to diesel-like operating curves, which, however, only reflect the different air flows necessary to overcome the power losses at the various engine speeds.

Admittedly, these remarks are only valid if the **throttle is located downstream of the charger**. The layout with the throttle upstream of the compressor will be discussed in Sect. 9.2.2.

Therefore, boost pressure control is of much greater importance for the supercharged gasoline engine than for the diesel engine. Naturally, this is especially valid for the exhaust gas turbocharged gasoline engine.

9 Charger control intervention and control philosophies for fixed-geometry and VTG chargers

First the most important definitions regarding control processes should be given. They are defined by the German Industry Standard DIN 19226.

- **Open-loop control**: One or several input parameters influence one or several output parameters in a system, based on regularities which are relevant to the system and are either measured or predefined.
- **Closed-loop control**: The parameters being controlled are constantly measured. The result is compared with the reference variables and then the parameters are adjusted correspondingly.

Each control loop contains at least one series connection of controller (governor) and control path. In a control loop, a distinction has to be made between command response and disturbance response.

- **Command response** is the reaction of the controlled parameter in the control loop to a **change in the command (target) value**.
- **Disturbance response** is characterized by the reaction of the controlled parameter to a **change in the disturbance variable**.

9.1 Basic problems of exhaust gas turbocharger control

The exhaust gas turbocharger is only thermodynamically coupled to the engine. This is an inherent advantage, but also a problem of turbocharger systems. Once the compressor and turbine have been selected, a specific boost pressure curve results in the entire speed range of the engine, dependent on the load only. The curve corresponds to the turbocharger main equation (equilibrium between turbine power and the sum of compressor power and friction losses).

Therefore, the boost pressure can only be controlled via the control of compressor or turbine power, as was discussed in Sect. 8.5.

At an airflow rate given or required by the engine, the **compressor power** can be controlled either by changes in the compressor efficiency or by changes in the air quantity, e.g., by means of a compressor bypass.

Several possibilities exist for controlling the **turbine power**:

- change of turbine volume flow by means of an exhaust gas waste gate;
- change of turbine inlet flow conditions by means of variable turbine intake geometry, i.e., the VTG turbine;

– change of engine parameters which influence the combustion efficiency, e.g., varying the air-to-fuel ratio or the combustion process (heat release) and thus the exhaust gas temperature.

In order to utilize these possibilities for a targeted and meaningful change in charger behavior and of the pressure and volume flows provided by the exhaust gas turbocharger, a control system is necessary. If transient engine operation is of special relevance, the need for control of the turbocharger intensifies. For the adaptation of all modern exhaust gas turbocharger systems it is therefore also necessary to know the possible or acceptable control systems and philosophies.

9.2 Fixed-geometry exhaust gas turbochargers

Control measures possible for exhaust gas turbochargers with fixed turbine inlet geometry will be discussed first. This covers both turbine volute housings for radial turbines and corresponding intake housings for axial turbines.

9.2.1 Control interaction possibilities for stationary operating conditions

All the possibilities mentioned are effective for turbocharged gasoline and diesel engines operating under steady-state conditions. They are: relief (or blowoff) valve at air intake side, exhaust gas waste gate, bypass valve at air intake side.

Air blowoff via a relief valve represents the simplest way to obtain a specified boost pressure. For today's mixture formation and engine control systems, which measure the amount of intake air, this design is not suitable. It is ineffective since engine power is wasted due to an unnecessary high exhaust gas turbine inlet pressure. Figure 9.1 shows a possible simple layout.

If the boost pressure is to be controlled dependent on load conditions, e.g., in a gasoline engine, this valve can be utilized for this task by connecting the intake manifold pressure to the spring chamber (see Fig. 9.2).

For **exhaust gas waste gate control**, a bypass with a valve is installed which allows to route a part of the gas flow around the turbine. Provided that the desired boost pressure is lower than the pressure achievable with fully closed waste gate, its power and thus the desired boost pressure for a specified load point can be adjusted by control of the amount of exhaust gas flowing through the turbine. Further, a waste gate layout has the ability to extend the virtual volume flow range of the turbine – by circumventing it.

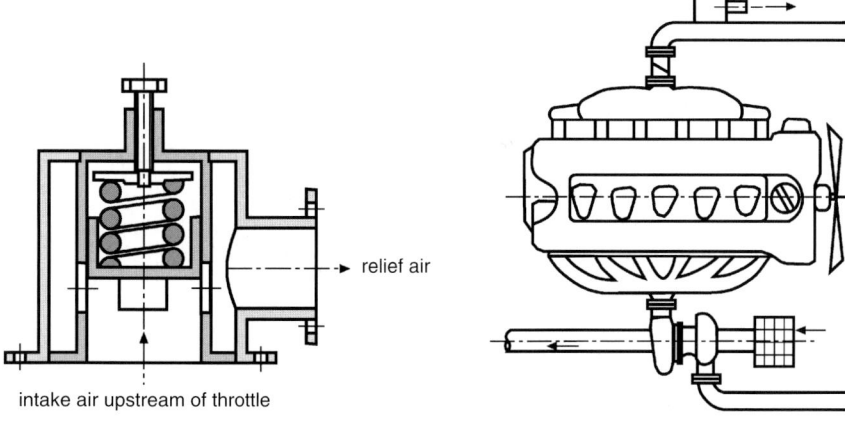

Fig. 9.1. Relief valve at air intake side

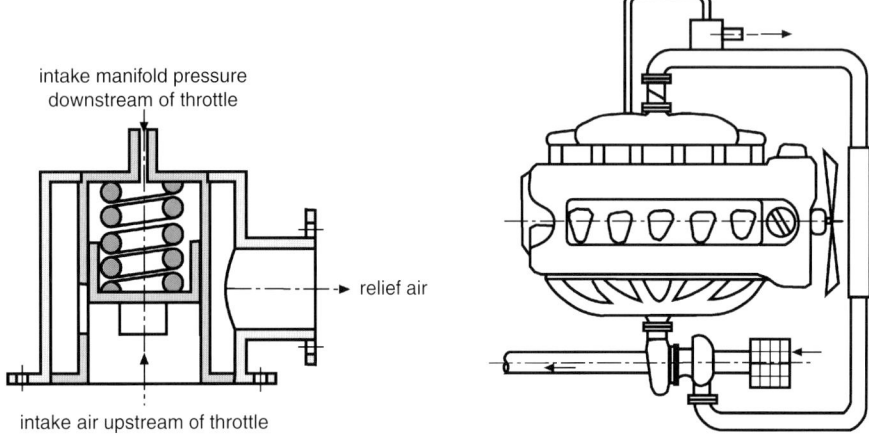

Fig. 9.2. Layout for boost pressure control via relief valve on the air intake side

However, it has the disadvantage that – in order to be able to achieve sufficient turbine power – the reduced amount of exhaust gas actually flowing through the turbine must have a higher turbine inlet pressure.

The waste gate, which is a blowoff valve, may be arranged separately in the exhaust manifold upstream of the turbine (Fig. 9.3). However, nowadays it is generally designed as a disc valve and integrated into the turbine housing. Figure 9.4 shows an example of the gas flow through the turbine and the waste gate, as designed by 3K-Warner. This valve is either controlled by the boost pressure

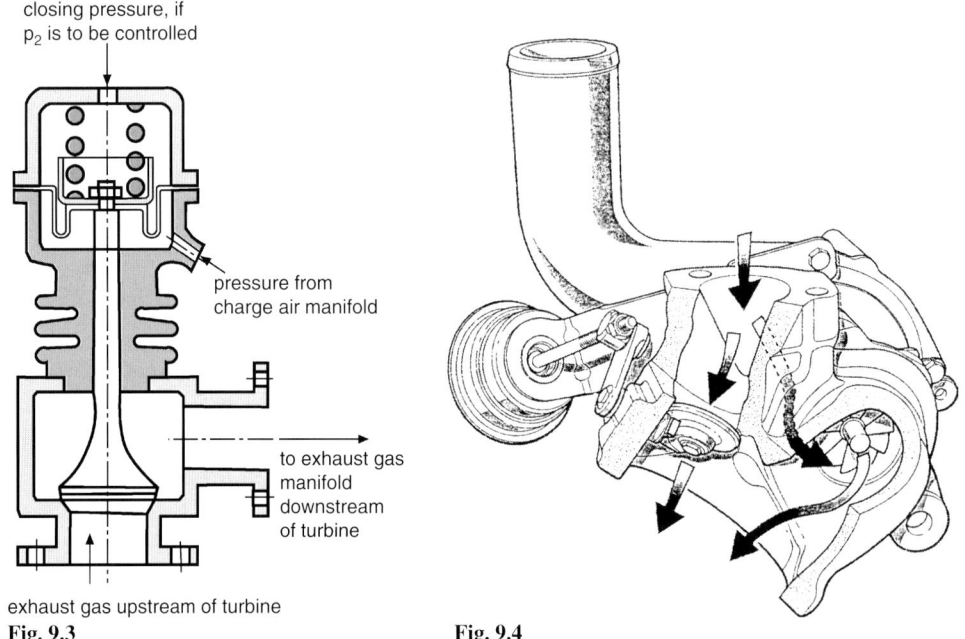

Fig. 9.3. Exhaust gas waste gate for separate arrangement

Fig. 9.4. Gas flow through turbine and waste gate

9.2 Fixed-geometry exhaust gas turbochargers

itself or – preferably – controlled electronically in accordance with a predetermined boost pressure target value.

If the spring chamber (Fig. 9.3) is also subjected to the boost pressure, the basic module for a very simple boost pressure control system is created (see Sect. 9.2.2).

Influencing compressor power by changing efficiency and air flow via a **bypass valve on the intake side** is frequently utilized in slow-speed engines. In these highly turbocharged engines, this control system can assure a sufficient distance between engine operating curve and surge limit, and

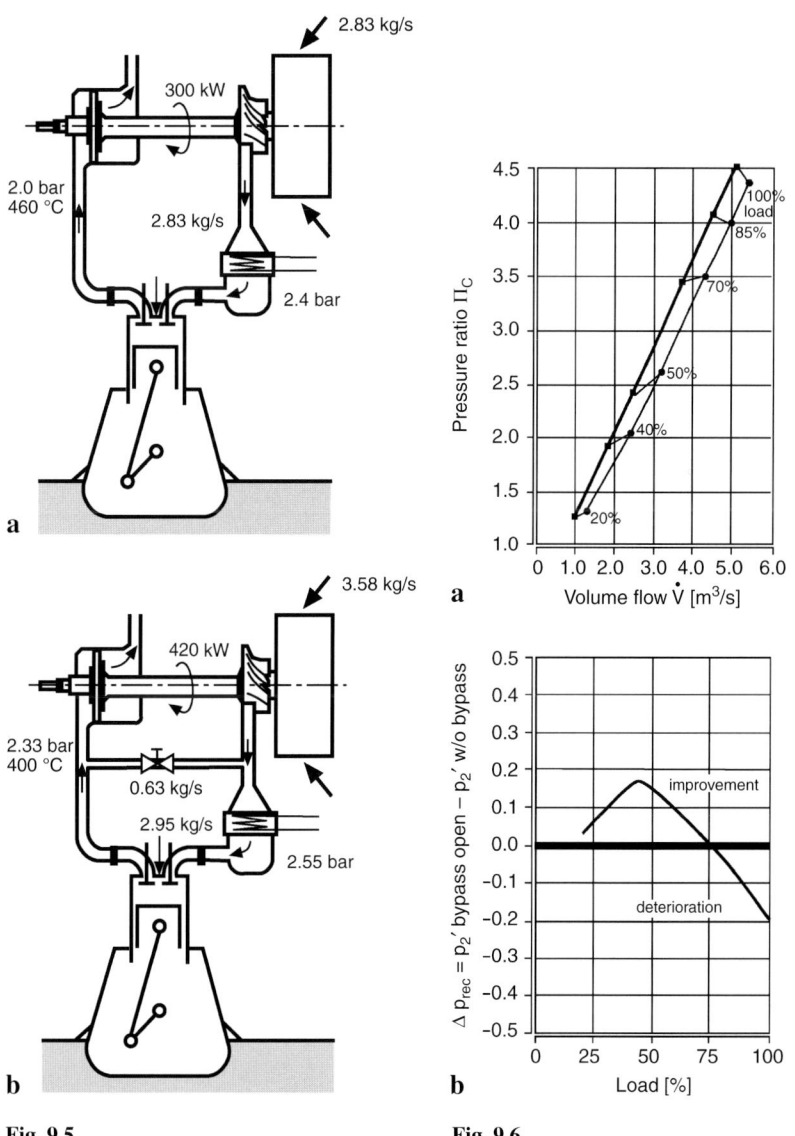

Fig. 9.5   Fig. 9.6

Fig. 9.5. Principal layout of an engine bypass line, including power, mass flow, pressure and temperature data, at 50% load, (**a**) without and (**b**) with bypass operation

Fig. 9.6. Volume flow changes in the compressor map (**a**), and the part-load advantages and disadvantages of bypass control (**b**) [54]

it can shift the engine operation line into a range with improved compressor efficiencies. Figure 9.5 shows the principal layout of such an engine, including power, mass flow, pressure, and temperature data at 50% load, with and without bypass. Figure 9.6a shows the changes in volume flow in the compressor map, and Fig. 9.6b shows the corresponding advantages and disadvantages at part-load.

9.2.2 Transient control strategies

For steady-state engine operation, boost pressure control is necessary for power and torque adjustment. As we have seen in Chap. 8, for transient engine operation, control of the transient behavior of exhaust gas turbocharged engines is absolutely necessary in order to achieve an acceptable load response. Partially, the control modules already described can be used for this task. However, and this will be discussed in more detail later, the control measures and strategies are changing.

General control options for turbocharged gasoline and diesel engines

Here, those possibilities will be discussed which can be applied independent of the type of the basic engine.

The waste gate described previously, which circumvents the turbine, can be utilized not only to control the boost pressure but also to improve the transient behavior of diesel engines. In doing so, the selected turbine must be as small as possible, accepting certain efficiency losses. This results in higher available turbine power and thus shorter charger speed buildup times in the lower engine speed range. Figure 9.7 shows this for an exhaust gas turbocharged gasoline engine. The disadvantage is the unavoidable increase in fuel consumption at high engine speeds.

Control options for turbocharged gasoline engines

In gasoline engines, the control options mentioned above assure knock-free combustion and compliance with the map limits of both compressors and turbines, by specifying a maximum possible or desired boost pressure value in predefined maps. In exhaust gas turbocharged gasoline engines, additional controls can improve the transient behavior further. Due to quantity control via the throttle, even at part load high boost pressures are generated and only the mixture mass flow is adjusted to the load.

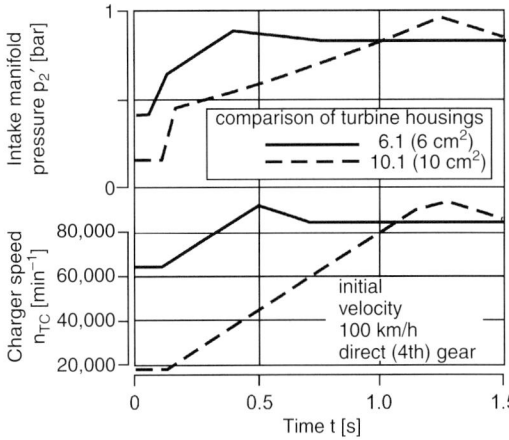

Fig. 9.7. Influence of turbine inlet area on torque curve

9.2 Fixed-geometry exhaust gas turbochargers

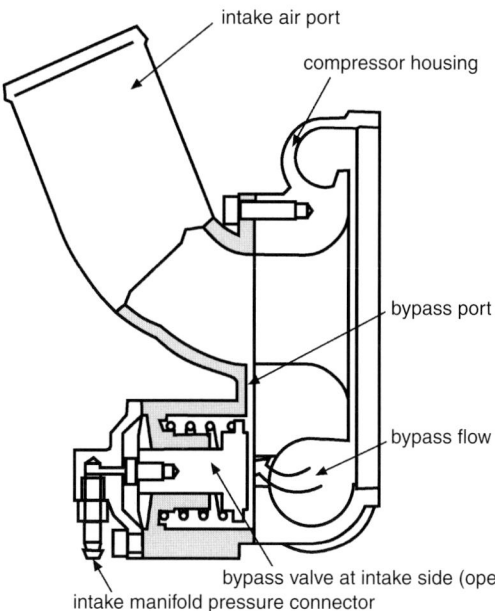

Fig. 9.8. Compressor bypass valve [KKK, now 3K-Warner]

The simplest measure is a bypass valve (Fig. 9.8), which serves several functions. On the one hand, by providing a virtual increase in compressor volume flow it allows a higher boost pressure at the surge limit. Thus, when the load is increased, a priori a higher charge pressure is available. On the other hand, it is absolutely necessary if the **throttle** is arranged **downstream of the compressor**. If this throttle is closed instantaneously during engine operation at higher load and speed, the air flow through the engine becomes negligible, but due to the inertia of the rotor assembly the exhaust gas turbine still provides excess power, at least for a short time. This would result in very high boost pressures at very low mass flow rates, i.e., operation of the compressor in the surge range.

To avoid this, the spring-loaded bypass valve (Fig. 9.8) opens – when the predetermined differential pressure between compressor outlet and intake manifold downstream of the throttle is reached and routes air back to the compressor intake. This reduces the boost pressure and again results in a virtual increase of the compressor volume flow, which prevents surging of the compressor.

A very effective measure to improve the load response is the arrangement of the **throttle upstream of the compressor**. Due to the quantity control of the engine described above, a corresponding amount of exhaust gas energy (and thus the same turbine power) is always available at part load, determined by exhaust gas mass and pressure.

The control of the mixture mass required for the actual part-load point is always obtained by throttling. This may be done upstream or downstream of the compressor. The **compressor** always delivers a **certain volume flow**, while the **engine** requires a particular **mixture quantity**. According to the equation

$$V = m/\rho, \qquad (9.1)$$

the volume of this mixture quantity depends on its density. Therefore, considering the state of the air, e.g., at compressor outlet, the actual mass flow corresponding to the volume delivered by the compressor can be determined. Figure 9.9 schematically shows the layout of throttles upstream and

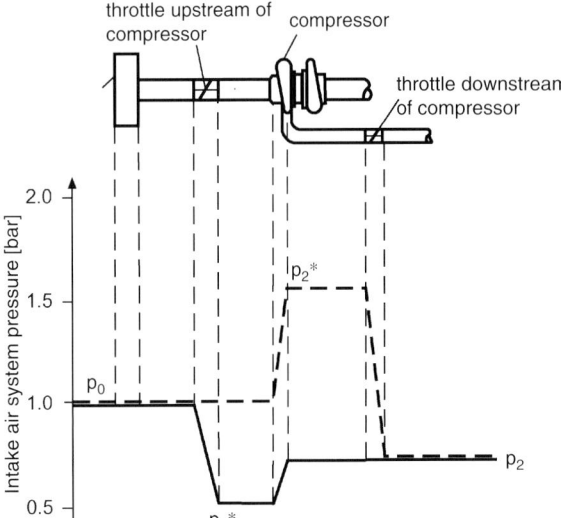

Fig. 9.9. Throttle arrangement upstream (solid line) and downstream (dash line) of the compressor and the resulting pressure conditions in the intake system

downstream of the compressor, and the resulting pressure traces in the intake manifolds. In both cases, ambient pressure p_0 prevails at the air filter. At low loads, this pressure is higher than the pressure p_2 in the intake manifold upstream of the engine. In case of a downstream arrangement of the throttle, the compressor aspirates the air at ambient pressure. Then the air is compressed to an interim pressure p_2^* downstream of the compressor, from where it is throttled to the required intake manifold pressure p_2.

In Fig. 9.10, corresponding typical operating points are plotted in a compressor map for bmep values of 2.4 and 6 bar at points A_1–A_3.

If the **throttle** is arranged **upstream of the compressor**, air is throttled from p_0 to an interim pressure p_1^* which is lower than the pressure p_2 in the charge air manifold (Fig. 9.9). Here the air has less density and a correspondingly larger specific volume (points B_1–B_3 in Fig. 9.10) at which it flows into the compressor and is then compressed from p_1^* to p_2. As Fig. 9.10 clearly shows, this occurs at better efficiencies, far away from the surge limit, and at higher compressor speeds. Therefore, it provides better takeoff conditions for sudden load increases. Figure 9.11 shows the differences between the two control strategies – throttle upstream or downstream of the compressor – at various loads, as well as the resulting compressor speeds, against engine speed. Compressor speed increases of up to 10,000 min^{-1} are achieved.

However, the discussed layout of the throttle upstream of the compressor is associated with two problems.

– For operation at **inlet pressures** significantly below ambient conditions, it requires a charger designed with an effective depression compressor housing **oil seal**. Even today this represents a challenging task and, of special significance, results in additional cost.
– If a charge air cooler with its pipes is added downstream of the compressor, the **charge air system volume increases** dramatically. This volume has to be drained at part load and filled at load increases, deteriorating the load response of the engine.

For steady-state natural gas engines, the layout with the throttle upstream of the compressor is state of the art today, since it allows the induction of the gas upstream of the compressor, which results in a very well homogenized mixture.

9.2 Fixed-geometry exhaust gas turbochargers

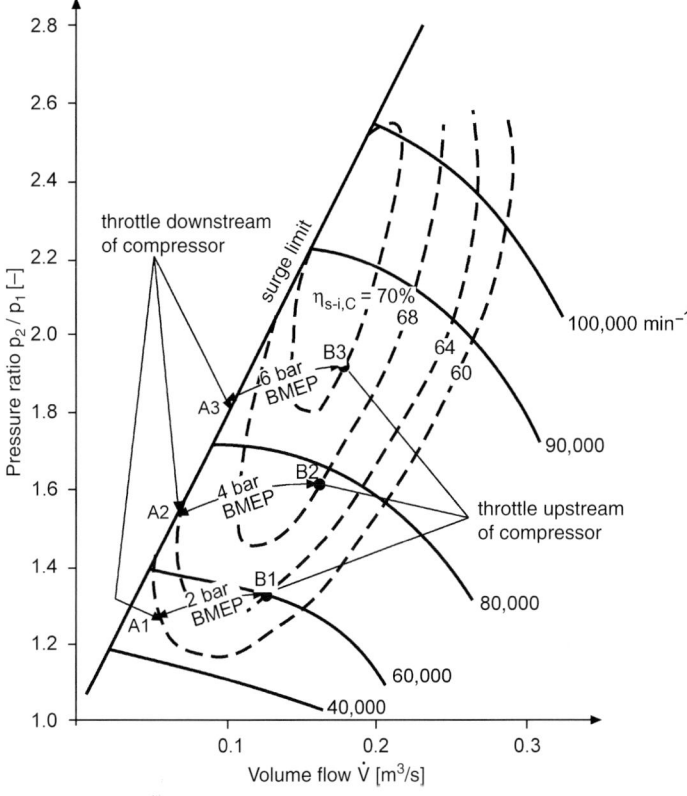

Fig. 9.10. Influence of the throttle layout on the location of the engine operating points in a pressure–volume flow map

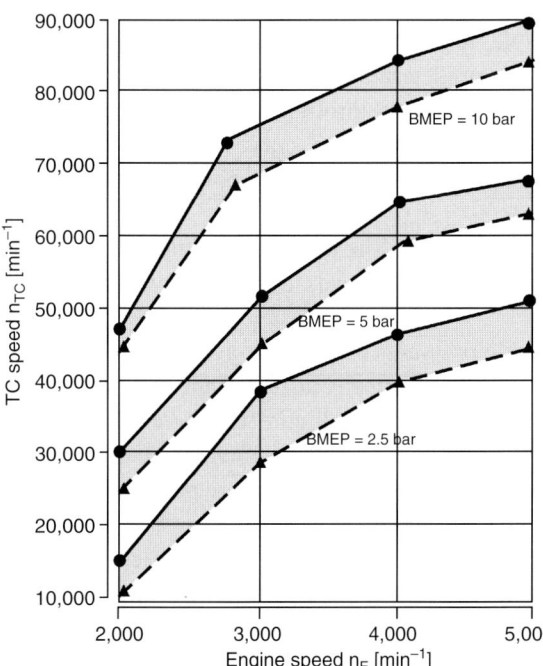

Fig. 9.11. Compressor speeds depending on throttle location upstream (solid line) and downstream (dash line) of the compressor

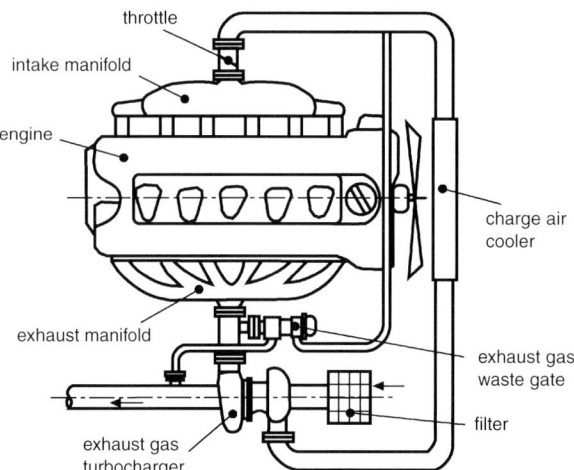

Fig. 9.12. Layout for maximum boost pressure limitation by waste gate control using the pressure upstream of the throttle

9.2.3 Part-load and emission control parameters and control strategies

Gasoline and natural gas engines

It is necessary to ensure that the full-load boost pressure limit is not exceeded. In an exhaust gas turbocharged gasoline engine with the control throttle located downstream of the compressor, the pressure upstream of the throttle must not increase too much at part load. The throttle controls the engine load exactly by throttling the boost pressure.

If the boost pressure and, thus, the pressure ratio at small mass flows is too high, the operating point drifts into the instable range of the charger map, to the left of the surge limit. This can be avoided by using the pressure upstream of the throttle as control pressure in the waste gate actuator, as shown in Fig. 9.12.

As long as sufficient turbine power is available, this control system limits the boost pressure upstream of the throttle to a constant maximum or predetermined value – also at part load. The advantage of this layout is the constant pressure upstream of the throttle with its good load response at sudden load increases. This results in engine operating characteristic curves as shown in Fig. 9.13. However, part-load fuel consumption is increased due to the high turbine power required for the load-independent supply of the full boost pressure.

Therefore, it may make sense to utilize the pressure downstream of the throttle – in a gasoline engine a direct indication for the load – as control variable. This leads to the layout with a part-load waste gate sketched in Fig. 9.14.

Pressures higher than desired and even unstable pressures can thus be avoided. However, the load response suffers somewhat, since the waste gate also opens at part load. By reducing the part-load compressor power, the fuel consumption is also reduced. Figure 9.15 shows the layout as well as the pressure sensors for the part-load blowoff control, and also the schematic layout of the electronic pressure control in the upper chamber of the waste gate via pressure pulse valves (pulse width modulation valves).

This allows an engine operation with a constant pressure gradient at the throttle, which may be advantageous for the control system. In order to achieve a buildup of boost pressure as fast and precisely as possible, and an exact control of the load-determining boost pressure at part load, it is worthwhile to combine all the measures discussed up to now in a **waste gate boost pressure**

9.2 Fixed-geometry exhaust gas turbochargers

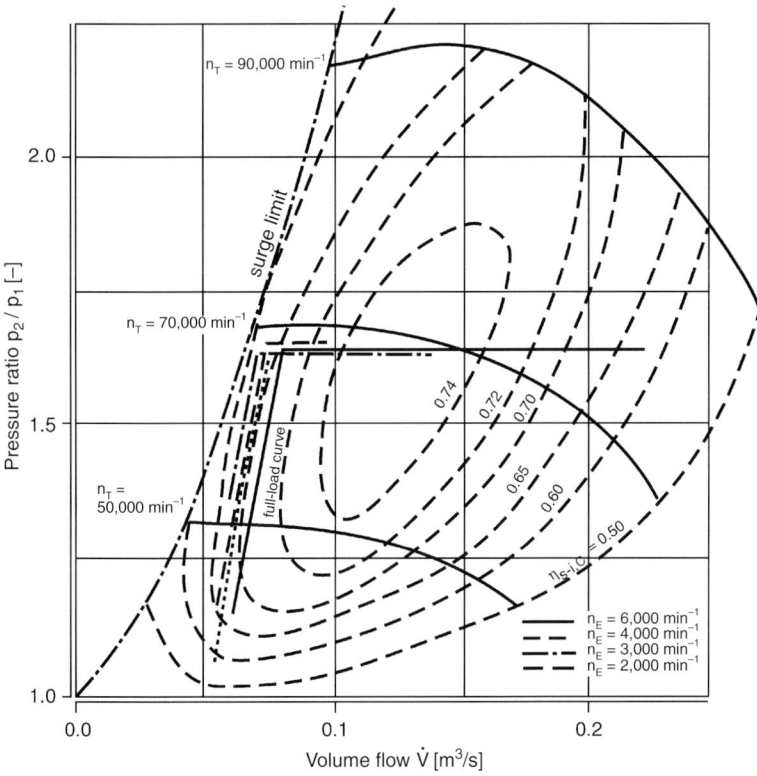

Fig. 9.13. Engine swallowing characteristics, control system according to Fig. 9.12

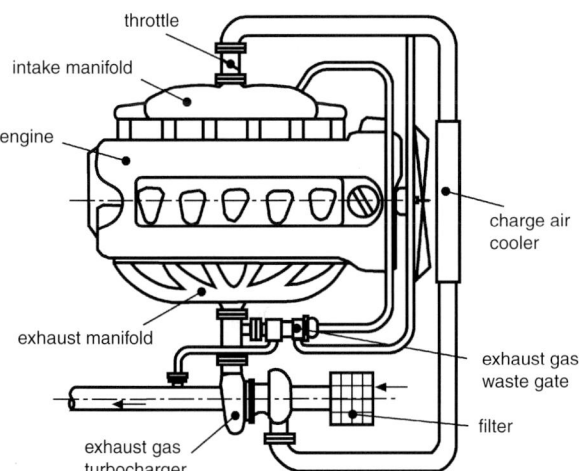

Fig. 9.14. Layout with full- and part-load waste gate control using the differential pressure up- and downstream of the throttle

control system. To do this, the spring chamber (Fig. 9.3) of the waste gate is connected to the boost pressure upstream of the throttle until the desired boost pressure downstream of the throttle – i.e., the load-determining boost pressure value – is achieved. This arrangement avoids a creeping opening of the valve at small differences between target and actual boost pressure, which would be associated with the disadvantage of exhaust gas pressure losses. The waste gate is fully closed.

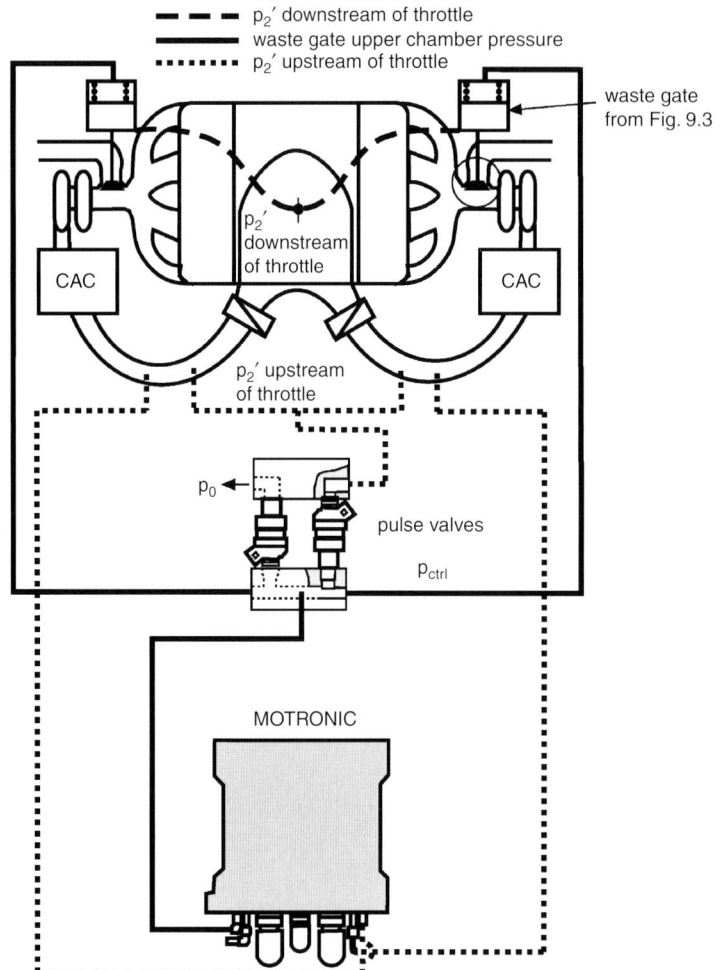

Fig. 9.15. Layout and pressure sensors for part-load blow off control, and schematic layout of electronic pressure control in the upper chamber of a waste gate via pressure pulse width modulation valves

On the other hand, with such a layout any desired value below the maximum possible boost pressure (waste gate fully closed) can be reached. This is also used to control the air-to-fuel ratio at part load of modern diesel engines. In gasoline engines it is used for load control in the entire boost pressure operating range.

However, this requires a suitable control layout, as described in Sect. 9.3.6.

Diesel engines

Besides full-load control, a supercharged diesel engine also needs control at part-load operation. On the one hand, this is aimed at reducing its exhaust emissions, on the other, at improving its load-response out of part load.

For example, the **part-load boost pressure** can be optimized for minimum emissions of one or several exhaust gas components. Then these values are entered into a map, which subsequently is used to control the waste gate for optimal boost pressure. Obviously, only a reduction of the maximum pressure achievable without activation of the waste gate is possible. To improve the boost pressure buildup and thus the load response of the engine under these operating conditions, the full closing of the waste gate, already described, must in any case also be possible.

The turbine size of a turbocharger with a waste gate is also a very important factor contributing to reduced emissions. The choice of a rather small turbine, associated with high waste gate gas flow rates, offers the possibility of generating a high exhaust gas turbine inlet pressure. This can be used for exhaust gas recirculation, which needs a driving pressure gradient between exhaust system and charge air manifold. As was the case with part-load boost pressure control, this measure may be associated with a – possibly substantial – increase in fuel consumption.

Special control interventions for engine braking

For engine braking, all fixed and waste gate charger systems must be equipped with an exhaust brake throttle downstream of the turbine. During engine braking, this causes high pressures in the exhaust system and an associated, essentially undesirable, reopening of the exhaust valves. It also creates high pressures in the turbine housing, with all the associated problems for the proper sealing of the charger bearings.

9.3 Exhaust gas turbocharger with variable turbine geometry

9.3.1 General control possibilities and strategies for chargers

In a charger with variable turbine geometry (VTG), the boost pressure can be easily changed during operation and thus controlled. This is achieved by adjusting the inlet blades of the turbine, i.e., by changing the gas entry angle into the rotor. Figure 9.16 shows such a charger. The advantage of the VTG charger is that the total exhaust gas quantity can always be utilized for power generation in the turbine. This has a very positive impact on turbine efficiencies since it significantly widens the usable flow rate range of the turbine (Fig. 9.17). The necessary exhaust gas turbine inlet pressure, particularly in comparison with waste gate control, is significantly reduced. In order to fully utilize this advantage, the position of the turbine inlet blades has to be controlled on the basis of suitable control parameters.

9.3.2 Control strategies for improved steady-state operation

Since gasoline engines with VTG chargers are not yet in mass production, the following remarks refer to engine quality control of the diesel engine. With VTG, for any engine operating point the

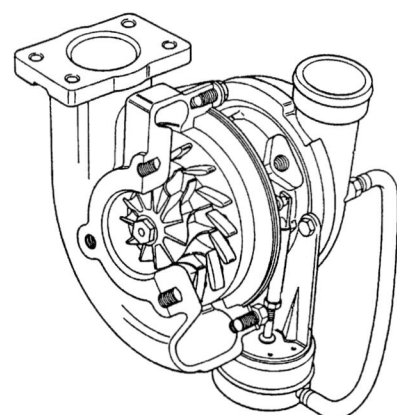

Fig. 9.16. Drawing and cross section of a VTG charger

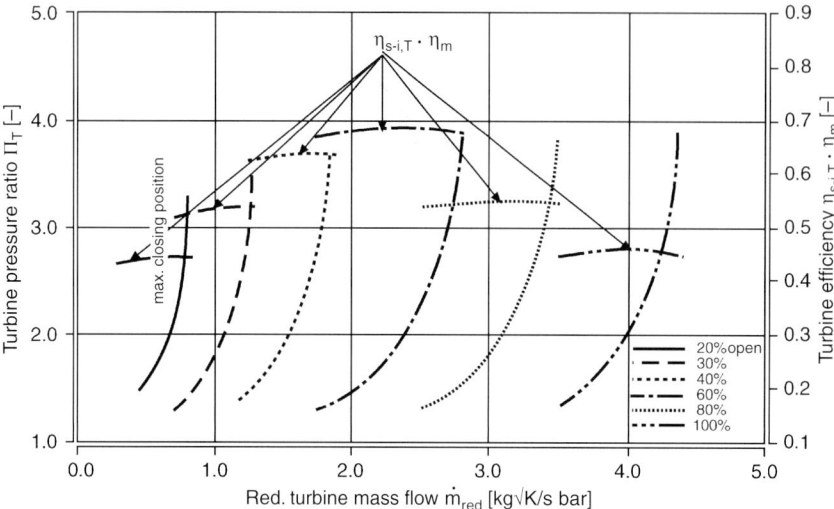

Fig. 9.17. Map range of a VTG charger

quantity and pressure of the charge air can be widely varied by changing the turbine inlet area, i.e., by changing the turbine power.

By choosing a particular angular position for the inlet blades, described here as relative blade position, that airflow rate can be chosen which, e.g., offers lowest fuel consumption. As shown in Fig. 9.18, a deviation from the optimum setting results in increased fuel consumption. This is caused by the following reasons: At greater relative blade positions – which in this context means a larger turbine area – the delivered air mass and thus the air-to-fuel ratio λ decrease, resulting in a

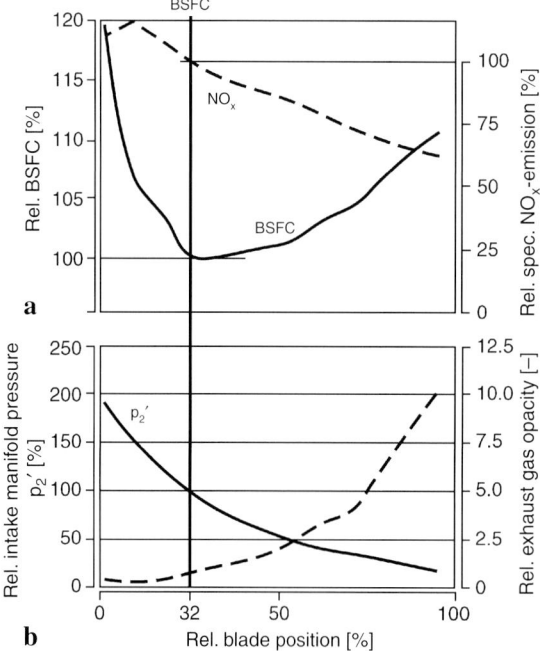

Fig. 9.18. Dependence of major engine data on VTG blade position

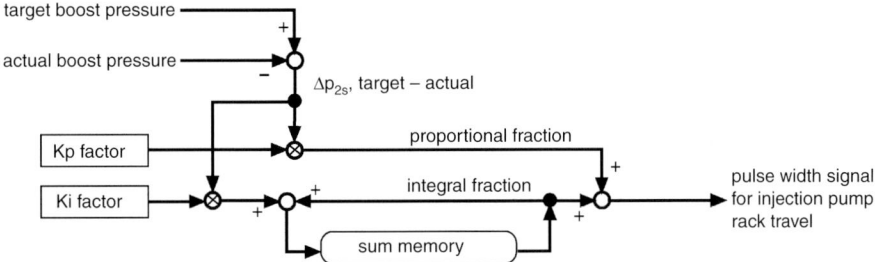

Fig. 9.19. Simple PI controller for the control of p'_2

slower combustion process, the combustion duration is extended and the wall heat losses increase. Evidence of this is an increased emission of particulate matter.

At smaller relative blade positions, i.e., at decreased turbine areas, the gas exchange losses, due to negative scavenging pressure gradients, will be increased. The air flow rate and thus blade angle resulting in the lowest fuel consumption is that which leads to the best compromise between air-to-fuel ratio and scavenging pressure gradient.

9.3.3 Control strategies for improved transient operation

In general, the two criteria mentioned above are also applicable under transient operating conditions. In early controller layouts as sketched in Fig. 9.19, simple PI controllers were used to control p'_2 in such a way that at a load change the new, higher boost pressure is achieved as fast as possible. Such controllers showed unsatisfactory results when tested under transient conditions (Fig. 9.20). Figure 9.20b shows the boost pressure curves for a fixed geometry charger and a VTG charger with p_{2s} control. With the VTG charger, the boost pressure increases much faster than with the fixed geometry charger.

However, as plotted in Fig. 9.20a this can result in a totally unacceptable characteristic of the increase in engine torque, which lags for about 2 s far below even that of the fixed-geometry charger.

The reason for this can be seen in Fig. 9.20c. When adjusting for the most rapid boost pressure increase, the turbine area is reduced to its predetermined minimum. As a result, the turbine inlet pressure significantly increases, which leads to high negative gas exchange mean effective pressures. These have to be compensated by the engine, resulting in a loss of torque available at the flywheel. Therefore, at a load increase, it is not sufficient only to adjust the turbine inlet blades to their minimum position until the desired boost pressure is achieved. Rather, it is important that in the process of adjustment to a new boost pressure value, at sudden load increases certain limits (Fig. 9.21) are not exceeded.

As shown in Fig. 9.21, a load increase can be subdivided into four phases. Starting from a steady-state part-load condition (1), e.g., 20% of full-load, the demanded load is increased instantaneously to 100% (shown in Fig. 9.21 by the accelerator pedal travel, i.e., load requirement, curve). In this phase (2), the VTG is closed to a specified minimum area (Fig. 9.21a). At the same time, it is important that the air-to-fuel ratio λ does not fall under a specified limit, primarily determined by the exhaust gas opacity value. In addition, a too large negative gas exchange scavenging pressure gradient ($p'_2 - p_3$) cannot be tolerated, at least not for a longer period. This is controlled by reopening the VTG in phase 3, although the desired boost pressure may not yet have been obtained.

The full boost pressure, λ, and the scavenging pressure gradient approach their final values only in phase 4. Numeric cycle simulations are also an excellent tool for the development of

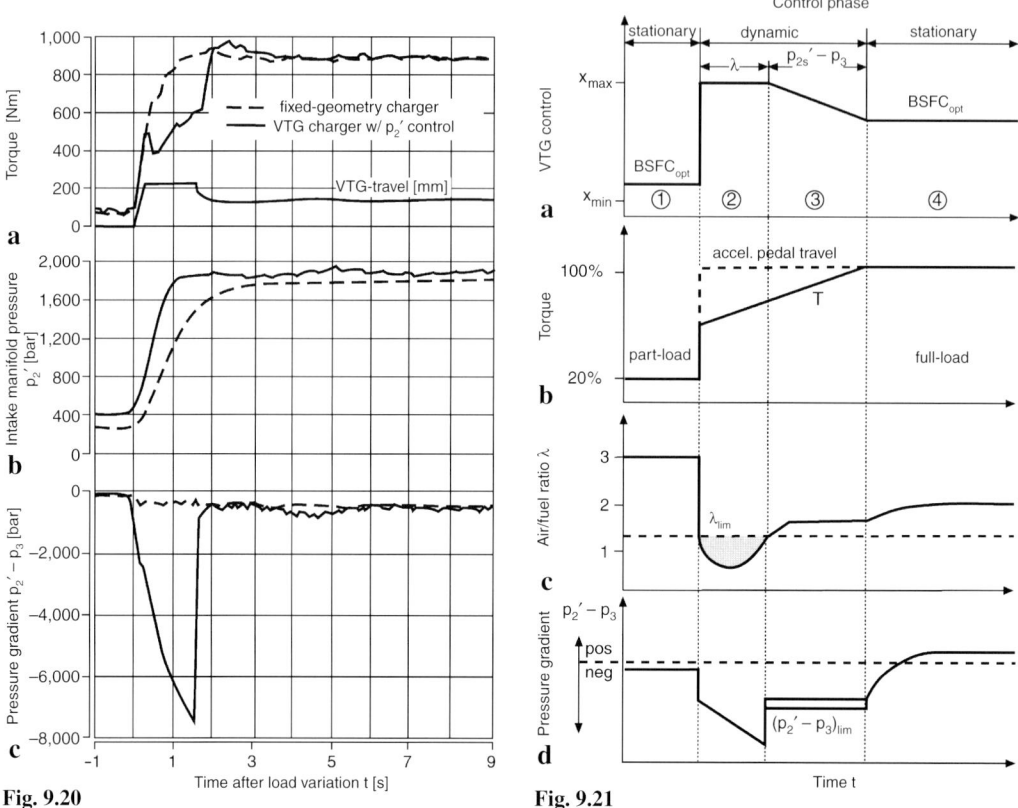

Fig. 9.20

Fig. 9.21

Fig. 9.20a–c. Load step results with simple PI controller

Fig. 9.21a–d. Basic control limits and parameters

optimized control strategies for such transient charging processes. (A corresponding example will be described in detail in Sect. 9.3.7.)

The desired VTG control strategy can be realized best if the two critical engine parameters, the scavenging pressure gradient limit $\Delta p_s = p_2' - p_3$ and the λ limit, can be directly processed.

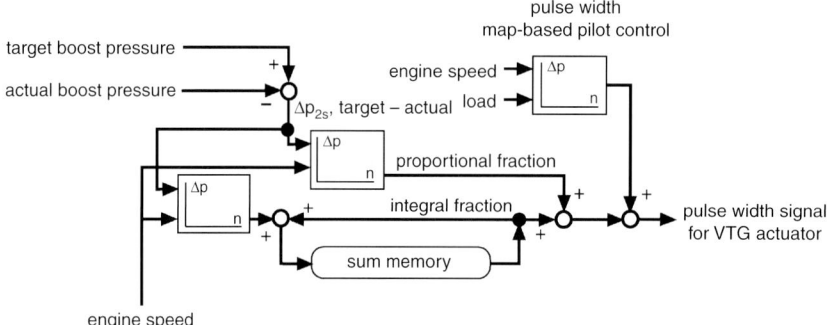

Fig. 9.22. Extended PI controller with load- and speed-dependent maps for the I and P fractions to be chosen, and with a map-based pilot control of the VTG [DC]

9.3 Exhaust gas turbocharger with variable turbine geometry

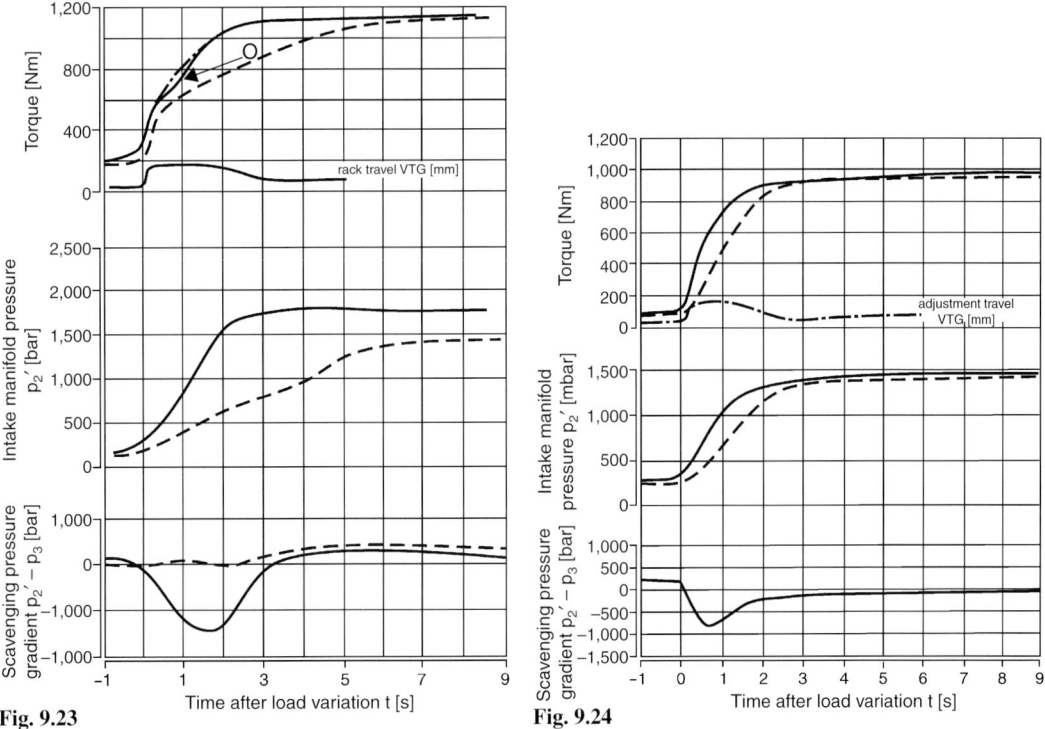

Fig. 9.23. Load response with an extended PI VTG controller at lowest full-load speed [DC]; dash line, waste gate charger; solid line, VTG with map-based pilot control; O, torque limited because of exhaust gas opacity

Fig. 9.24. Load response with an extended PI VTG controller at high full-load speed [DC]; dash line, waste gate charger; solid line, VTG with map-based pilot control

However, currently it is not yet possible to measure these parameters directly with sensors of sufficient durability and low cost. A possible solution is to expand the PI controller, which was mentioned before, via load- and speed-dependent maps for the integral and proportional fractions to be selected, combined with a VTG actuator with pilot control (Fig. 9.22). With this improved VTG controller, satisfactory pressure and speed buildup times can be obtained. Figure 9.23 shows this for a full-load application at low engine speed, and Fig. 9.24 at high engine speed.

A further improvement is possible for vehicle applications, especially for trucks. During gear shifts, due to insufficient exhaust gas mass flow the fixed-geometry charger significantly drops in speed, and thus the boost pressure. If at that point the VTG inlet guide blades are closed, the speed drop, and boost pressure loss, is significantly reduced. In addition, the air-to-fuel ratio λ which occurs when load is reapplied after the gear shift is improved. This is important for transient exhaust emission tests. Figure 9.25 shows the effect of such a gearshift VTG control, resulting in a significantly reduced boost pressure drop during the shift, an improved load response after the shift, and thus improved vehicle acceleration. Further details regarding controller layout will follow in Sect. 9.3.6.

9.3.4 Special control strategies for increased engine braking performance

Due to advancements in supercharging technology, desired power and torque values can be realized with ever decreasing engine displacements. When such modern high-power engines are utilized

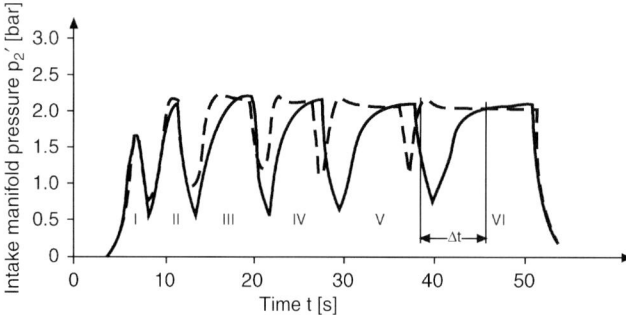

Fig. 9.25. Influence of a gearshift VTG control on the boost pressure drop during the gearshift interruption [DC]. Waste gate charger (solid line); VTG with pilot control (dash line)

in trucks, an additional problem arises: The small-displacement engine has to generate higher specific braking power. Further, such higher engine braking power is required at relatively low engine speeds, in order to enable braking for adjustment of the vehicle speed to the traffic flow without gear shift down. VTG technology can offer solutions to this problem, provided the following conditions are met:

– Even when the VTG area is small, it still must assure sufficiently high turbine efficiencies.
– These small turbine inlet areas must be controlled accurately and with smallest hysteresis. The VTG must be able to mechanically withstand the resulting high exhaust gas turbine inlet pressures (including the sealing of the turbine and bearing housings).

During braking, very high pressure ratios occur at smallest turbine areas (Fig. 9.26), since only in this way the turbine power necessary for high air flows during engine braking can be provided.

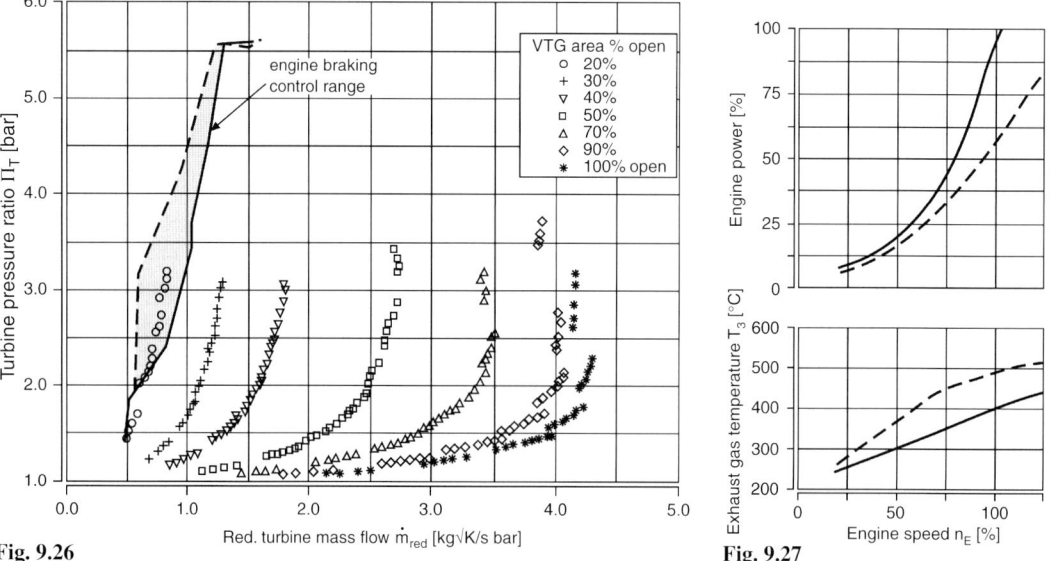

Fig. 9.26. VTG-pressure–volume flow-turbine diagram, with swallowing curve during engine braking

Fig. 9.27. Engine braking power and exhaust gas temperatures with VTG for increased braking power. VTG and constant throttle (solid line); constant throttle and exhaust throttle flap (dash line)

The higher air flow rates which can be achieved in this way during engine braking result in a significant increase in the braking performance (Fig. 9.27), as desired, even at a lower thermal engine load, due to lower exhaust gas temperatures.

9.3.5 Special problems of supercharged gasoline and natural gas engines

Additional problems arise if VTG charger technology is applied to gasoline engines. On the one hand, the high exhaust temperatures during full-load and during part-load operation, and on the other hand, the significantly increased demands on control accuracy and speed, caused by the gasoline engine's quantity control are critical. The complete control strategy for an exhaust gas turbocharged gasoline engine poses high demands on the layout and architecture of an electronic control system, due to
– throttling at low load,
– controlled exhaust gas recirculation in the lower- and medium-load range,
– the need for precise load control by controlling the boost pressure in the higher-load range (otherwise high gas exchange losses would occur, resulting in increased fuel consumption),
– cold-start and warmup problems in connection with the necessary λ control and the control of the catalyst temperature.

9.3.6 Schematic layout of electronic waste gate and VTG control systems

In general, of course we have to distinguish between the **diesel engine's quality control** and the **gasoline engine's quantity control**.

Gasoline engines

The controller architecture becomes very complex for exhaust gas turbocharged gasoline engines. However, due to ever improving simulation tools supporting the controller layout (Sect. 9.3.7), no unsolvable difficulties remain. Figure 9.15 shows a principal diagram of an electronic control system for an exhaust gas turbocharged gasoline engine with waste gate charger. Here, for an engine equipped with two turbochargers, the full boost pressure upstream of the throttles is measured and fed into the control unit as an input variable. Additionally, this measured signal is routed to a pulse valve assembly which generates a control pressure via pulse width modulation valves for two-directional flows.

This control pressure is measured and controlled by the control unit via pulse timing in such a way that a force equilibrium is achieved in the waste gate control assembly (e.g., a membrane pressurized on both sides and under spring pretension), resulting in the waste gate position which corresponds to the desired boost pressure p_2.

Diesel engines

The advantages of VTG chargers described above – e.g., widely adjustable boost pressure, better load control in case of gasoline engines, mutual tolerance adjustment of charger and engine – can only be utilized in combination with a powerful electronic control system. Here, in any operating point the actuator must be able to set the turbine inlet blades to any position. The characteristics of the actuator and the control loop are decisive for the **dynamics** of boost pressure buildup and the **control performance**.

Since the system and thus the system characteristics depend only on the exhaust gas flow, a non-linear control path results for the actuation of the inlet blades.

Additionally, other limiting relations influence the control system and thus may result in control oscillations. These limitations include the air-to-fuel ratio λ, particulate emissions, exhaust backpressure, and control functions for a harmonic torque rise of the engine. The control unit must be able to handle these non-linearities and to assure a stable boost pressure or boost pressure buildup in the complete load and speed range of the engine, with good control dynamics. As the discussion in chapters 9.3 and 9.4 has shown, this cannot be achieved with simple PI or PID control units. In addition, the control unit has to compensate for tolerances in the charger itself as well as for deteriorations in charger performance within its lifecycle.

For these reasons, VW applied for their VTG diesel engines, e.g., the TDI engine rated at 81 kW, a PDI control unit with adaptive parameter selection and additional disturbance stabilization in order to further improve the dynamics and stability of the control loop. As research has clearly proven [75], an adjustment of the control parameters to each individual load point is absolutely necessary. Figure 9.28 shows the structure of this boost pressure control unit. The control parameters are determined in engine bench tests, then defined for all operating ranges and stored in the control unit. A main adaptation parameter is the fuel consumption, calculated from injection quantity and engine speed. To avoid an overshoot of the boost pressure at full load, a disturbance stabilization is added, utilizing the values used for the calculation of the fuel consumption – and augmented by the ambient pressure and the charge air temperature.

If the VTG charger is to become established in truck engines, further control parameters have to be included, e.g., controlled exhaust gas recirculation, or even VTG-controlled engine braking.

As an example, Iveco introduced its Cursor-9 diesel engine with VTG charger utilizing the control structure as shown in Fig. 9.29.

Other manufacturers are also engaged in the development of VTG chargers for truck applications. Such VTG chargers will allow not only to control or even increase the braking

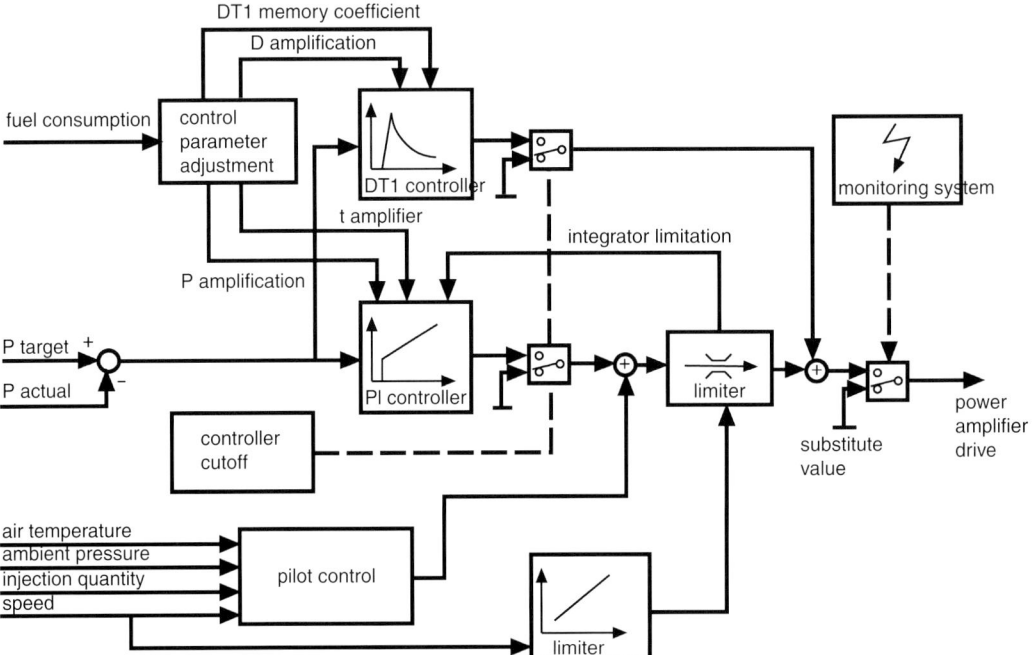

Fig. 9.28. Structure of a boost pressure control unit for a passenger car diesel engine with VTG

9.3 Exhaust gas turbocharger with variable turbine geometry

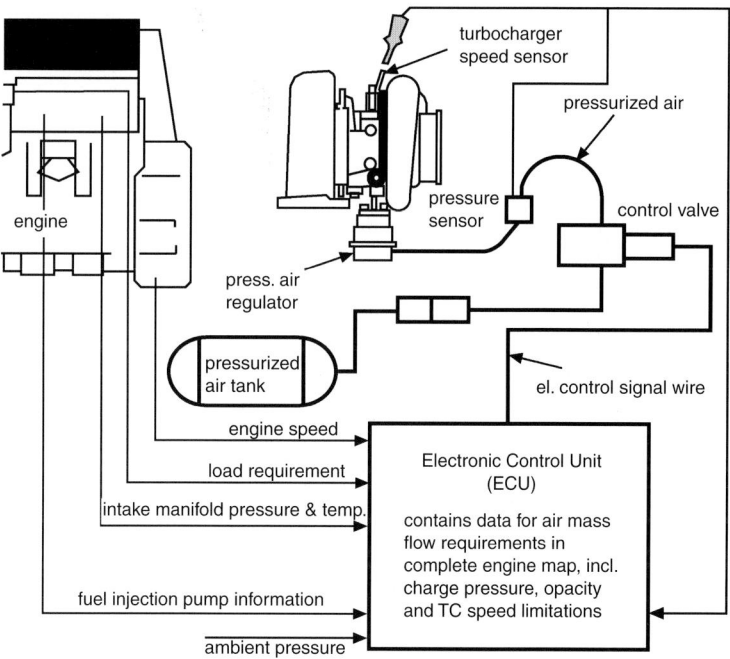

Fig. 9.29. Hardware components of a boost pressure control unit for a truck diesel engine with VTG

performance, but also to control exhaust gas recirculation systems.

9.3.7 Evaluation of VTG control strategies via numerical simulation models

As an example for numeric simulation, we will discuss the control process for a VTG turbine using the 6-cylinder passenger car engine introduced in Sect. 5.5. For such a study a simulation model verified with extremely precise measurements should be used (Fig. 9.30).

Within the scope of numeric simulations, the turbine position can be optimized in each phase of the load increase. Target parameters may include, e.g., the charger response time, the air-to-fuel ratio, the cylinder peak pressure, the gas exchange work, and the specific fuel consumption. In the example discussed here, the first version of the charger control system shows the VTG control

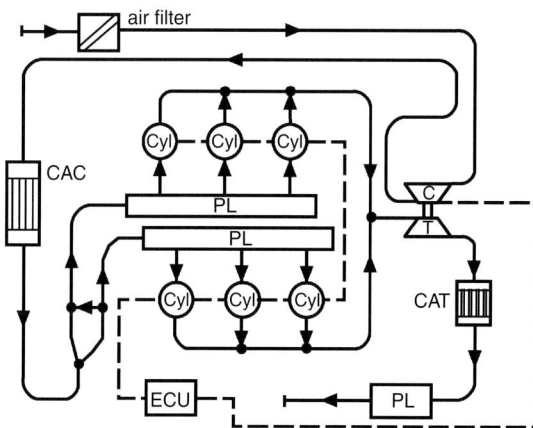

Fig. 9.30. Simulation model of a 6-cylinder passenger car engine, with ECU simulation for the control of its transient behavior

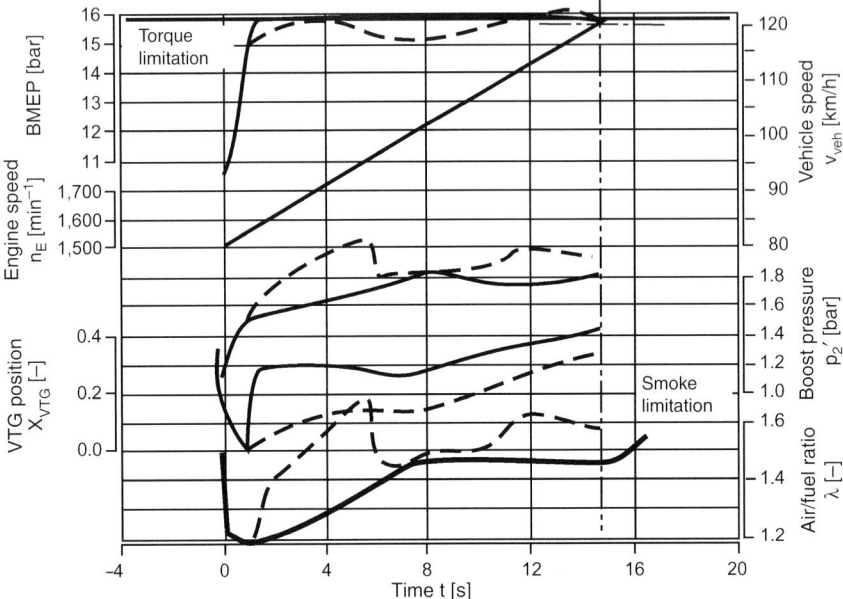

Fig. 9.31. Simulation of the influence of the turbine control strategy on engine operating behavior during a vehicle acceleration (from 80 to 120 km/h) with conventional boost pressure-guided control (dash line), $t = 14.9$ s, and with optimized smoke-limited control (solid line), $t = 14.6$ s

strategy discussed above, which is only seemingly target-oriented: Immediately after the sudden load increase the turbine is operated at minimum swallowing capacity, in order to generate a compressor drive power as high as possible. Primarily as a result of the mechanical inertia of the system, a delay in turbine adjustment may occur.

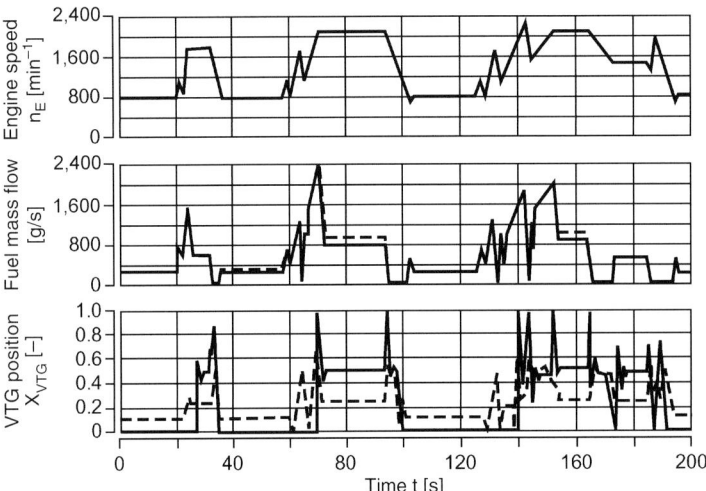

Fig. 9.32. Simulation of the effect of a boost pressure control strategy with smoke and gas exchange work as governing parameters (solid line) in the MVEG driving cycle for a vehicle of 2,100 kg weight equipped with 2.5 liter HSDI diesel engine; dash line, conventional strategy with boost pressure as governing parameter

9.3 Exhaust gas turbocharger with variable turbine geometry 183

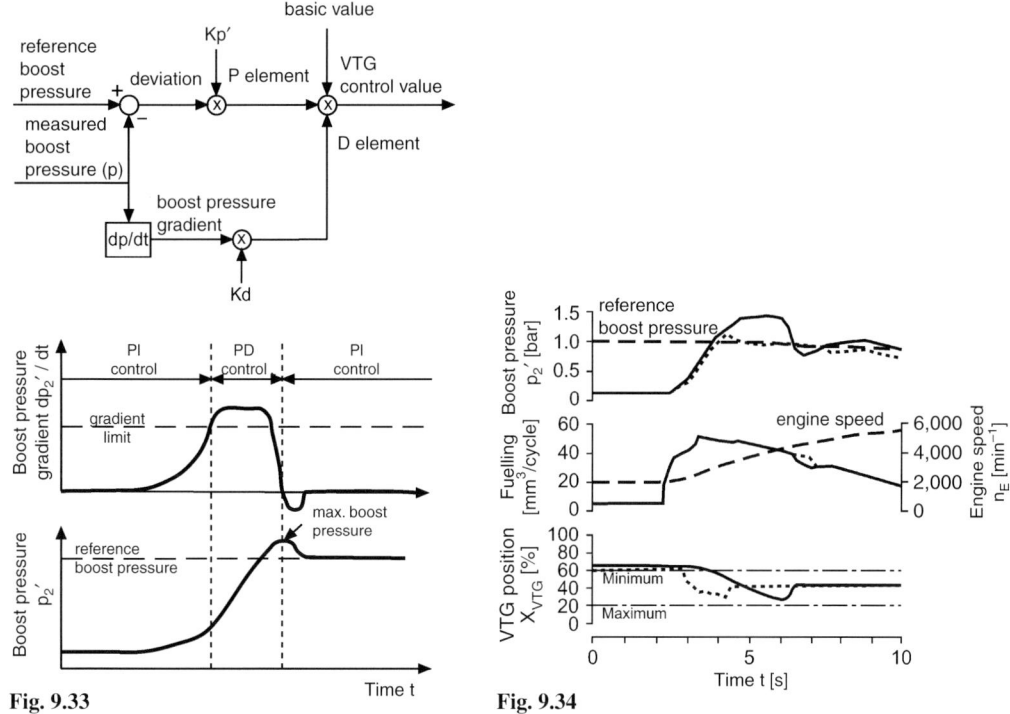

Fig. 9.33. Boost pressure control strategy of a model-based PID control system [29]

Fig. 9.34. Boost pressure control process during a load increase. Comparison between standard PI control unit (solid line) and model-based PID control system (dot line) [29]

Once the limitation by either mean effective pressure or peak pressure or transmission torque is reached, a further boost pressure increase would result in increased fuel consumption (retarded ignition timing to limit peak pressure, high gas exchange work due to high exhaust backpressure). Therefore, in this phase of the load change, the inlet blades of the variable-geometry turbine can be further opened. Thus, also the air-to-fuel ratio does not increase into a range which would be associated with higher fuel consumption. This approach is accompanied by improved specific fuel consumption, e.g., a fuel quantity reduction of 3.6% for the example shown in Fig. 9.31, which simulates a vehicle acceleration in 5th gear from 80 to 120 km/h.

With this control strategy, driving cycle simulations resulted in a fuel consumption improvement of about 1.5% in the European MVEG cycle (Fig. 9.32).

In addition to an optimum control strategy, the layout and optimization of the control algorithms is as important for the achievement of best engine operating conditions. Reference 29 describes such a layout of a control algorithm, especially with regard to the minimization of boost pressure overshoot, common to VTG chargers (as described in Sect. 9.3.6; Fig. 9.33).

By switching from a standard PI control loop to a model-based control algorithm, these pressure overshoots can be significantly reduced (Fig. 9.34). This is important not only for the exact adjustment of the air-to-fuel ratios during transient processes, but also in order to stay within given peak firing pressure limits of the engine, which could be exceeded in case of extreme overshoots of boost pressure.

10 Instrumentation for recording the operating data of supercharged engines on the engine test bench

The importance of verification of simulation models and results with measured data was mentioned in Chap. 9. Various measurement parameters are considered for such verifications, e.g., pressures, temperatures, mass flows, power, and speed. In this chapter we present a very compact overview of the preferred measuring techniques and devices used for data acquisition. For a more detailed study of this subject we refer to the pertinent literature [110].

Prior to more detailed presentation of the measurement techniques and methods, the parameters to be measured have to be defined. Figure 10.1 shows a typical setup of the measuring points for a bench test of a supercharged diesel engine, identifying these parameters.

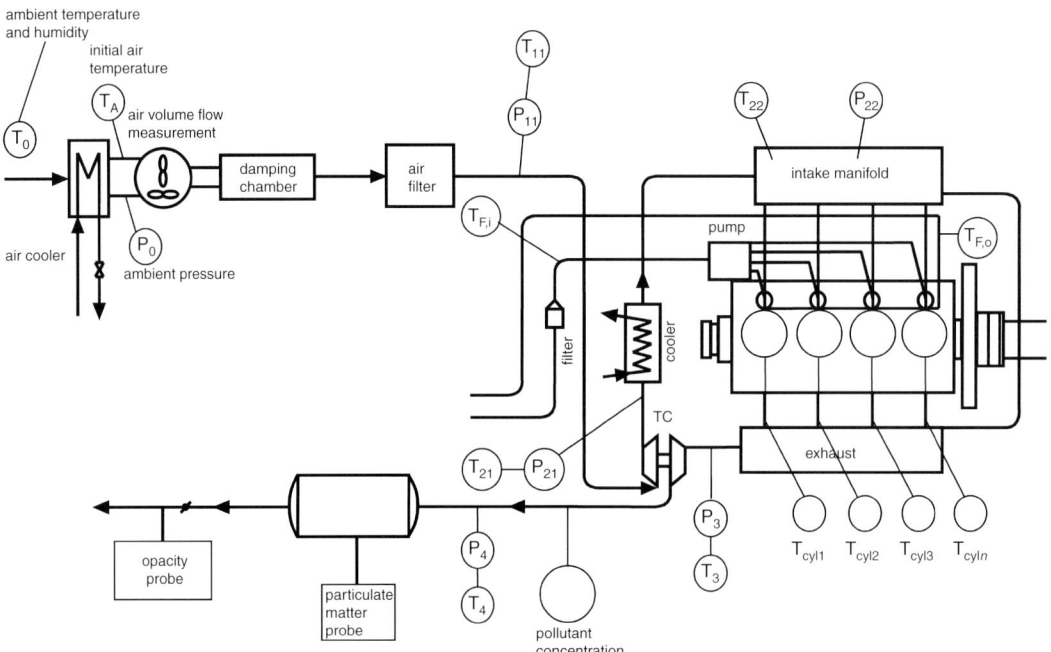

Fig. 10.1. Measurement setup for a bench test of a supercharged diesel engine with charge air cooling

10.1 Measurement layout

At the air inlet into the intake system, the ambient temperature T_0 and the relative humidity PHI are measured. The subsequent gas meter (e.g., by RGM Messtechnik GmbH) measures the volume flow. For that, besides the ambient pressure p_0, also the air temperature T_A at the intake into the volume flow measurement device is important, since the air mass flow is calculated with these parameters. The pressure and temperature change up to the compressor (e.g., in the air filter) is recorded via the pressure and temperature parameters T_{11} and p_{11}. Downstream of the compressor, the compressor efficiency can be determined by measuring the pressure p_{21} and the temperature T_{21} (here the measuring point has to be carefully selected). To evaluate the efficiency of the charge air cooler as well as to calculate the volumetric efficiency of the engine, reduced to intake manifold conditions, the measurement of p_{22} and T_{22} in the intake manifold is necessary.

On the exhaust gas side, it is advantageous to measure the individual cylinder exhaust gas temperatures $T_{cyl,1}$–$T_{cyl,n}$ in order to be able to compare the uniformity of the individual cylinders. Pressure p_3 and temperature T_3 (indices 31 and 32 in case of a twin-flow turbine) should be measured as close as possible to the inlet of the turbine. This allows an assessment of the turbine operational characteristics, and the wall heat losses of the exhaust manifold.

Downstream of the turbine, besides the temperature T_{41}, the pressure p_{41} is of special importance for the determination of the backpressure caused by catalyst and muffler. The exhaust gas sample probes, which are necessary for the measurement of the raw emissions including particulate matter, are collected in the exhaust system downstream of the turbine.

On the engine itself, besides speed n, torque T, and the inlet and outlet temperatures of the coolant $T_{W,int}$ and $T_{W,out}$, the lubrication oil inlet temperature T_{oil}, and the oil pressure p_{oil} also have to be measured continuously. Further important parameters which must be continuously recorded are the fuel mass flow $\dot{m}_F$ and the blowby volume flow $\dot{V}_{blowby}$.

Measurement parameters especially linked to the supercharging system are the speed of the turbocharger, the position of the variable blades (compressor preswirl, diffuser blading, and turbine guide blades), position of relief, blowby, and waste gate valves and pressure indications in the cylinders and at carefully selected locations in the intake and exhaust systems.

Table 10.1 shows a summary of an extended measurement data set for a typical engine development test bench.

10.2 Engine torque

Engine torque represents a control measuring parameter for any engine test. Under steady-state operation, it can be derived from the brace torque of the engine test bench. The bracing force acting via a lever arm is measured via strain gage load cells or via precision balances. Depending on the intended use and engine-power class of the supercharged engine to be tested, various test benches may be utilized. The most common passive systems are eddy current brakes (Fig. 10.2) and hydraulic brakes (Fig. 10.3). Eddy current brakes are utilized in wide power (10 to 1,000 kW) and speed (up to 20,000 min^{-1}) ranges. Hydraulic systems are favored for very large, medium- and slow-speed engines.

In addition, the development of modern engines mandates that the transient behavior of the engine can be already tested on the test bench. Accordingly, for this purpose electric 4-quadrant test benches were developed, including control units, which enable both braking operation in generator mode and electrically powered operation. In this case, torque can be transferred to the engine

Table 10.1. Extended measurement data set for a typical engine development test bench

Parameter	Equipment	Preferred measuring principle
Engine torque	Dynamometric brake	Reaction torque via strain gage load cell
Engine speed	Optical angle marker sensor	IR-transmitted-light/reflected-light photo sensor
Air volume flow	Gas meter	Rotary piston speed correlated to volume flow
Blowby mass flow	Blowby meter	Volumetric flow measurement via pressure drop at calibrated orifice
Fuel consumption	Fuel balance	Gravimetric mass measurement
TC speed	Optical marker sensor	IR-reflected-light photo sensor
Static and dynamic pressure	Pressure sensor	Strain gage and piezoelectric strain measurement
Temperature	Temperature sensor	Resistance thermometer, thermocouple
Emission	Exhaust gas analyzer	CO, CO_2: NDIR absorption of CO CO_2: paramagnetic effect of O NO_x: chemoluminescence at NO_2 formation HC: HFID via HC gas chromatography PM: filter mass measurement

Fig. 10.2   **Fig. 10.3**

Fig. 10.2. Eddy current brake for power range up to 300 kW

Fig. 10.3. Hydraulic brake for large engine tests

(coasting mode), as it occurs under driving conditions when coasting or during gear shifts. Figure 10.4 shows a cut-away view of such a 4-quadrant power brake.

The latter power brakes are even used for the development of F1 racing engines. To execute the actual speed gradients, coasting and braking powers of 1,000 kW must be covered. The speed range of these power brakes extends up to 22,000 min^{-1} (utilizing a reduction stage). The maximum possible speed gradient is about 30,000 min^{-1}/s.

10.3 Engine speed

Engine speed is often measured electrically inside the power brake of the test bench. On the other hand, corresponding measuring devices can also be fitted to the engine itself. Optical measuring instruments, either reflecting or absorbing infrared light signals from a measuring disc equipped

10.4 Turbocharger speed

Fig. 10.4. Cut-away view of an electric 4-quadrant power brake (up to 800 kW)

with angle or trigger markers, are favored. The change in light intensity is transformed by the infrared sensor into a pulse signal. After digitalization of the signal, the speed can be determined from counting the pulse signals per time unit. The angle marker signal can additionally be used to control measured parameters which must be resolved by angle, e.g., pressure indications. Figure 10.5 shows such an angle transmitter for speed measurement of rotating components.

10.4 Turbocharger speed

Measuring devices such as the angle transmitter using discs with markers are not suited for the measurement of turbocharger speed. The very sensitive rotor dynamics of the turbocharger would be severely disturbed by the slightest changes in the rotor assembly. Therefore, optical methods are usually used for turbocharger speed measurement. Such sensors, specially developed for turbocharger speeds up to 200,000 min^{-1} without affecting the charger, are suited for measurements under both steady-state and transient engine operating conditions. A laser beam is targeted on the compressor impeller, where it is reflected once per revolution from a reflecting marker. The sensor receives the signal and converts it into an output value consisting of a periodic sequence of voltage signals which then can be utilized for further signal processing (e.g., AVL Trigger Box TB350 for

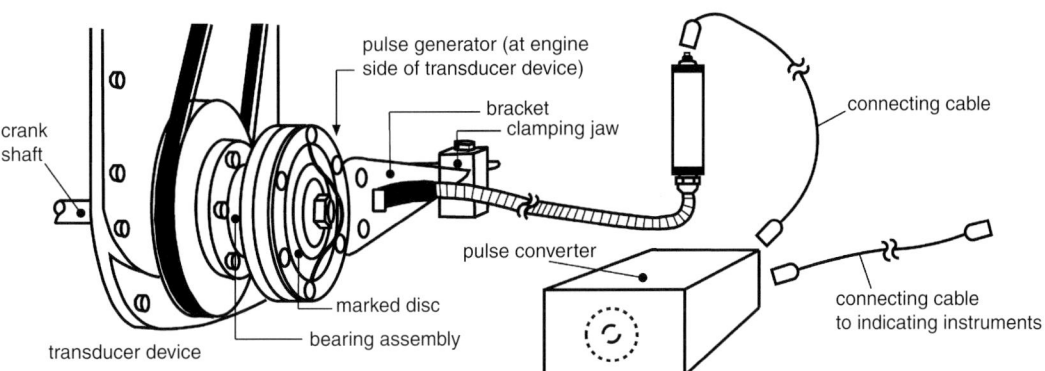

Fig. 10.5. Angle transmitter for speed measurement of rotating components

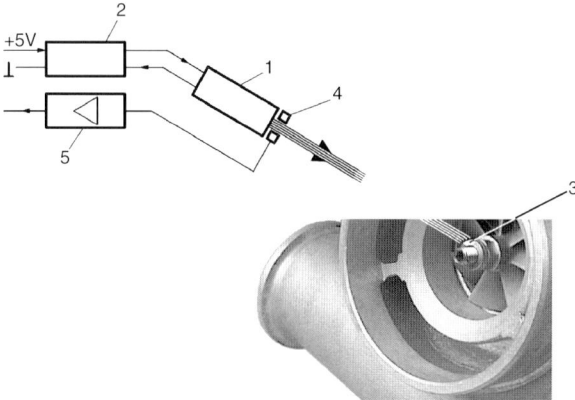

Fig. 10.6. Mode of operation and component description of a TC speed sensor. 1, Laser collimator (consisting of laser diode, monitor photodiode, and lens); 2, electric power supply component; 3, color marking at the compressor impeller of the turbocharger (retroreflecting color); 4, light detector; 5, amplifier circuit

connection to the test bench, oscilloscope, counter; Fig. 10.6). The large optical range of the sensor enables measuring without an impact on the intake air flow.

10.5 Engine air mass flow

A further essential measuring parameter is the engine air mass flow. Usually, it is determined indirectly via a volume flow measurement, combined with a determination of the density of the intake air. Experience shows that sound and exact results are obtained with rotary piston gas meters, especially if a damping chamber is added as a buffer volume just downstream of the gas meter. Figure 10.7 shows an example of such a measuring device.

The perfect suitability of these devices for the volume flow measurement of the engine air flow is emphasized by their very minor measuring error (<1% between 5 and 100% of the maximum measuring range) and their small pressure losses (up to 5 mbar at maximum flow rate).

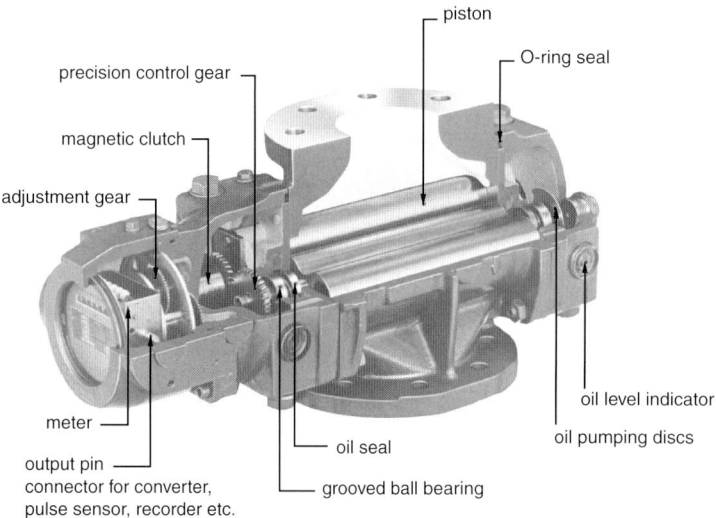

Fig. 10.7. Rotary piston gas meter for the volume flow measurement of engine intake air [2]

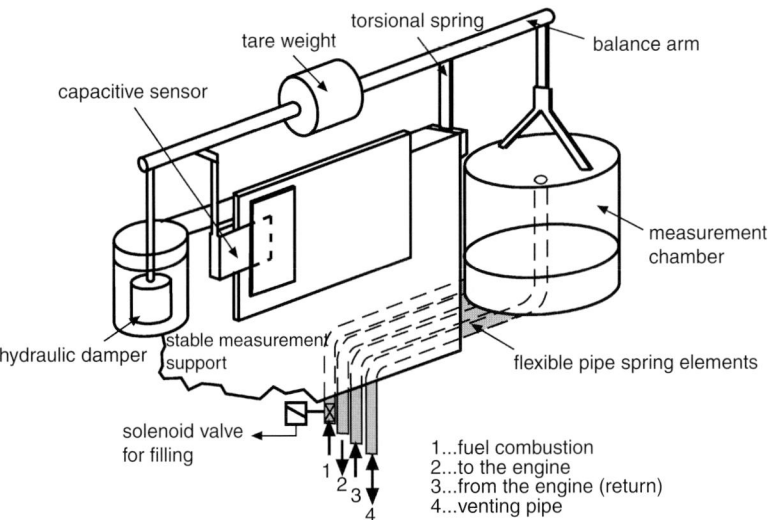

Fig. 10.8. Fuel balance for gravimetric fuel consumption measurement

10.6 Fuel mass flow

The fuel mass flow has to be measured very precisely since it directly influences the corresponding specific fuel consumption of the engine. In principle, this measurement can also be performed indirectly on the basis of a volume flow measurement. Due to the additional inaccuracy in the determination of the fuel density, this way of measurement could include a relatively large margin of error. Therefore, the measurements should preferably be performed using the gravimetric principle. Here, the fuel mass consumed per time unit, i.e., the flow rate, is directly determined by a balance (deflection according to change in mass). With known characteristics of the balance, this assures a consistent accuracy of the measurement in the complete measuring range, i.e., also at very low absolute fuel flow values. Figure 10.8 shows a sketch of a fuel balance operating to this principle.

10.7 Engine blowby

A further mass flow to be determined during bench tests is the engine blowby. As experience has proven, an indirect measurement is most appropriate, where the pressure loss at a calibrated orifice associated with a particular volume flow is measured. Figure 10.9 shows the schematic layout of such a measuring device.

10.8 Pressure and temperature data

Especially for supercharged engines, the measurement of the pressure and temperature data in the intake and exhaust systems is important. Generally, this means the time-averaged data, i.e., the mean values during the engine cycle. Due to the cyclic behavior of piston engines, the instantaneous values, resolved by crank angle, deviate substantially from these averaged data. This was shown in Sect. 5.5.3 by means of measured and calculated intake and exhaust manifold pressures. Thus, in addition to the average values, the variations in time have to be measured via pressure indications. These indications are especially important in the combustion chamber, since they enable a direct analysis of the engine high-pressure cycle.

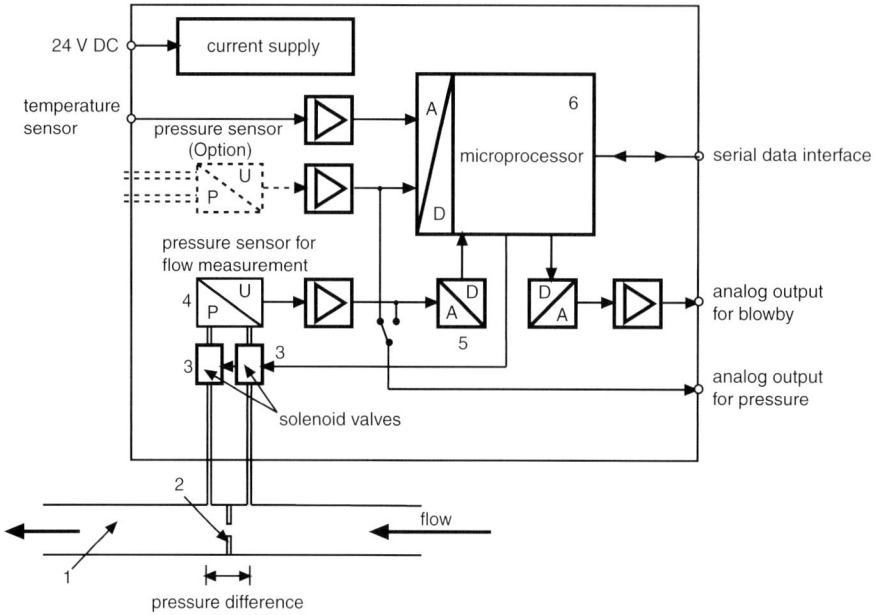

Fig. 10.9. Indirect engine blowby determination via pressure difference measurement at a calibrated orifice

For these measurements, angle resolutions of 0.1 to 0.2° crank angle should be chosen. On the one hand, this will result in a sufficient density of the signal data allowing, if needed, data filtering. On the other hand, for the analysis of the cylinder high-pressure signal (e.g., determination of the combustion characteristics) it will provide exact information regarding the pressure gradients.

For the determination of the average values (in the intake and exhaust systems), the static pressures are acquired via pressure bores in the pipe wall. The signal is converted by a pressure transducer (usually based on a strain gage sensor) into a voltage signal which is proportional to the pressure. Depending on the sensor characteristics and calibration, the signal can then be linearized and analyzed. The transient (dynamic) pressure indication is performed via special pressure pickups, often piezoelectric sensors (pressure application leads to a charge change in a crystal) or strain gage load cells. The location of these sensors must be very carefully chosen, since flow disturbances (separation effects) may lead to significant errors in the measurement signal. Figure 10.10 shows possible installation locations in the combustion chamber as well as a cross section of such a sensor.

Up to temperatures of 300 °C, mean temperatures are measured via Pt-100 sensors. Mean temperatures up to 1,000 °C can be measured using NiCr-Ni thermocouples type K. The Pt-100 sensors are based on the principle of resistance thermometers, which show a direct proportionality between the temperature and the resistance of the sensor. In contrast, the NiCr-Ni sensors operate on the basis of the principle of thermocouples, which – at the joining of different metals – generate an increasing voltage when the temperature is increased. Figure 10.11 shows typical characteristics of resistance thermometers and thermocouples.

With temperature measurement data, as well as the corresponding simulation data, it generally has to be considered that the instantaneous temperatures fluctuating around the mean value – especially pronounced in the exhaust system – also occur at differing mass flows. This also

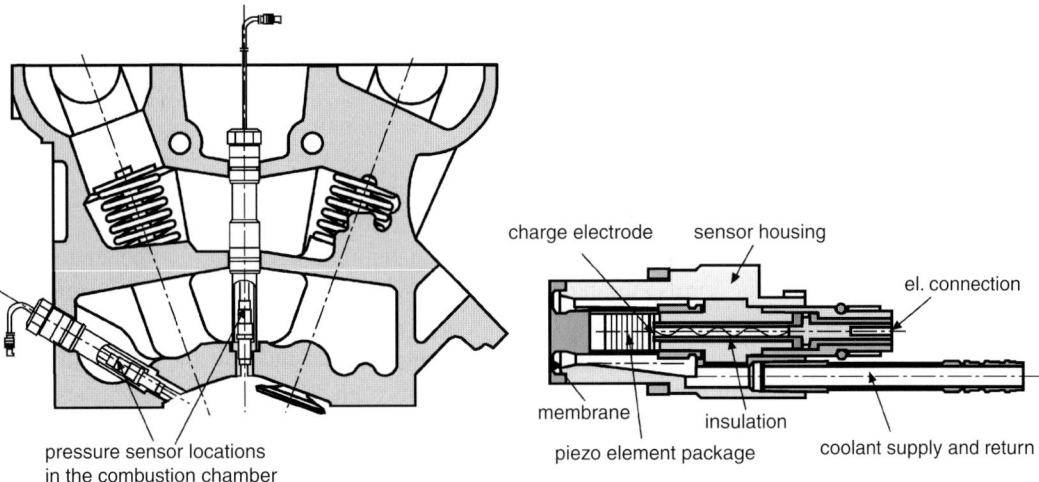

Fig. 10.10. Installed locations and cross section of a piezoelectric pressure sensor for cylinder pressure indications

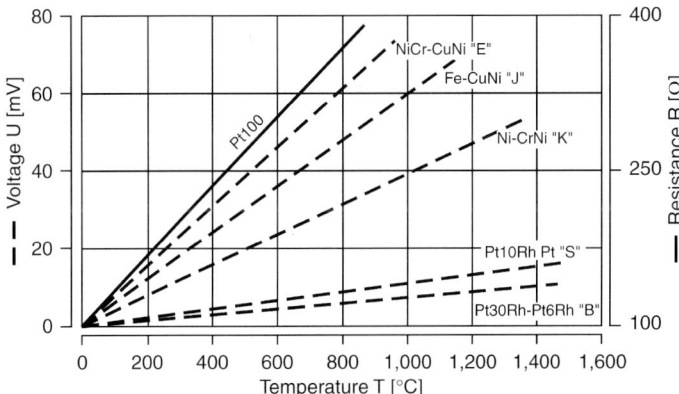

Fig. 10.11. Characteristics of resistance thermometers and thermocouples for gas temperature measurement

influences the heat transfer to the thermocouple, resulting in a higher weighting of those temperatures which occur at high mass flow rates and thus flow speeds. Experience shows that the measured mean temperatures essentially correspond to a mass flow-weighted value:

$$T_{\text{med}} = \frac{1}{720\dot{m}} \int_0^{720} T(\varphi)\dot{m}(\varphi)\,d\varphi. \tag{10.1}$$

Therefore, the simulation values for the corresponding measurement points should be weighted by Eq. (10.1) before they are compared to the actually measured data.

10.9 Emission data

Finally, for a full assessment of the engine operating behavior it is also necessary to measure and analyze the engine's emission data. A distinction is made between the gaseous exhaust emissions and the soot and particulate matter emissions.

In the broadest sense, the group of gaseous emissions covers the following:

all unburned hydrocarbons
carbon monoxide
all nitrogen oxides NO_x (sum of NO and NO_2)
fraction of non-methane hydrocarbons
methanol and formaldehyde fractions

The latter two components relate to special rules and regulations in the United States and have to be measured only for corresponding applications. When measuring the gaseous emission components, all or a part of the engine exhaust gas is mixed with a calibrated dilution air mass flow. The emission concentrations in the total gas flow are related to those of the known dilution mass flow. Additionally, the dilution simulates after-reactions of the exhaust gas in the atmosphere and prevents water condensation during the measurement. Subsequently, the gas mixture is either collected in exhaust gas collection bags – whose contents are measured after the test via gas analyzers for CO, CO_2, and NO_x concentrations – or is routed to a continuously operating exhaust gas analyzer. In contrast to this, the HC measurement is performed via a separate, heated, probe, which routes the exhaust gas directly to a continuously measuring HFID (heated flame ionisation detector).

Soot and particulate matter emissions are determined via the black smoke number and exhaust gas opacity, and as emissions of particulate matter. The first is measured via the calibrated optical density of a filter paper, under both steady-state and transient operating conditions. Opacity measurements are based on a (calibrated) reduction of the translucence of the exhaust gas by soot, fuel, lubricating oil, and water vapor particles. It enables a continuous measurement, allowing a very good resolution of dynamic processes. Particulate matter is measured gravimetrically, by weighing the mass of soot and its attached soluble (fuel and lubricating oil) and insoluble (sulfur compounds, water, ash and wear particles) components deposited on a filter.

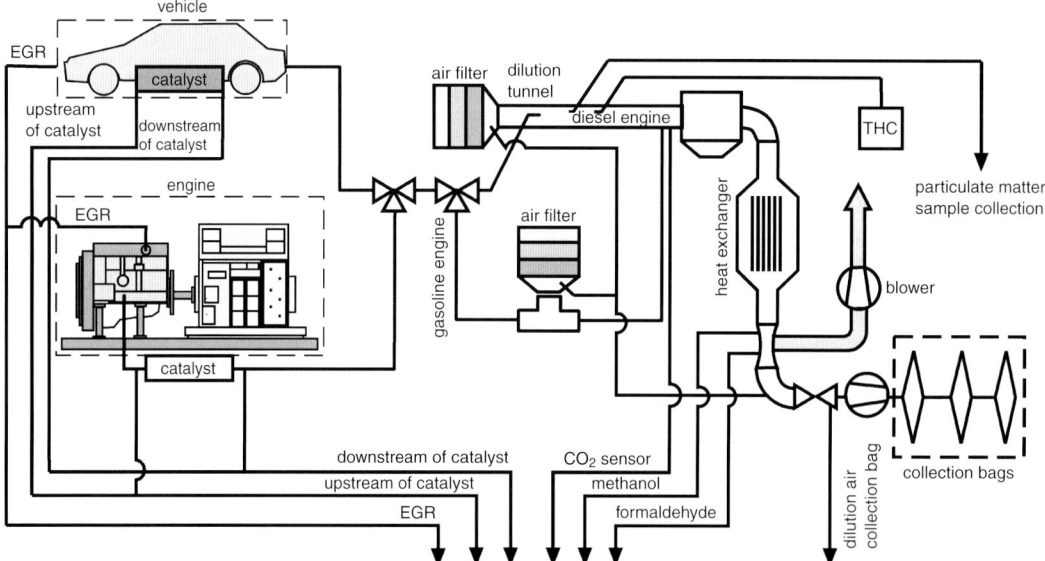

Fig. 10.12. Layout principle of an exhaust emission measurement system for engine test benches and vehicle measurements

10.9 Emission data

Finally, particulate matter analysis methods should be mentioned which are capable of measuring particulate emissions – under steady-state and transient engine operating conditions – in respect to the number of particles and their size, in form of size spectra (particle size as a function of particle size class). Mobility spectrometers are capable to perform such measurements (DDMPS, dual differential mobility particle spectrometer, or TDMPS, transient differential mobility particle spectrometer).

Figure 10.12 shows the layout principle of such an exhaust emission measurement system.

11 Mechanics of superchargers

This chapter will cover especially the mechanical and production engineering-related topics of superchargers.

11.1 Displacement compressors

Displacement compressors particularly are piston compressors of different shapes (e.g., rotary piston compressors), screw-type compressors and spiral compressors. As an example, Fig. 11.1 shows a spiral charger (Ecodyno) for two delivery volumes. This again shows clearly that the overall size of a displacement compressor increases about linearly with increasing volume flow. Figure 11.1a shows the model S-100/35, rated at a geometric delivery volume of about $700\,cm^3$. Figure 11.1b shows the model S-100/50 with a geometric delivery volume of about $1,000\,cm^3$, about 40% more than the S-100/35. If the effective compressor length of the smaller compressor is defined as 100%, the length of the larger compressor is increased by about the same percentage.

11.1.1 Housing and rotors: sealing and cooling

With the exception of the spiral charger, all displacement compressors are rotary piston chargers, either with or without internal compression. Today, housings and rotors of these are mostly manufactured by pressure die casting.

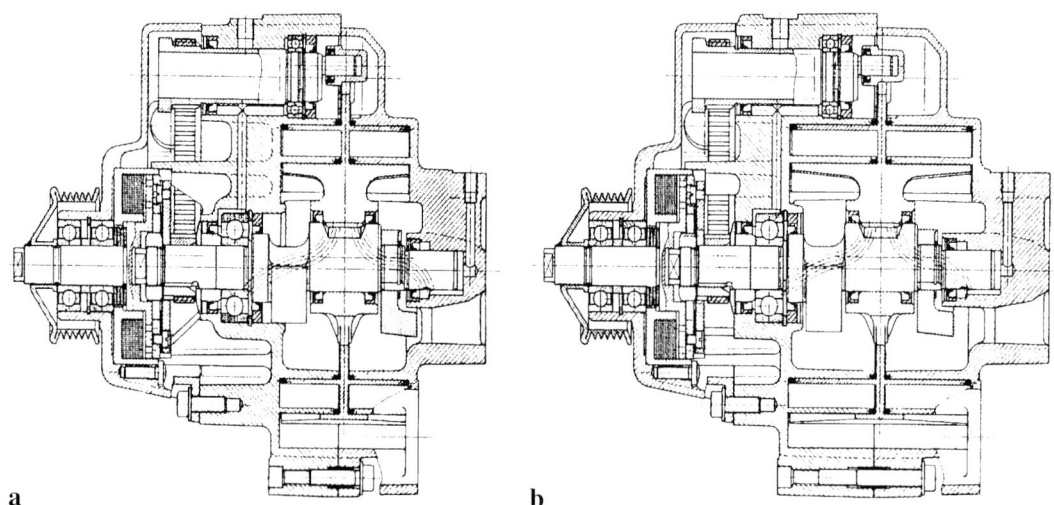

Fig. 11.1. Size comparison of two spiral chargers (Ecodyno): **a** model S-100/35 with geometric delivery volume of about $700\,cm^3$; **b** model S-100/50 with geometric delivery volume of about $1,000\,cm^3$

The machining of the charger rotors poses problems primarily for two reasons. First, in order to reduce noise emissions, the blade configurations in most cases are highly twisted and partially interlocked. Second, the thermal expansion of the housing and of the rotors is different. This makes it difficult to achieve a good sealing between rotor and housing, and between the rotors themselves, which is a mandatory requirement for high efficiencies and small gap losses. To increase its rigidity, the housing itself is designed with ribbings, which additionally complicates the goal of similar heat dissipation of housing and rotors in order to obtain similar component heating and thus similar component expansion.

Today, sealing between housing and rotor as well as between the rotors is mostly accomplished by applying an abrasive lubricating layer (graphite-containing paste) on the rotors and/or housing. In the initial break-in phase of the charger, abrasion of the sealant layer achieves a uniform and very narrow gap between housing and rotor, and from rotor to rotor. This play adjusts itself corresponding to the maximum expansion differences and the highest temperature differences occurring during compressor operation, but only stays constant as long as no excessive component heating occurs, e.g., by overloading the charger. Thus, the narrowest gaps are optimally adjusted only for the most adverse operation case, and all other operating points must manage with these sealing gaps.

The charger type therefore also has to be selected under consideration of the number and size of sealings in the frontal and circumferential areas of the charger, and the maximum glide speeds occurring during operation. Today, maximum glide speeds of about 60 m/s are allowed.

11.1.2 Bearing and lubrication

For rotary piston and spiral chargers, only ball bearings are used today. On the one hand, this is due to their extremely small bearing width, which also influences the possible sealing gap size, on the other hand, because of better lubrication conditions.

11.2 Exhaust gas turbochargers

In the discussion of exhaust gas turbocharger designs and their production methods, one has to distinguish between the small chargers as used in passenger cars and trucks, which are produced in large numbers and must be very inexpensive [98], and the large (heavy-duty) chargers produced in small quantities and developed with the goal of highest efficiencies and pressure ratios, which are used in medium-speed and slow-speed heavy-duty engines.

11.2.1 Small chargers

11.2.1.1 Housing: design, cooling and sealing

Today, in mass production, the housing of exhaust gas turbochargers consists of three parts: the actual bearing housing with the mounted compressor and turbine wheels and the housings for compressor and turbine attached to it.

Compressor housing

The compressor housing surrounds the impeller and additionally contains the inlet for the air into the charger, the diffuser (generally without blades), the volute, which is the air-collecting plenum, and possibly additional air and recirculation channels for flow-stabilizing measures. Nowadays, most compressor housings are cast from aluminum or magnesium.

Turbine housing

The turbine housing correspondingly surrounds the turbine wheel, mostly a centripetal rotor. Additionally, it contains the exhaust gas intake and, for fixed geometry chargers, the turbine inlet cone integrated into the housing, as well as (in most cases) the waste gate. For VTG chargers, the turbine inlet blades, including their adjustment mechanism, are mounted on a special disc-shaped shield which is connected to the bearing housing. Turbine housings are cast. For special applications, e.g., marine, where the surface of the complete exhaust system may not exceed a specified temperature limit, water-cooled housings are used (see bearing housing, below).

The materials to be used have to be selected depending on the actual exhaust gas temperatures. For temperatures up to 750 °C, GGGX-SiMo51 is used today; this covers most diesel engine applications. For higher temperatures up to about 850 °C, mostly GGG NiCr 20 2 (D2) is used. For highest exhaust gas temperatures up to a maximum of 1,050 °C, i.e., gasoline engines, GGG NiCr 35 5 2 (D5) is used. If a compressor impeller or turbine rotor bursts, the debris must not penetrate the corresponding housing. This mandates sufficiently dimensioned wall thicknesses of both housings.

Compliance with this requirement is tested in a so-called **containment test**. Here, the rotor is accelerated to the point of bursting. Then the containment safety of the housings is analyzed. With the rotor materials mentioned above, the bursting speed is approximately 50% above the maximum acceptable operating speed.

Bearing housing

The bearing housing contains the bearings for the rotor assembly, the corresponding lubricating and cooling oil circuit, the shaft seals towards compressor and turbine housings, and the thermal protection shields of the bearings. In most cases the bearings are located inside, i.e., between the compressor and the turbine. Due to the small distance between the bearings and the hot turbine housing, substantial heat flows occur to the bearing next to the turbine – increased by the heat flow inside the charger shaft. With insufficient **thermal protection**, this can lead to oil coking and, consequently, failure of lubrication and/or solid material friction, in any case in increased bearing wear. Depending on the highest occurring temperatures, the following design variants help to avoid such problems.

With an appropriate bearing housing design (Fig. 11.2), the bearing block located next to the turbine has to be thermally decoupled by maximizing the length of the heat conduction path. For example, the connection between the bearing block and the bearing housing can be designed in such a way that it is routed towards the compressor side and behind the oil intake. Further improvement is achieved by a so-called heat shield arranged in the back of the turbine rotor, which to a large extent prevents direct contact between the hot exhaust gases and the bearing housing. An additional oil-jet cooling on the hot side of the charger shaft reduces the heat input by the shaft into the bearing.

Cooling

For gasoline engines, where the exhaust gas temperatures are 200 to 300 °C higher than for diesel engines, charger housings mostly are equipped with additional water cooling (Fig. 11.3). In this case, the bearing housing is integrated into the cooling circuit of the engine. If problems with heat accumulation occur, e.g., when the engine is stopped immediately after operation at high load, a small thermostat-controlled auxiliary water pump has to be utilized to flush the water-cooled bearing housing also after the engine and, thus, the main water pump were stopped.

The thermal stability of a charger is verified by a **start–stop test** in which the charger is equipped with temperature sensors at typically critical locations. Then the engine is operated at full-load and immediately stopped, which is repeated for a specified number of cycles. The test is passed when

11.2 Exhaust gas turbochargers

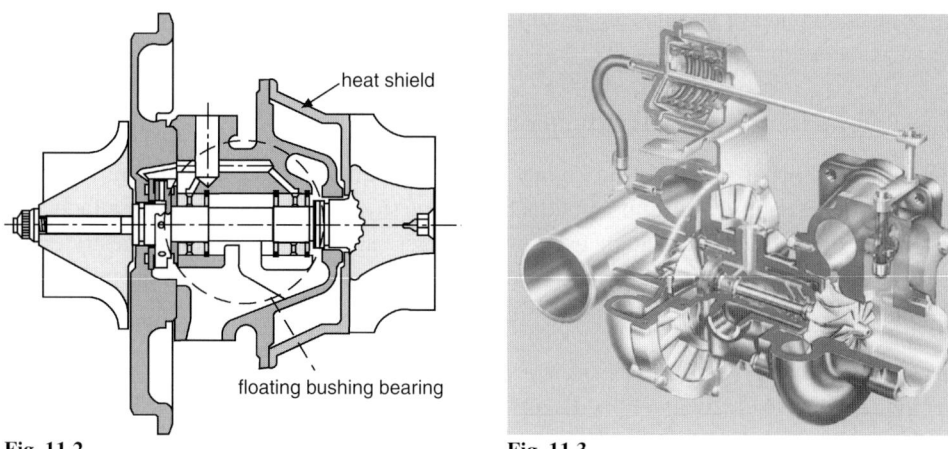

Fig. 11.2 **Fig. 11.3**

Fig. 11.2. Design of a bearing housing for maximum thermal insulation and with floating bushing bearing

Fig. 11.3. Waste gate turbocharger for gasoline engines, with water-cooled bearing housing

neither the maximum acceptable component temperatures are exceeded nor noteworthy amounts of oil carbon are found.

Sealing system

At the shaft passage, the bearing housing has to be sealed off, on the one hand to prevent oil losses both on the compressor and on the turbine side, on the other hand to prevent the hot exhaust gases from flowing from the turbine into the bearing housing. Mass-produced chargers are equipped with one piston ring each, which sits in a corresponding groove of the rotor shaft (Fig. 11.4). These piston rings do not rotate but are rigidly held in the bearing housing and thus form a kind of noncontact labyrinth seal.

In special turbocharger applications, e.g., for natural gas engines, the throttle, as the load-controlling device, is often located upstream of the compressor. The reason for this is an improved homogenization of the gas in the compressor, which results in better mixture formation. In this case, substantial pressure drops below ambient conditions can occur in the compressor housing, which cannot be handled adequately by the piston ring labyrinth seal described above. Here, an additional seal, generally a carbon slip ring, has to be applied.

There are functional tests for all of these sealing systems. On an actual engine, the charger performance is checked in the entire map. For the test of the sealing ring on the compressor side, the pressure at the inlet to the compressor is lowered to a value which could occur in case of a soiled air filter. To test the sealing on the turbine side, the engine crankcase pressure is increased

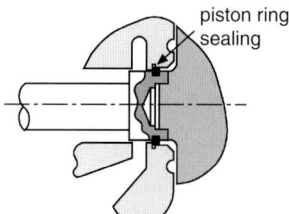

Fig. 11.4. Charger shaft piston ring sealing system

such that a plugged crankcase ventilation system is simulated. In both cases, no oil should leak towards the corresponding rotors.

11.2.1.2 Rotor assembly: load and material selection

Independent of the turbocharger's size and layout, the rotor assembly is the critical component. It consists of the shaft, the compressor impeller attached to one side, and the turbine rotor attached to the opposite side. For small turbochargers, used in passenger car and truck engines, mostly radial compressor and turbine rotors are used; more recently, also mixed-flow turbines are utilized, a design combining axial and radial flow characteristics.

Compressor impeller

Aluminum cast alloys are exclusively utilized as material for the compressor impeller. This includes not only today's passenger car and truck mass-produced chargers, but also larger high-speed engines with several chargers. The acceptable circumferential speeds reach about 550 m/s. If reduced durability can be tolerated, higher circumferential speeds are possible. Further increases are possible by applying measures to reduce stress peaks in the rotor, e.g., reinforcing the rotor back in the vicinity of the hub (Fig. 11.5). Fully machined rotors made of forged aluminum are only utilized in special cases.

Turbine

Up to a diameter of about 150 mm, turbines are today exclusively designed as radial or mixed-flow tangential turbines. With decreasing diameter, this design shows better efficiencies than comparable axial turbines. Additionally, in combination with turbine housings without inlet guide blades, it can be produced at low cost. Due to the high turbine inlet temperatures, the materials used must still offer sufficient strength to carry the stresses caused by the circumferential speeds as required for the compression process. Nowadays, for mass-produced turbines essentially two materials are utilized:

- GMR 235 for exhaust gas temperatures up to approximately 850 °C at the turbine inlet, i.e., primarily for applications with diesel engines,
- Inconel 713 (73% Ni, 13% Cr) for exhaust gas temperatures up to 1,050 °C, i.e., for turbocharged gasoline engines.

The main components of both castable alloys are nickel and chromium.

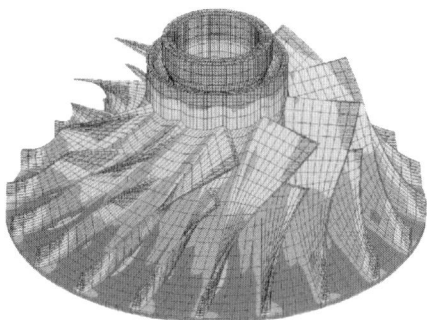 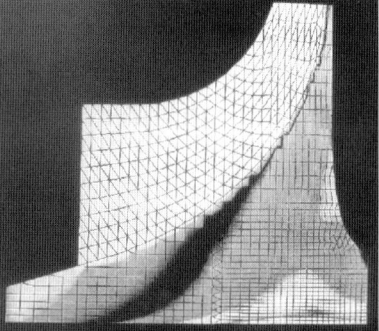

Fig. 11.5. Reduction of stress peaks by reinforcing the rotor back in the vicinity of the hub, shown on a compressor impeller

11.2.1.3 Bearing, lubrication, and shaft dynamics

The rotors of the mass-produced turbochargers described here reach speeds of up to $200,000\,\text{min}^{-1}$. At the same time, their durability should be guaranteed for up to 1 million km (truck applications). Only specially developed floating bearing systems can satisfy these stringent requirements reliably and at low cost.

Radial bearing with floating bushing

In this design, floating bearing bushings are arranged between the compressor- and turbine-side bearing blocks in the bearing housing. Within these bushings, the shaft can rotate, without wear, in an oil film. The bushing itself also rotates in an oil film between bearing block and bushing, in such a way that the bushing, made of brass, rotates at about half of the shaft speed, i.e., it floats. Thus, the friction speed in the bearing can be halved. In addition, the double oil film acts as an improved damper, resulting in more stable dynamics of the rotor shaft (Fig. 11.2).

With the proper choice of the **lubrication gap widths** between bearing block and bushing and between bushing and shaft, the hydrodynamic load capacity of the bearing and its damping behavior can be optimized. Here, the lubrication gap width between shaft and bushing is dimensioned under the aspect of load capacity, while the gap between bushing and bearing block is dimensioned under the aspect of optimized damping. Increasing lubrication gap widths increase the damping and reduce the load capacity.

Single bushing bearing

Single bushing bearings are regaining importance in mass-produced turbochargers. Here, the rotor shaft rotates within a single, long and fixed bushing, which on its outside is surrounded by oil (Fig. 11.6). In this case, since there is no rotation, the outer gap of the bearing can be especially optimized for bearing damping. The bushing is recessed on both sides in its center part. This design makes a smaller bearing span in the bearing block possible, which results in a shorter total length of the turbocharger. The assembly is easier, resulting in lower production costs.

Neither the floating bushing nor the single bushing bearing carry forces in axial direction. However, as a rule, compressor and turbine rotors are subject to different gas forces, which would lead to a lateral movement of the rotor assembly. To prevent this, a further bearing, the axial bearing, is necessary.

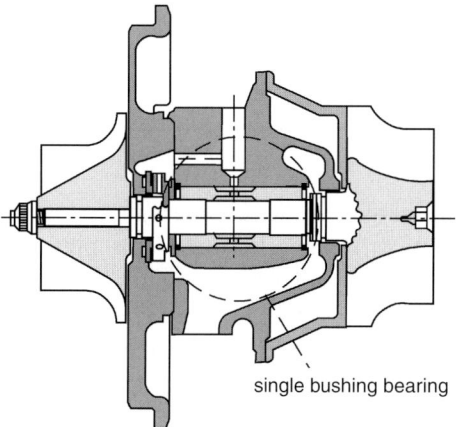

Fig. 11.6. Design example of a floating single bushing bearing

Axial bearing

The axial bushing carries forces in axial direction. Today, it is mostly designed as a wedge surface friction bearing. Two discs serve as stop face. These are rigidly connected to the shaft, while the axial bearing itself is mounted in the housing. An oil deflection plate prevents too much oil from getting close to the shaft seal.

Lubrication

Today, all turbochargers are lubricated with engine oil and integrated into the engine's oil circuit. Thus, the necessary lubricating oil enters the bearing housing at a pressure of about 4 bar, where a throttle reduces the pressure to about 2 bar before it is routed to the bearings. Oil drainage occurs at ambient pressure. To assure an unobstructed oil drainage, the drainage pipe must be dimensioned correctly and the oil feedback position into the engine must be located above the engine's oil level.

Balancing

Rotor assemblies – such as those in a turbocharger – rotating at very high speeds naturally have to be balanced particularly well. The components – the turbine rotor with shaft and the compressor impeller – are individually prebalanced. After assembly, they are fine-balanced in the completed turbocharger (also see Sect. 11.2.1.4).

Rotor dynamics

During operation, the rotor assembly of any turbocharger is influenced in its rotary motion by the pulsating admission of exhaust gas into the turbine, its own residual imbalance, and by the mechanical vibrations of the combustion engine. This may excite vibrations in the rotor assembly.

Nowadays, these rotor dynamics and the associated **shaft shift paths** of the rotor assembly are carefully simulated and also measured. The goal is to avoid excessive deflections of the rotor assembly in the bearings, which – especially in combination with low lubricating-oil pressures and high oil temperatures – may lead to instabilities and metallic contact, resulting in increased bearing wear.

Turbine blade vibrations

The excitations mentioned above may cause similar problems for the turbine blades as were discussed for the rotor dynamics. This can lead to undesired blade resonance vibrations, endangering both the charger's reliability and durability. The dynamic loads of the turbine blades occurring under full-load engine operating conditions are therefore measured via suitably arranged strain gage load cells; then they are analyzed, and the blades are optimized in regard to their vibration behavior.

11.2.1.4 Production

On the one hand, function and operating conditions of turbochargers are demanding high production precision and quality. On the other hand, the complete charger is subject to strong cost pressure. Since turbochargers are produced in ever larger quantities, their production processes have been continually improved and refined. At the same time, the production costs decreased significantly. Therefore, within the scope of this book, the production sequence can only be covered briefly and with few examples. The special production know-how that has rendered today's turbochargers into cost-efficient, durable, and reliable products, essentially is the manufacturers' domain.

Production of the turbine rotor and connection with the shaft

As previously explained, turbine rotors are made of high-temperature nickel alloys which must be melted and cast under vacuum. The necessary ceramic shell forms are produced utilizing the lost-wax casting process. First, the wax models necessary for each rotor are combined, at 1 : 1 scale, in cast clusters. Then, by dipping them several times into a ceramic slurry and subsequent sanding, they are covered with a 6–10 mm thick, fireproof ceramic shell. After the drying and setting of the shell, the wax models are melted out and the resulting casting molds can be baked. The actual raw parts are cast in these molds, which are heated during casting. After cool-down, the shell can be removed, the castings can be cut off the cluster, and they can be machined.

The cleaned and occasionally reworked turbine rotor is connected to the charger shaft via friction welding. The shaft is coupled to a flywheel mass and accelerated to a speed of about $1,000\,\text{min}^{-1}$. Then the drive and flywheel mass is disconnected and the shaft is pressed against the stationary rotor with a predetermined force. The resulting friction heats up the shaft and rotor to such an extent that both components are welded to each other.

After friction welding, the next steps are stress-free annealing and subsequent machining of the shaft. The bearings are hardened and the complete rotor is subjected to further heat treatment. The final dimensions are achieved by grinding, where especially strict requirements for dimensional accuracy and roundness have to be met. Then the outer contour of the turbine rotor is ground and the grooves for the sealing piston rings are machined. Finally, utilizing special machines, the completed turbine–shaft combination is balanced by countersinking the back of the turbine.

Production of the compressor impeller

The production of the compressor impeller starts with a master form produced on a 5-axis milling machine. On the basis of the master form, a master mold – a hollow form – is produced from which the so-called work models – made of rubber – are extracted. Mounted on a worktop, the rubber work model is used to produce hard plaster models. After drying, the rubber model must be extracted from the plaster model. This is done by a rotary motion of the rubber model, where the model must not be damaged.

This production manner of course influences the form of the elastic rubber model and accordingly that of the possible compressor impeller and its contours. The hard plaster models are then clamped between plates, where they act as casting molds for the aluminum die cast of the impellers. The machining of the compressor impeller starts with the insertion of the center bore where concentricity between impeller molding blank and bore is assured by special clamping fixtures.

Subsequently, impeller back and contour are machined. When finished, the compressor impeller is also balanced.

Production of compressor and turbine housings

The molding blanks of both housings are cast from the respective materials chosen. Subsequent machining encompasses lathing, drilling, and milling, performed on CNC machines (computer numerical control). Final machining, which includes the milling of the flange faces as well as insertion of bores and screw threads, is performed using circular transfer machines. Since the compressor housings are mostly made of aluminum die cast, due to their high cast accuracy and good surface quality, they only need minimal final machining. At the end of machining, the turbine housings are deburred and preserved to prevent corrosion.

Machining of the bearing housing

Similar to the turbine housing, the machining of the bearing housing – which is cast in gray iron – is performed on multiple-spindle CNC machines. The connecting flanges on both turbine and compressor sides have to be lathed, and the bearing bore has to be drilled and either ground or honed. The bearing housing is protected against corrosion as well.

Assembly

The assembly process is divided into the core assembly and the final assembly. In between, the complete rotor assembly is again fine balanced at high speed in the core assembly. This core assembly consists of bearing unit, rotor assembly and compressor back wall. During assembly, first the bearings are installed into the bearing housing, then the turbine rotor–shaft combination including the sealing piston rings is inserted into the bearing. The axial bearing is installed and the compressor back wall is attached with bolts. Subsequently, the compressor impeller is slipped on the still-open shaft end and braced with a shaft nut. In mass production, the complete core assembly is performed automatically.

In the following final assembly, at group-work stations the compressor and turbine housings are assembled and, if needed, the waste gate or VTG mechanisms attached and tuned. This finalizes the production of the turbocharger and it is delivered as a supplier component to the engine manufacturer.

11.2.2 Large chargers

In applications with medium-speed and slow-speed heavy-duty engines, rated between 1,000 and nearly 10,000 kW per cylinder, the exhaust gas turbocharger – today exclusively used here – is an essential design element which significantly influences power, fuel consumption, and installed size. Both its conceptual design and its production are, therefore, subject to considerations totally different from those for the previously discussed mass-produced charger. Between these two extremes, naturally many intermediate stages exist. Within the scope of this book, only the other extreme, the truly large charger, will be discussed.

11.2.2.1 Design, housing, cooling, sealing

Design

While the assembly of small chargers generally involves four components – the rotor assembly, bearing housing, turbine and compressor housings – the design of a large charger is much more complex. Fully machined compressor impellers and bladed diffusers are used, as well as axial turbines with external intake and central exhaust gas outlet, which necessitates a totally different design for the bearing housing (see below). As an example for a large charger, Fig. 5.39 shows a NA/S charger by MAN.

Bearing

The bearing is also quite different to that of a small charger. While small chargers are designed exclusively with "inside" bearings, large chargers predominantly use "outside" bearings. Figure 11.7 shows a layout principle. Further, in most cases roller bearings with their own oil supply are used, for the following reasons:

lower power losses at normal speeds
high short-term overload capability of the roller bearings

11.2 Exhaust gas turbochargers

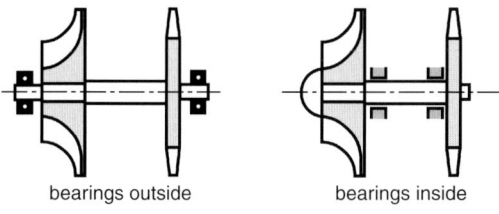

bearings outside   bearings inside

Fig. 11.7. Layout principle of the bearings, outside and inside

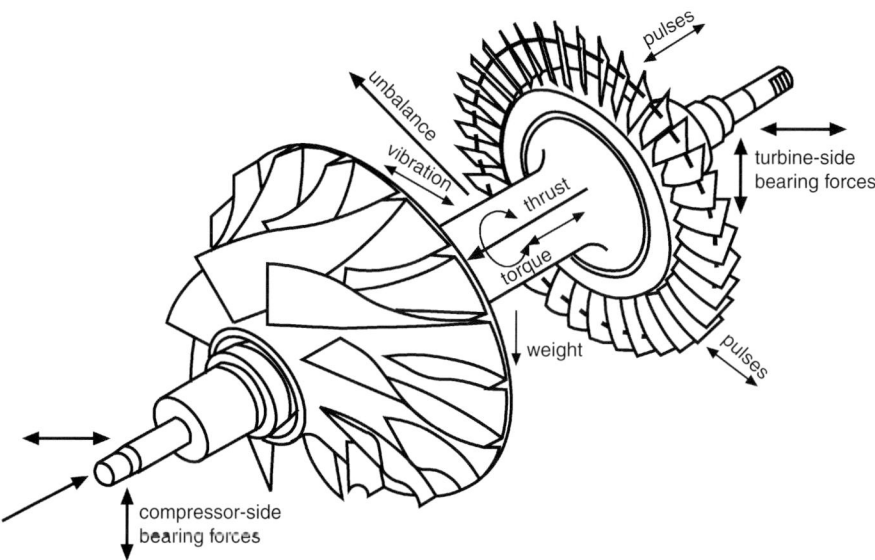

Fig. 11.8. Rotor assembly bearing forces and their causes

high insensitivity against short-term fluctuations in oil supply
oil supply independent of engine lubrication circuit
no oil contamination by the engine
possibility of choosing special oil qualities and viscosities

The forces acting on the rotor assembly of a turbocharger are shown in Fig. 11.8. Looking at these forces and considering the durability requirements of slow-speed engines, it can be easily seen that significantly higher effort has to be invested in the bearing design of these chargers than for the small chargers.

As an example, Fig. 11.9 shows a compressor-side double-ball roller bearing with dampening intermediate layer at the outer rings and its own splash-disc oil pump. Besides this design, so-called four-point bearings are also utilized, as shown in Fig. 11.10 in combination with an oil-jet pump, or regular double-ball bearings (Fig. 11.11) in combination with a hollow-shaft oil pump and a gear-oil pump.

Compressor housing

The compressor housing is the first especially complex component of large chargers to be discussed here. Usually, it consists of a ring-shaped and strongly tapered air intake to the compressor. On the intake side, a coaxially arranged lamella muffler is attached to dampen the compressor noise and to stabilize the intake air flow (Fig. 11.12).

Fig. 11.9

Fig. 11.10

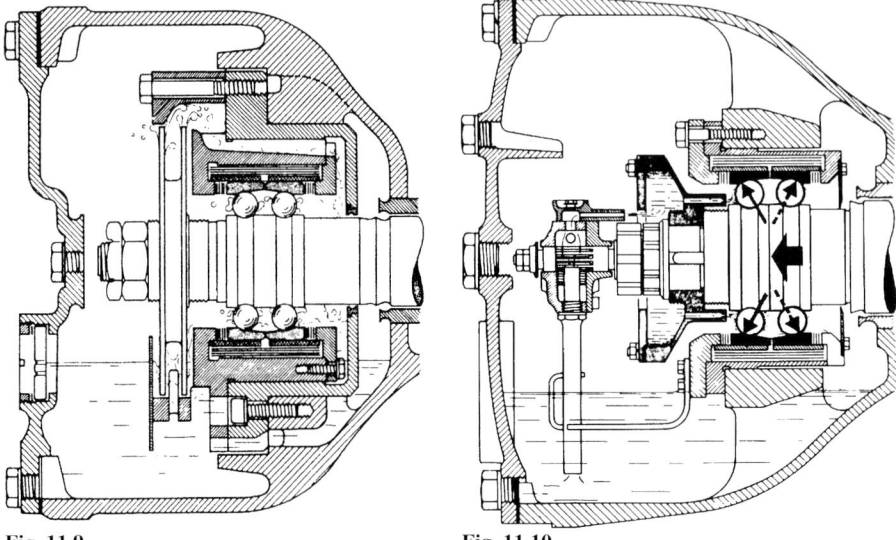

Fig. 11.9. Double-ball roller bearing of a large charger

Fig. 11.10. Four-point roller bearing with oil jet pump

Fig. 11.11

Fig. 11.12

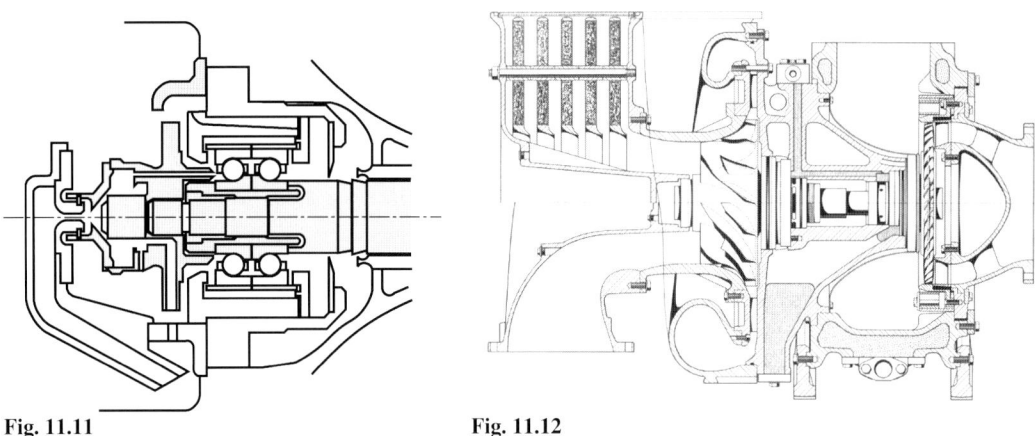

Fig. 11.11. Roller bearing with hollow-shaft oil pump (left) and gear oil pump (right)

Fig. 11.12. Compressor housing (left), exhaust gas plenum housing (center) and turbine housing (right) of a large charger with intake muffler

Inside the housing, which is reinforced by molded support struts, the compressor-side bearing, dampened at the outer ring, is installed close to the compressor impeller; the housing also includes the integrated oil pump and the oil tank. In most cases a double-ball bearing is used, which also carries the axial forces of the rotor assembly. The generally bladed outlet diffuser with subsequent air plenum is a further component which is additionally used as a connection to the exhaust plenum.

Exhaust gas plenum

The exhaust gas plenum (Fig. 11.12) collects the exhaust gases flowing from the axial turbine towards the compressor housing and usually routes the gas upwards into the exhaust gas manifold. Together with the compressor housing, it represents the load-bearing element of the charger. It is water-cooled in order to avoid thermal distortions, since those would negatively influence the viable gap dimensions between compressor and compressor housing as well as turbine and turbine housing. Additionally, distortions could lead to bearing alignment problems. The exhaust gas plenum has an inner protective liner with flange, concentric to the charger shaft. It is directly attached to the turbine housing to assure bearing alignment. This arrangement protects the penetrating charger shaft from excessive heat loading. The turbine housing is located on the other side of the exhaust gas plenum.

Turbine housing

Also the turbine housing (Fig. 11.12) has a concentric, towards the axial turbine tapered (cone shaped), exhaust gas intake. The inlet guide blade ring and the outer turbine ring are attached to it. Via a concentric neck it assures an aligned connection to the compressor housing. Thus it also closes the manifold to shield it from exhaust gas.

The turbine-side roller bearing is arranged inside the turbine housing. Usually, it is an axially movable and also dampened roller bearing, in order to accommodate the thermal expansion of the charger shaft. It also features its own lubricating oil supply including pump and reservoir. The complete housing is water-cooled. A double labyrinth seal on the charger shaft protects against exhaust gas entering from the turbine into the bearing housing.

Charger shaft

The charger shaft is also much more complex than that for small chargers. Due to the outside bearings, which enable a larger distance between the bearings, the shaft must be carefully calculated and optimized regarding bending and torsional vibration modes and natural frequencies. Since high power must be transmitted into the compressor impeller, the connections have to be dimensioned carefully. On the turbine side, the shaft carries the rotor wheel for the axial blades, which are usually supported in the wheel via self-centering fir-tree dovetail connection.

Cooling

Theoretically, ample cooling of all hot components seems to be extremely desirable to limit distortion and assure proper bearing alignment, as well as to guarantee cool bearings and oil supply. However, considering the fact that the use of heavy oil is mandatory for large low-speed engines and this is usually associated with the danger of coking and contamination, attention has to be paid that the contamination-endangered components remain at sufficiently high temperatures to avoid coking. Therefore, the cooling system must be strategically designed to result in a desired and particular coexistence of insulation and cooling measures.

11.2.2.2 Rotor assembly

Compressor impeller

In order to optimize efficiency, the flow guiding components as well must be machined as precise as possible. Production costs are of much lower importance. Also in this case, all impellers are radial and mostly manufactured from titanium; they are machined from solid metal by means of

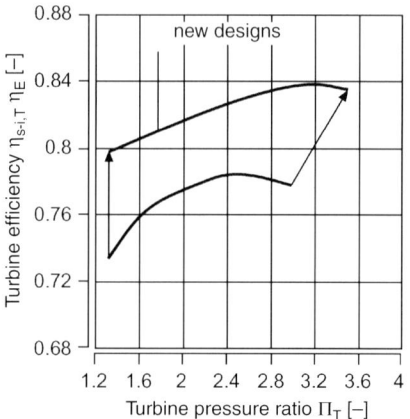

Fig. 11.13. **Fig. 11.14.**

Fig. 11.13. Compressor impeller, fully machined from solid metal

Fig. 11.14. Effective turbine efficiencies of large modern chargers against the expansion ratio [124]

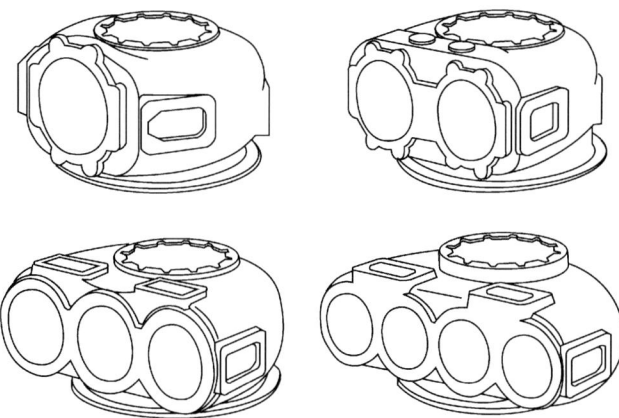

Fig. 11.15. Exhaust manifold connection to the turbine housing

5-axis milling machines. Strongly twisted and backswept blades with reset intermediate blades are state of the art (Fig. 11.13).

Circumferential speeds of up to about 600 m/s are possible. The described high-power compressor impellers are more and more combined with profile-bladed diffusers for further efficiency increase. The compressor efficiencies, thus, achieve values of up to 88%.

Turbine

The turbine of large chargers is for the most part designed as an axial turbine. This design exhibits explicit advantages in efficiency in comparison to a radial turbine at diameters of more than 300 mm. The turbine can be optimally adjusted to particular supercharging requirements with relatively simple changes in blade length. Today, turbine efficiencies of up to 85% can be achieved (Fig. 11.14) in combination with an upstream guide ring equipped either with molded blades or – more and more – with variable-geometry blades, as well as downstream exhaust diffusers. Various design variants for attaching exhaust manifold branches to the turbine housing are shown in Fig. 11.15.

11.2.2.3 Production

Large chargers described in this chapter are always manufactured in very small numbers. They are produced in the most modern machining centers, where the highest demands regarding production accuracy, low tolerances, and reproducibility can be met.

12 Charge air coolers and charge air cooling systems

As shown in Chap. 2, charge air cooling plays a decisive role in the design of supercharged engines with high power density, low fuel consumption, and low emissions. Therefore, special attention has to be paid to the layout of charge air cooling systems and their components for specific engine concepts and applications. For this purpose, knowledge about different cooler designs and suitable charge air cooling systems is as important as knowledge about their respective characteristics.

12.1 Basics and characteristics

The density of the air aspirated by the engine depends on its pressure and temperature ($\rho = p/RT$). Therefore, the objective is to achieve high boost pressures at a temperature increase as low as possible.

Real compressors feature efficiencies $\eta_{\text{s-i,C}} < 1$ (as compared to the ideal isentropic compression process). Therefore, the actual compression of the air results in a significant temperature increase. This increase depends on the pressure ratio selected and on the compressor efficiency, as shown in Fig. 12.1. It can be described by the following equation:

$$\Delta T = T_2 - T_1 = (T_2 - T_1)_s / \eta_{\text{s-i,C}} = T_1\left[(p_2/p_1)^{(\kappa-1)/\kappa} - 1\right]/\eta_{\text{s-i,C}}. \tag{12.1}$$

As can be seen, even at very good efficiencies of $\eta_{\text{s-i,C}} = 0.8$, the temperature is increased at a pressure ratio of $\Pi = 3$ by approximately $135\,°C$.

Consequently, a density increase ρ_2/ρ_1 of only about 2 can be achieved (Fig. 12.2) at the selected pressure ratio.

If the density is to be increased as much as possible, the charge air must be cooled down to achieve a density recovery. Air and water can be utilized as coolants. The intercooler efficiency is the characteristic used to assess the quality of the charge air cooling. It is defined as the ratio between the actual and the maximum possible heat removal:

$$\eta_{\text{CAC}} = (T_2 - T_2')/(T_2 - T_C), \tag{12.2}$$

where T_2 and T_2' describe the charge air temperature up- and downstream of the charge air cooler and T_C describes the coolant temperature.

As an example, again assuming a compressor efficiency of $\eta_{\text{s-i,C}} = 0.8$, a pressure ratio of 3, and an ambient, i.e., coolant, temperature T_C of $20\,°C$, an increase in the density ratio up to 2.7 is possible (Fig. 12.3) depending on the intercooler efficiency η_{CAC}.

12.2 Design variants of charge air coolers

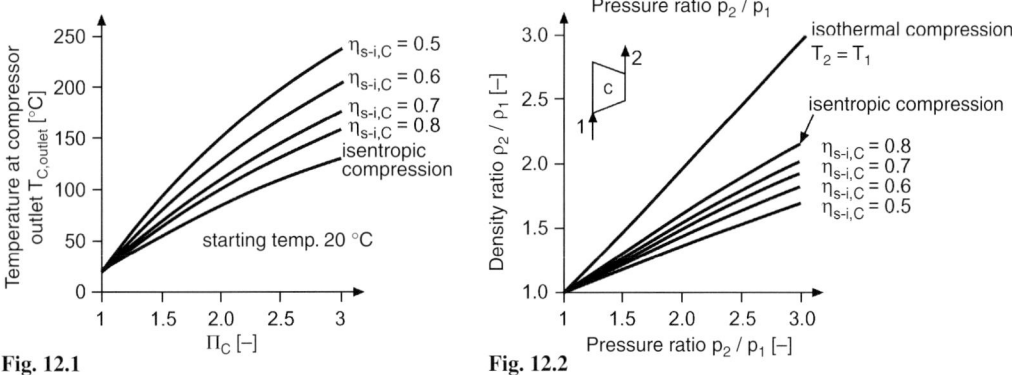

Fig. 12.1. Charge air temperature increase depending on pressure ratio and compressor efficiency

Fig. 12.2. Charge air density increase depending on pressure ratio and compressor efficiency

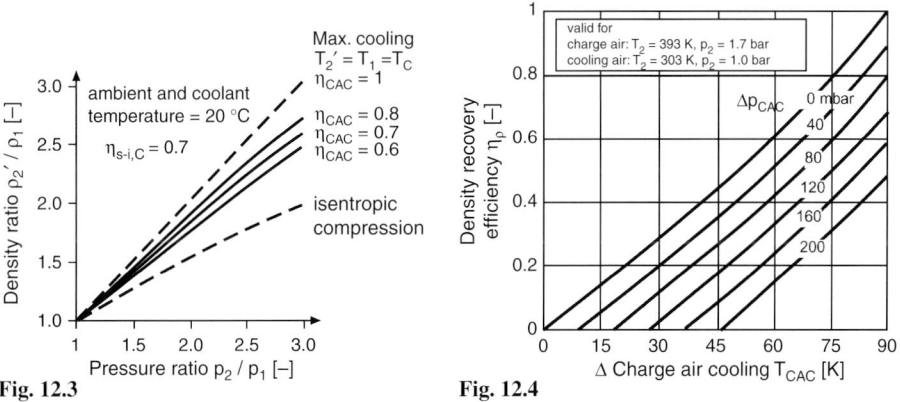

Fig. 12.3. Charge air density status before and after cooling depending on pressure ratio and intercooler efficiency

Fig. 12.4. Achievable pressure ratio depending on pressure loss, intercooler efficiency and temperature drop

It is also possible (and nowadays preferable) to define an efficiency of density recovery η_ρ:

$$\eta_\rho = \Delta\rho/\Delta\rho_{max}. \tag{12.3}$$

Figure 12.4 shows this efficiency for specified pressure losses Δp_{CAC} in the intercooler, against the charge air temperature drop ΔT_{CAC}.

12.2 Design variants of charge air coolers

In principle, charge air coolers (intercoolers) consist of a number of heat transfer areas which route the charge air and coolant flows in such a way that no mixing occurs between the media and the heat transfer is as good as possible. To achieve this, fins are arranged in the respective flow channels to increase their surfaces and thus increase the heat transfer. The heat transfer occurs directly via the described walls and fins. Therefore, they have to be made of a material that effectively conducts

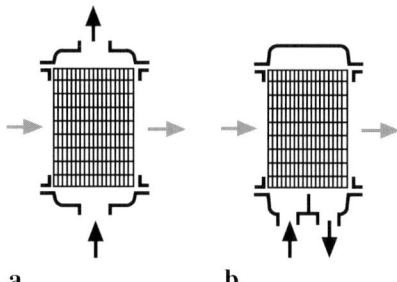

Fig. 12.5. Design of charge air heat exchangers: (**a**) single-cross-flow heat exchanger, (**b**) double cross-counter-flow heat exchanger

heat; in general, metal is used. To minimize the design complexity, the flows of the media are routed according to either the cross- or the cross-counter-flow principle (Fig. 12.5). The charge air cooler should transfer, i.e., conduct, heat at a transfer rate as high as possible. The heat transfer rate is calculated as follows:

$$\dot{Q} = k \cdot A_C \cdot \Delta T, \tag{12.4}$$

where k describes the heat transfer coefficient and A_C the cooling surface. Both factors, k and A_C, depend on the intercooler size and design, while ΔT exists as a boundary condition.
kA_C is termed the heat transfer value. It depends on the following factors:

- α_{in} heat transfer coefficient on the coolant side
- α_{out} heat transfer coefficient on the charge air side
- A_{in} cooling surface on the coolant side
- A_{out} cooling surface on the charge air side
- δ thickness of the transferring wall
- λ thermal conductivity of the transferring wall

The relationship is as follows:

$$\frac{1}{kA_C} = \frac{1}{\alpha_{in} A_{in}} + \frac{\delta}{\lambda} + \frac{1}{\alpha_{out} A_{out}}. \tag{12.5}$$

The equation shows the influences of charge air, coolant, and the walls in relation to the heat transfer. The magnitude of this value is determined by the summands in the denominator, which themselves depend on the intercooler:

size of the cooling surfaces A_{in} on the coolant side and A_{out} on the charge air side
heat transfer coefficients α_{in} and α_{out}
quotient δ/λ of wall thickness and thermal conductivity of the wall

From Eq. (12.5) it can also be derived that the product αA has to be as large as possible to achieve a high value of kA_C, i.e., a high heat transfer value. This fact determines the intercooler design and size. Accordingly, due to very differing properties of the possible coolants, air or water, the design and layout of corresponding intercoolers is quite different.

Due to the similar order of α values on the coolant and charge air sides in air-to-air intercoolers, the cooling surfaces A_{in} and A_{out} also must be about the same size.

12.2 Design variants of charge air coolers

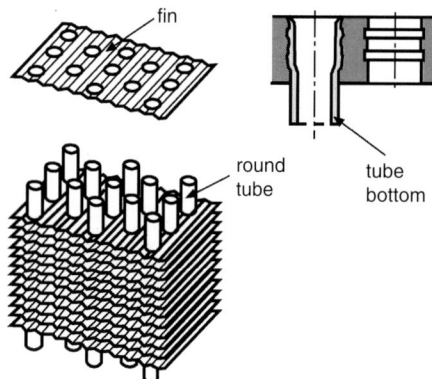

Fig. 12.6. Round-tube intercooler [115]

For air-to-water intercoolers, the ratio of the α values between water and air is about 10 : 1. Thus, the cooling surface on the air side always represents the limiting critical value. No fins are necessary on the water side.

Therefore, depending on the choice of coolant, very differing intercooler designs are possible. Particularly if the operational influences of air and water as coolants are considered, especially contamination and corrosion.

12.2.1 Water-cooled charge air coolers

Depending on the design, a distinction is made between round-tube and flat-tube intercoolers.

Round-tube intercoolers

The core of the round-tube intercooler consists of a multitude of fin plates which are crossed by the tubes (Fig. 12.6). The tubes are connected to the plate flanges such that conduction of heat is supported as good as possible. The connection is produced either by hydraulically or mechanically expanding the tubes or by soft-soldering the plate flanges to the tubes. The heat transfer quality of these two methods is not significantly different. However, there are major differences in view of possible material combinations. When hydraulic expansion is chosen, nonsolderable combinations can also be connected with sufficient heat conductivity, e.g., stainless steel, copper, brass, or titan (tubes) with fins of copper or aluminum. Round-tube intercoolers are mostly used if the coolant is untreated water or ocean water (most severe operating conditions). In this case, the coolant at least has to be filtered before entering the intercooler, and the intercooler has to be cleaned at regular intervals. Further, due to contamination and corrosion, minimum and maximum water flow rates have to be observed and maintained, and occasional erosion damages at the tube inlets have to be monitored.

In regard to their efficiency factor $\eta_{\text{cyc,CAC}} = \dot{Q}/\Delta p_{\text{Loss}}$, round-tube intercoolers reach certain limits. These can be significantly extended if flat-tube intercoolers of the same dimensions are applied.

Flat-tube intercoolers

Due to the flow efficient shape on their charge air side, flat-tube intercoolers generate less pressure loss there. Thus, they can be equipped with a higher fin density. However, the relatively narrow tube channels are not suited for contaminated water and thus should only be used in closed-loop cooling systems (Sect. 12.3). Two designs exist for flat-tube intercoolers.

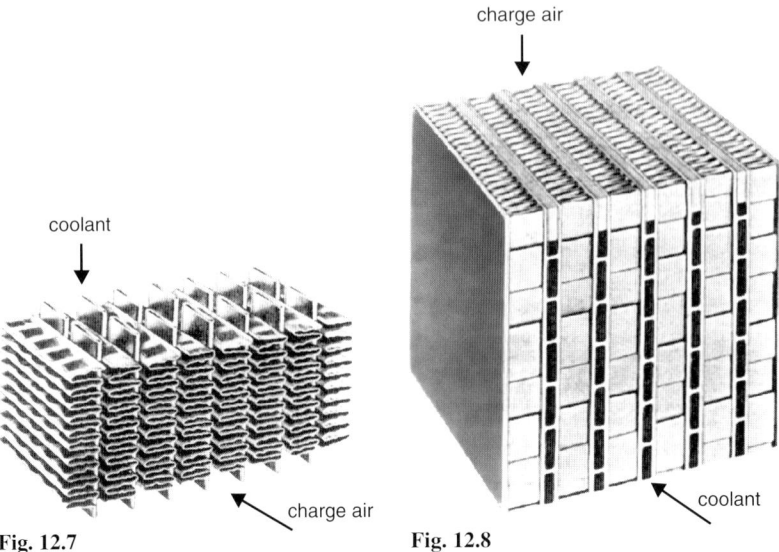

Fig. 12.7. Flat-oval-tube lamella intercooler without interior fins [115]

Fig. 12.8. Flat-tube intercooler in rod-sheet design [115]

In a combination of flat-oval tubes with thin-walled tube bottoms (Fig. 12.7), the tubes (with or without internal fins) are soldered to the air lamellas, side frames and tube bottoms. At the ends of the intercooler cores, water plenums are welded or bolted on. This design requires special stamping tools for the tube bottoms. Therefore, it can only be used economically in mass production.

In a combination of tube wall sheet metal and terminal ledges (Fig. 12.8), the completely brazed intercooler block consists of sheet metal, rods and support lamellas, which together constitute the water channels, as well as air-side lamellas and their side frames. The water plenums are welded to the block ends. Since no expensive model-specific tools are necessary, this design is also suited for low-volume production.

12.2.2 Air-to-air charge air coolers

For applications in aircraft or on-road as well as off-road vehicles where water is not directly available as coolant, air-to-air charge air coolers have to be utilized. In these, in contrast to the water-cooled intercoolers, the charge air flows through flat tubes, which in most cases are finned on the inside. Cooling-air lamellas are arranged between these tubes. Figure 12.9 shows such an intercooler in rod-sheet design with interior-finned flat tubes; Fig. 12.10 shows one in flat-oval-tube lamella design, also with interior-finned flat-oval tubes.

12.2.3 Full-aluminum charge air coolers

In order to further optimize the capacity, weight and production costs, intercoolers in increasing numbers are made fully of aluminum. Here too, two differing designs can be chosen:

- intercooler design with small block depth (about 30 mm) and tubes without internal fin, for medium charge air mass flows;
- design with block depth of about 50 to 100 mm, with turbulence inserts in the tubes, for especially high-capacity requirements at high compactness.

12.3 Charge air cooling systems

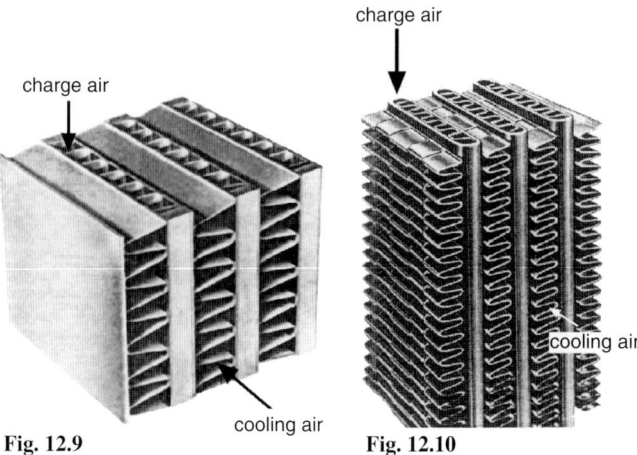

Fig. 12.9. Air-to-air intercooler in rod-sheet design [115]

Fig. 12.10. Air-to-air intercooler in flat-oval-tube lamella design with interior fins [115]

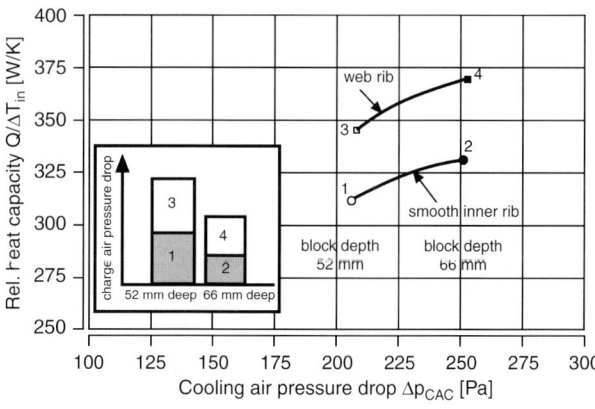

Fig. 12.11. Behr diagram for charge air cooler layout [85]

Figure 12.11 shows the capacities of various tube–fin combinations for different block depths. With these designs, it is also possible to combine several heat exchanger components in one unit – a so-called monoblock.

12.3 Charge air cooling systems

The cooling systems always consist of the charge air cooler itself, the corresponding piping or connections, and any additional components which may be needed for charge air heat management. Two systems are primarily utilized: direct charge air cooling, with air or water as coolant, or indirect charge air cooling with corresponding intercooler combinations.

Figure 12.12 shows a layout principle of a **direct charge air cooling** system. If air-to-air cooling is applied, the achievable temperature reduction and, thus, the achievable density recovery mainly depend on the intercooler size and on its efficiency. ΔT values of about 15–20 °C above the ambient, i.e., coolant temperature, can be achieved. In systems with untreated water or ocean water cooling, values of 5–10 °C above coolant temperature are even possible. A charge air heat management system (e.g. cooling only at high loads or similar controls) can only be achieved with significant additional effort. On the other hand, the basic system is simple, robust, and low-cost.

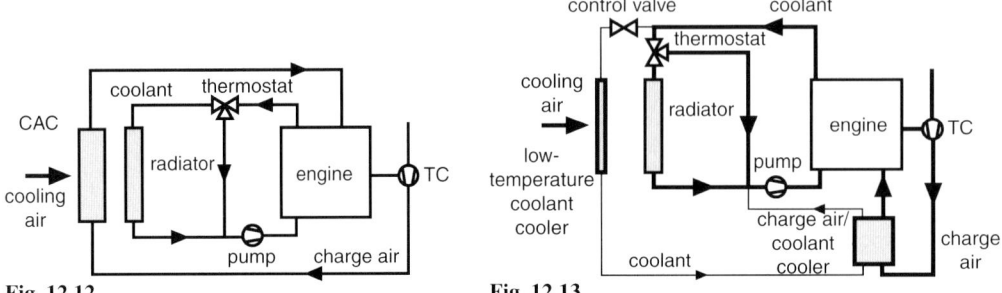

Fig. 12.12. Principle layout for direct charge air cooling [85]

Fig. 12.13. Principle layout for indirect charge air cooling [85]

Figure 12.13 shows the **indirect charge air cooling** system. The actual charge air cooler is designed as air-to-water charge air cooler, in a layout as compact as possible. It is cooled via a second cooling circuit. Thus, with limited cooling surfaces in a vehicle, a higher temperature reduction with improved density recovery becomes possible, accompanied by a smaller pressure loss in the charge air system. Further, the charge air temperature can be effectively influenced via a valve-controlled connection to the engine coolant circuit. This enables a very flexible engine heat management system. The necessity of an additional coolant pump is a clear disadvantage. Altogether, the system is more complex, heavier, and more expensive.

13 Outlook and further developments in supercharging

This chapter will examine potential future developments – based on the actual state of the art of supercharging system design – using scenarios which are as realistic as possible.

13.1 Supercharging technologies: trends and perspectives

As mentioned in the historic overview, with the exception of inexpensive engines, i.e., very small engines and the passenger car gasoline engine, practically all internal combustion engines are supercharged today.

It is especially important for supercharging designs to be cost-effective. In the near future, dramatic changes may occur. More stringent legislative boundary conditions and regulations for internal combustion engines will limit noise and pollutant emissions, and will require improved fuel economy.

The situation with medium- and heavy-truck, but also modern direct-injection passenger car diesel engines may serve as an example. Without supercharging and charge air cooling, both of these engine types could no longer comply with the pollutant emission standards imposed by law.

On the vehicle side, new features and functions are demanded, e.g., trucks with highly increased engine braking power and passenger cars with further improved transient behavior. All this will lead to a shift of the cost away from the basic engine, which must become less expensive, to more powerful, multifunctional supercharging systems. Moreover, the constant cost pressure will lead to higher degrees of supercharging. The passenger car gasoline engine will also have to follow the trend towards supercharging, especially since here the potential for reduced fuel consumption is by far greater than for diesel engines.

13.2 Development trends for individual supercharging systems
13.2.1 Mechanical chargers

Mechanical supercharging, i.e., supercharging by means of a displacement compressor mechanically powered by the engine, is presently experiencing a remarkable renaissance in the classic gasoline engine. On the one hand, the reasons for this can be found in the inherent problems of exhaust gas turbocharging the gasoline engine, as was discussed at length. On the other hand, in most cases the supercharged engine only represents the peak-power version of an engine family, so that the supercharching system must be designed as an add-on. This can much more easily be achieved with a supercharging system which does not have an impact on the hot side of the engine, as compared to exhaust gas turbocharging.

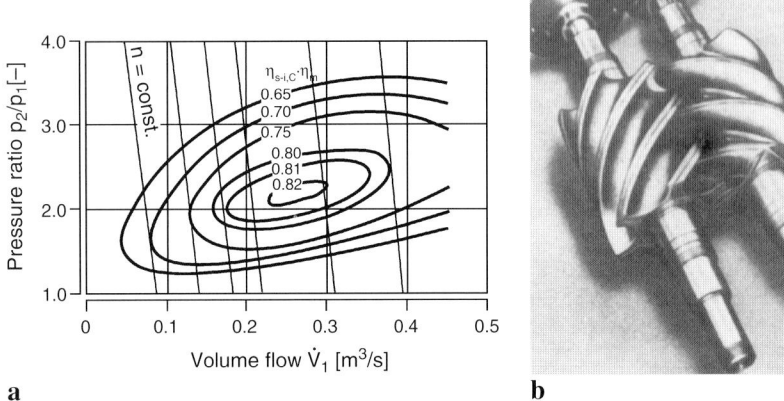

Fig. 13.1. Map (**a**) and rotor pair (**b**) of a modern screw-type (Lysholm) compressor

Further, the mechanical charger can feature advantages with regard to its installation.

Marketing is an additional aspect. In the emotionally influenced passenger car market, a component branding "Compressor" seems to be more attractive than the widely used exhaust gas turbocharger, which is linked to truck and passenger car diesel engines.

In any case, supercharging via a mechanically powered displacement compressor makes sense for small gasoline engines if the displacement compressors are further developed in view of their natural boost pressure curve against engine speed, and their achievable pressure ratios and efficiencies. Furthermore, another open problem of mechanical supercharging systems is the noise emission.

All open challenges of mechanical supercharging could be effectively solved, e.g., by utilizing a well developed screw-type compressor (Lysholm compressor). On the one hand, it can achieve high pressure ratios, on the other (due to its high internal compression and its quasi-continuous delivery) high total efficiencies (Fig. 13.1a), and this at low noise levels.

Figure 13.1b shows the rotor pair of such a compressor. Its success will depend on whether such a **precision charger** can be produced with sufficient durability, with small tolerances, and at acceptable costs. With further development regarding cost and tolerances, the so-called spiral charger (Fig. 11.1) may indeed be a solution for future small supercharged gasoline engines.

We do not yet know, whether the current Roots blowers will need a variable speed connection to the engine, via a variable belt drive or at least a shift gear, to improve their boost pressure curve against engine speed. This certainly will never be an inexpensive solution.

13.2.2 Exhaust gas turbochargers

As mentioned several times, the exhaust gas turbocharger is now applied for nearly 100% of the market for all diesel engine sizes down to those for light trucks, i.e., to displacements of about 3–4 liters.

Its development continues in a way previously deemed impossible. The passenger car diesel engine is now practically exclusively turbocharged, and only mass-produced gasoline engines, up to now rather rarely turbocharged, await this development.

13.2 Trends for individual supercharging systems

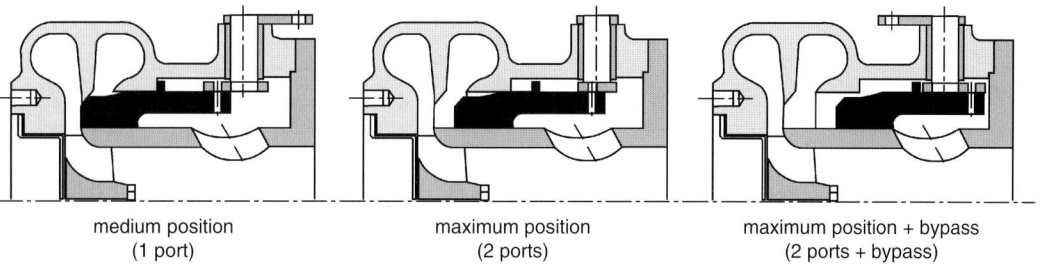

| medium position | maximum position | maximum position + bypass |
| (1 port) | (2 ports) | (2 ports + bypass) |

Fig. 13.2. Sliding ring VTG by 3K-Warner

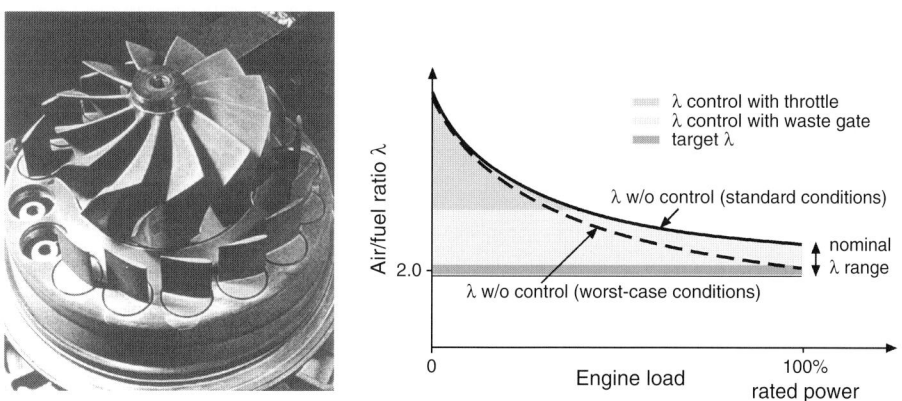

Fig. 13.3. Large engine VTG for natural gas operation [ABB]

The further development of variable turbine geometry (VTG) will certainly play an important role with regard to both its function and its cost. As an example, Fig. 13.2 shows a low-cost sliding-ring VTG design developed by 3K-Warner.

This is also true for medium-speed engines and slow-speed engines. Figure 13.3 shows a VTG charger for a medium-speed natural gas engine, and a two-stage VTG axial turbine of a large charger is sketched in Fig. 13.4.

A similarly important role will be played by compressor-related measures to increase the pressure ratio with simultaneous extension of the usable map. For example, Fig. 13.5 shows the advances in possible maximum compressor pressure ratios from 1946 to 1990.

More advances will also be made regarding the permissible exhaust gas temperatures. Here, ceramic turbine materials are promising. Figure 13.6 shows a possible connection between a ceramic turbine rotor with a metallic turbocharger shaft.

Sheet-metal turbine housings with small heat capacity are under development for the purposes of reduced response times of charger and catalyst.

13.2.3 Supercharging systems and combinations

In the future, more specific charging systems will be utilized in order to even better satisfy the differing requirements for the application of supercharging in truck and passenger car engines.

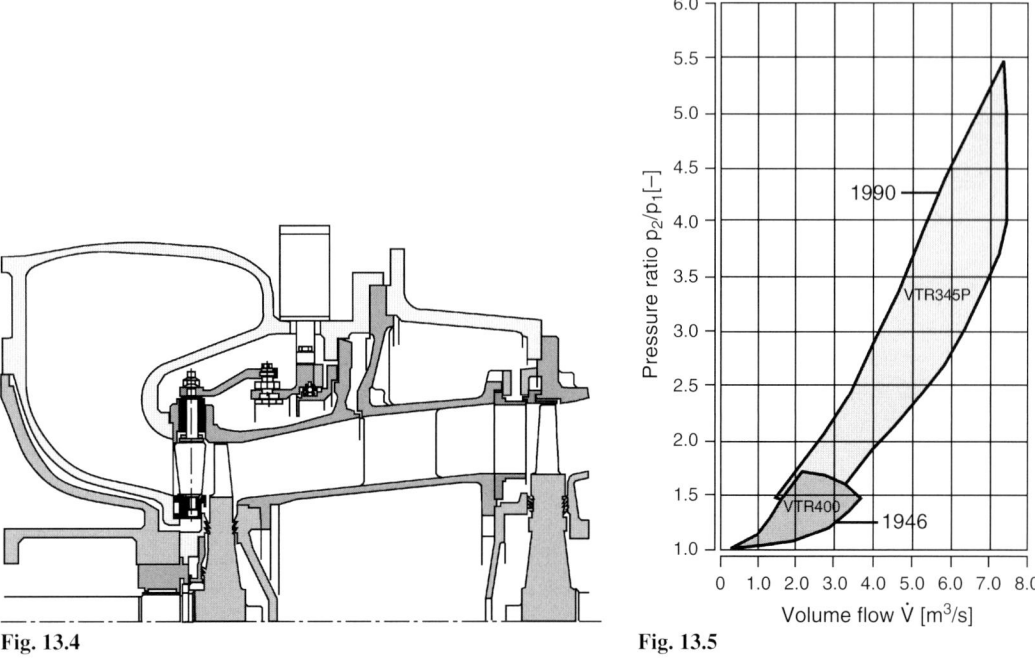

Fig. 13.4.

Fig. 13.5.

Fig. 13.4. 2-stage axial turbine for slow-speed engines, with VTG in first stage [159]

Fig. 13.5. Advance in possible compressor pressure ratios of radial compressors [75]

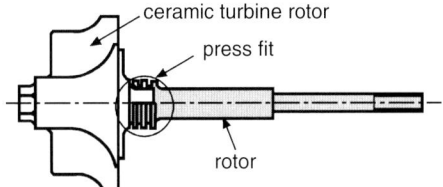

Fig. 13.6. Connection between a ceramic turbine rotor and a metallic charger shaft [KKK, now 3K-Warner]

Register charging

Register charging is commonly used in slow-speed engines already. But it also could gain significance for truck applications if the VTG charger cannot meet the durability demands required under these conditions. However, in this case new operating strategies have to be considered.

In vehicle applications, charger switching in the main operating range of the engine is difficult for safety reasons (power loss in critical driving situations). A possible switching strategy could be that the start-up range is covered using only one charger – with much higher torque – and under all other driving conditions two chargers are used. Furthermore, the operational speed range of an engine can be extended with this charging system.

For modern 4-valve engine designs, in combination with variable valve control, it might be desirable to provide a separate exhaust port branch for each valve to optimize the gasdynamic conditions – twin-flow and cylinder combinations – for both chargers.

13.2 Trends for individual supercharging systems

Two-stage controlled supercharging

Two-stage controlled high-pressure supercharging has at least the same chance for future success as register charging. It can achieve not only very high boost pressures at low engine speeds but also improved load response and – due to the higher possible air excess figures – reduced pollutant emissions. Especially large high-speed engines can take advantage of this in the future, since in this engine category already today highly developed and correspondingly expensive charging systems are utilized.

Two-stage exhaust gas turbochargers

Currently, designs are under development which combine two-stage chargers in one housing, powered by one turbine. This would result in significant installation advantages.

Turbocompounding

Turbocompound operation is sporadically applied today for heavy-duty truck engines – i.e., a smaller segment of the entire automotive powertrain market. However, its further implementation strongly depends on the one hand on future fuel cost, as well as on further efficiency increases of the flow components, and on the other on future emission levels and reduction strategies, because the negative pressure gradient between intake and exhaust manifold of a compound engine makes exhaust gas recirculation much easier. Furthermore, at least in the truck sector, the rule of thumb regarding the application of all these efficiency-increasing exhaust gas energy recovery systems is that the additional cost must be amortized by the customer within one year. A further prospect can be seen in single-stage compound turbocharger operation, which will be described below.

For slow-speed engines, compound operation is also of great significance. Here, economy is the decisive factor, i.e., the lowest possible operating costs, especially fuel costs. However, mechanical solutions have recently been abandoned in favor of electrical energy recovery (i.e., the secondary turbine directly powers a generator, efficiently feeding the electric supply system on board at low price). A new problem, caused by the further increased efficiencies of the basic engine, has arisen: When a secondary turbine is used, the remaining exhaust gas energy is too low to preheat the heavy oil sufficiently.

Supporting the exhaust gas turbocharger

At present, much research is directed toward a significant improvement and extension of the functionality of turbocharging systems by supporting the exhaust gas turbocharger. Here, a distinction must be made between applications, i.e., passenger car, truck, stationary engines and slow-speed engines.

For **applications in passenger cars**, the primary goal is to eliminate the "turbolag" of turbocharged engines which exists even when a VTG charger is used. Naturally, at start-up only the aspirated torque of the basic engine is available, since boost pressure can only be generated after an increase in the demand for torque, i.e., an increase in the amount of fuel injected into the engine.

A possible solution could be an additional charger which can be inexpensively integrated into the intake and charge air system. Electricity from an onboard battery would be sufficient to power such an auxiliary charger. It would operate for short periods only. At low engine speeds, as soon as vehicle acceleration is demanded by the driver, the additional charger would be powered, thereby increasing the boost pressure. Pressure ratios of about 1.4 to 1.6 would be sufficient to double

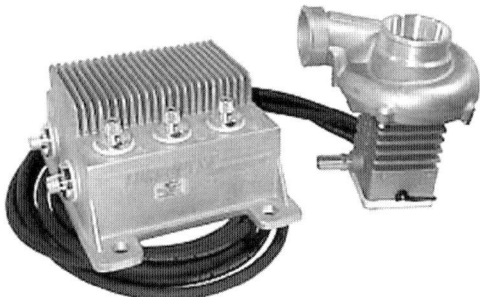

Fig. 13.7. Electrically powered Garrett turbo compressor by Honeywell

the start-up torque. Turbodyne/Honeywell was developing such a start-up system under the name Turbopac (Fig. 13.7).

Further advantages may be achieved with such a system during the cold-start and warm-up phases of diesel and gasoline engines. It would preheat the intake air. In a gasoline engine this could lead to a significant reduction or even abandonment of start-up enrichment. In a diesel engine it could result in improved cold-start behavior and possibly lead to a reduction of the compression ratio necessary for cold starts. Thus, also reduced peak firing pressures could be achieved during normal operation.

For **applications in trucks**, the problem of start-up necessitates the use of engines with sufficient displacement to start on an incline with only the aspirating torque. Thus, an additional charger could provide a significantly improved start-up characteristic. But there are even further interesting application aspects for trucks, supporting an additional drive mechanism for the exhaust gas turbocharger which is in any event necessary. If the exhaust gas turbocharger can be mechanically or electrically coupled to the total system in the entire load and speed range of the engine, this will possibly improve its load response, as a single-stage compound operation, and its engine braking behavior.

Independent of the exhaust gas energy available, the turbocharger is accelerated during **positive load steps** by an electric motor – either arranged on the charger shaft or connected to it via a clutch – to a speed which enables it to generate the boost pressure desired in the actual operating point (e.g., start-up of a fully loaded truck at an incline). The time during which this support is needed will usually be very short, since once boost pressure is available, the turbine power increases rapidly and can cover the power requirements of the compressor.

Besides utilizing an electric motor located on the turbocharger shaft as an additional drive (also for the purpose of increasing the boost pressures in the low-speed range), in **single-stage compound operation** this unit can be used also as a generator. Under all those operating conditions in which, e.g., a waste gate is used to control the boost pressure, the charger speed (and thus the boost pressure) can be reduced via the generator, and the recovered electric power can be fed back into the onboard electric system. However, for this suitable controls, power electronics and energy storage components are necessary. Further, the turbines utilized have to meet this additional requirement. Even under the changed pressure–mass flow conditions they must feature high efficiencies since the turbocharger speed significantly deviates from the freewheeling speed. Theoretically it would also be possible to reroute the electrical energy back to the drivetrain, e.g., via an electric booster motor or a crankshaft starter generator. The success of all such systems depends on the efficiencies of the components, especially the electric components. Figure 13.8 shows a possible layout for such an electric compound system.

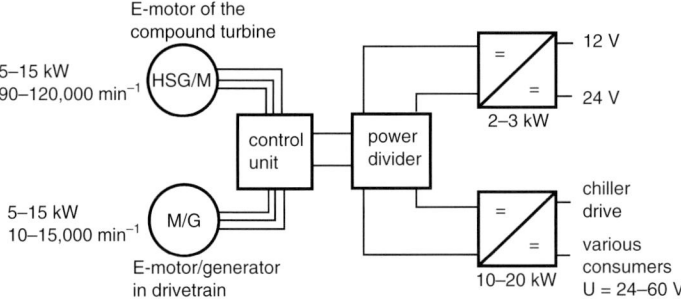

Fig. 13.8. Layout principle of a single-stage turbocompound system

If the described self-sustaining electric charger drive is also used under **engine braking conditions**, the engine braking power can be significantly increased since a far higher airflow through the engine is obtained. The airflow can be transferred into higher braking power, e.g., via a constant throttle, as it is now utilized in mass production already. Further, the electric power of the generator turns into additional braking power. Additional functions of such a system are feasible, e.g., a preheating of the charge air at cold-start via prestart air circulation. A similar system may also be designed mechanically, but this will not be further discussed here.

Mechanical additional charging

Systems combining a mechanical charger with an exhaust gas turbocharger have already been introduced by DaimlerChrysler and Volvo for truck engines. Recently VW has introduced such a system for its 1.4 liter TSI direct-injection gasoline engine; see Chap. 14.1 (Figs. 14.22–14.24). However, such combinations represent a very complex system and, possibly, may be less reliable with two different charger designs. Therefore, for various reasons including cost, they will only be used in special cases.

The situation is totally different for two-stroke engines. Especially in medium-speed and low-speed engines, this combination of mechanical scavenging auxiliary supercharger and turbocharger (Fig. 14.59) is state of the art today.

13.3 Summary

Supercharging of reciprocating internal combustion engines has become established as the most straightforward way to increase their power density and efficiency while also reducing their pollutant emissions. It will also significantly impact the gasoline engine market when more stringent requirements for **lower emissions**, and at the same time for **increased efficiency**, are demanded by society and politics. As soon as today's minor disadvantages of exhaust gas turbocharging under dynamic operation are eliminated, it will further promote a meaningful downsizing of engines especially for highly dynamic vehicle applications.

14 Examples of supercharged production engines

14.1 Supercharged gasoline engines

The history of supercharged gasoline engines started with automobile racing applications. As early as in the 1920s, but especially in the 1930s, remarkable specific power output levels of about 120 kW/l were achieved via mechanical supercharging of Auto-Union and Mercedes-Benz racing engines.

As turbocharger technology advanced, it also was first applied to racing engines. One of the best-known examples is the legendary Porsche type 917. In 1975, Porsche introduced into mass production a 2 liter, 4-cylinder engine with exhaust gas turbocharging for their model 924. The K26 charger built by 3K-Warner (formerly KKK) was equipped with an integrated bypass valve at the intake side; on the exhaust side the turbine mass flow was controlled via an external waste gate, as can be seen in Fig. 14.1.

Following Porsche, in the 1970s, among others, Saab (2 liter, 4 cylinders), Audi, and BMW introduced turbocharged engines. Figure 14.2 shows a 5-cylinder engine of which the exhaust manifold to the turbine warrants special mention. Due to the specific ignition sequence of the 5-cylinder engine, this manifold has a triple-flow design, in such a way that the routing of the exhaust gas up to the flange of the turbine is separate for cylinders 1, 2, and 5, and cylinders 3 and 4. As an example of an early mass-produced 6-cylinder inline engine, the 3.2 liter BMW engine is shown in

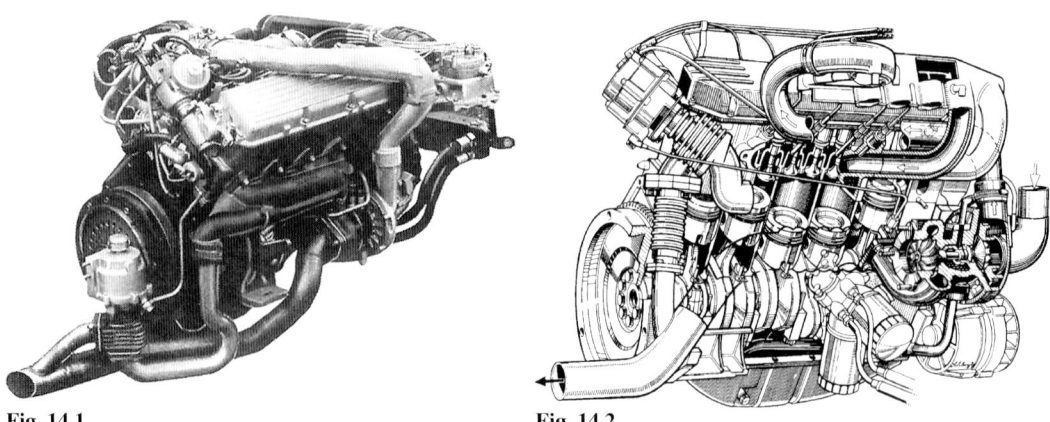

Fig. 14.1. View of the 2 liter, 4-cylinder turbocharged engine of the Porsche model 924-Turbo [159]

Fig. 14.2. Sectional view of the 2.14 liter, 5-cylinder turbocharged engine of the Audi model 200 [159]

14.1 Supercharged gasoline engines

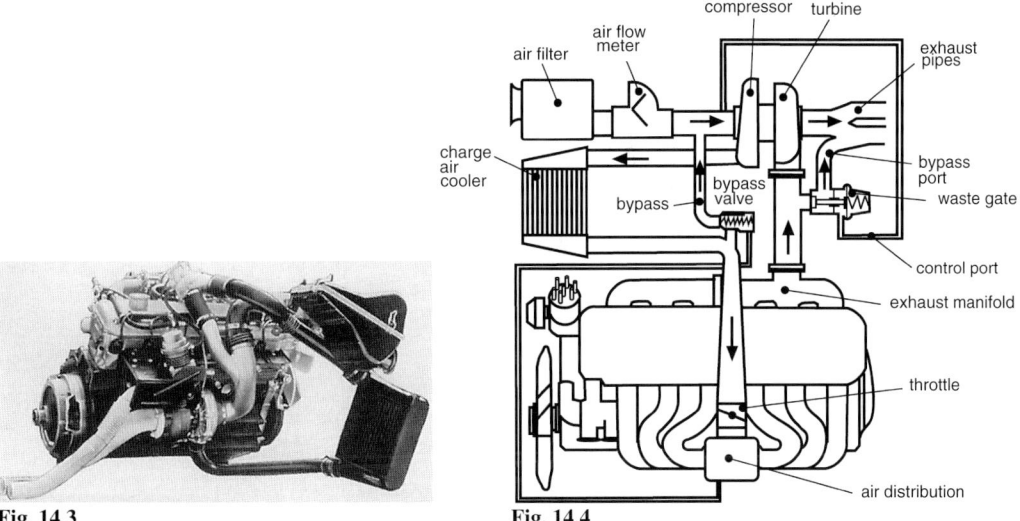

Fig. 14.3. View of the BMW 3.2 liter, 6-cylinder inline turbocharged engine with charge air cooling [159]

Fig. 14.4. Schematic of the charge air and exhaust gas routing of the BMW 3.2 liter, 6-cylinder inline turbocharged engine [159]

Fig. 14.3. Besides turbocharging, its power density and torque were further increased by utilizing charge air cooling and individual pulse charging manifolds, as can be seen in Fig. 14.4.

In the 1990s, the trend towards turbocharged gasoline engines increased. By means of improved turbocharger technology, the infamous turbo response lag could be largely eliminated, significantly improving its acceptance by the drivers. In the following, some examples of these modern supercharged engines will be described in more detail.

In 1994, Audi introduced its 5-valve gasoline engine series in its 4-cylinder turbocharged version (Fig. 14.5). The most important characteristics and performance data are summarized in Table A.1 in the appendix.

The charger unit consists of a turbocharger by 3K-Warner, series K03, and a downstream charge air cooler. Boost pressure control is achieved with a waste gate integrated into the turbine housing. A bypass control valve (short-circuit of the compressor cycle) is located on the compressor side,

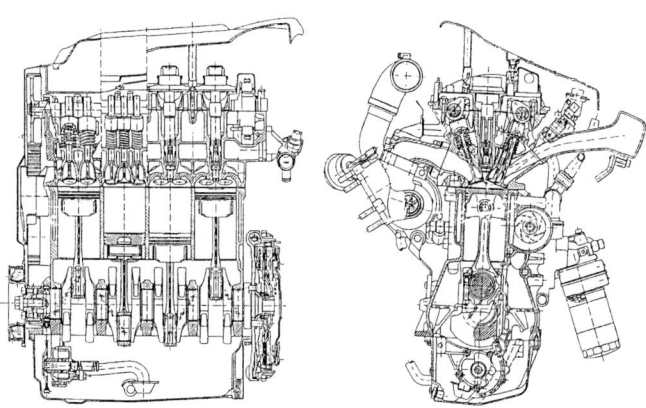

Fig. 14.5. Cross section of the Audi 5-valve turbocharged gasoline engine in its 4-cylinder version [117]

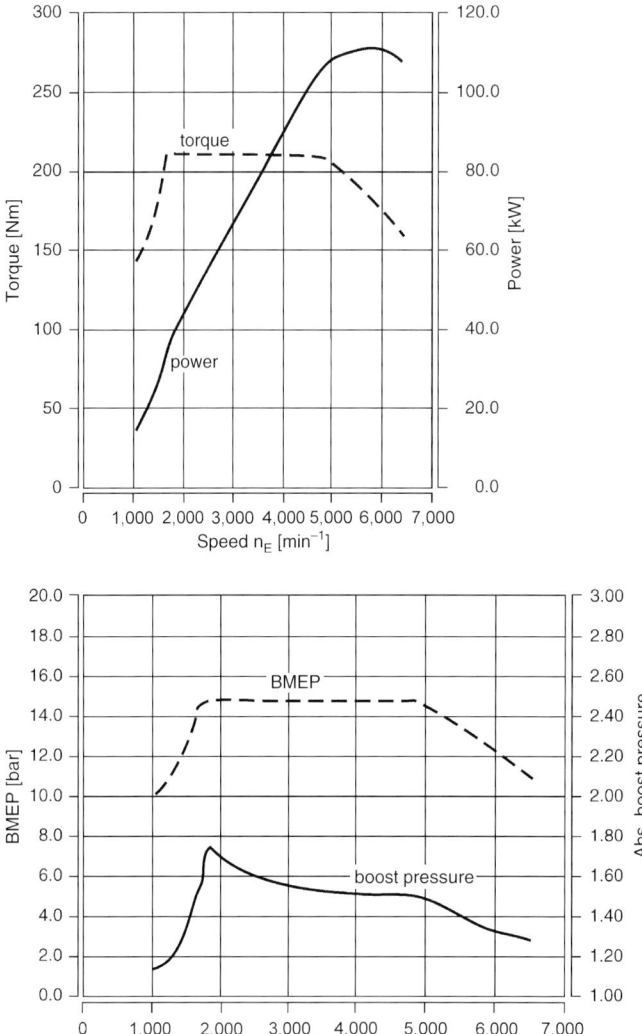

Fig. 14.6. Full-load operating data of the 4-cylinder version of the Audi turbocharged gasoline engine [117]

assuring that at rapid load changes, when the engine throttle is suddenly closed, the compressor does not operate beyond the surge limit.

Under steady state full-load conditions, this layout of the turbocharger reaches its maximum boost pressure at about 1,750 min^{-1}. Correspondingly, beyond this engine speed range, the rated torque of the engine, 210 Nm (corresponding to about 15 bar BMEP), is available (Fig. 14.6).

Due to its extreme engine compartment limitations, the MCC Smart passenger car requires a very compact layout of the complete engine (Fig. 14.7), especially its charger unit. In the case of the turbocharged M160 3-cylinder engine, the solution was to integrate the turbine housing into the exhaust gas manifold (Fig. 14.8), as was shown before already by Opel in its turbocharged 2-liter, 4-valve gasoline engine. By means of turbocharging, this engine with 0.66 liter displacement provides a torque of 80 Nm above 2,000 min^{-1}. Considering the total engine weight of 59 kg, this engine, which is rated at 40 kW, reaches a specific weight-to-power ratio of 1.48 kg/kW.

14.1 Supercharged gasoline engines

 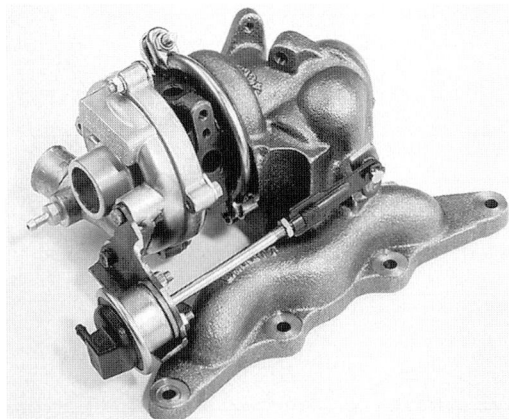

Fig. 14.7. **Fig. 14.8.**

Fig. 14.7. Smart 0.66 liter, 3-cylinder engine by Daimler Benz [82]

Fig. 14.8. Integration of the turbine housing into the exhaust gas manifold of the Smart 0.66 liter, 3-cylinder engine by Daimler Benz [82]

The highest power densities were achieved in F1 racing, utilizing turbocharged 1.5 liter gasoline engines. First they were used by Renault in actual races in 1977. The end of this development was marked by the Honda-RA168E-engine, which won the F1 championship with 15 of 16 possible victories during the last year of racing regulations allowing this engine type.

In this last year of F1 turbocharged engines, 1988, the boost pressure was limited to 2.5 bar, while in the year before a maximum pressure of 4 bar was allowed. Power levels of about 740 kW could be achieved from 1.5 liter displacement, i.e., an impressive figure of 495 kW/l displacement.

While the first turbocharged F1 engines utilized a single charger, in 1979 Renault adopted a bi-turbo arrangement, where each cylinder bank had its own turbocharger (Fig. 14.9). At a boost pressure of 2 bar, the engine shown in Fig. 14.9 generated a maximum power of 470 kW at 11,000 min^{-1} rated speed in its version for the 1982/83 racing season.

The previously mentioned culmination of the development of F1 turbocharged engines is represented by the example of the Honda-RA168E (Fig. 14.10). The impressive torque and power

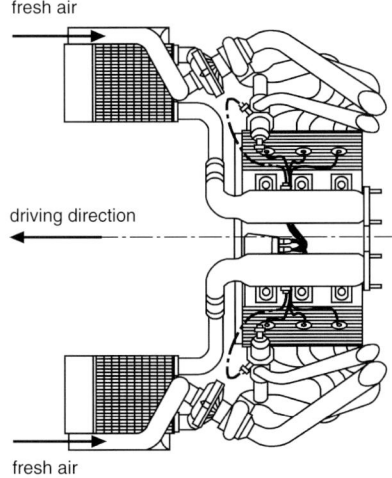

Fig. 14.9. Bi-turbo layout of the 1.5 liter, 6-cylinder F1 racing engine by Renault [72]

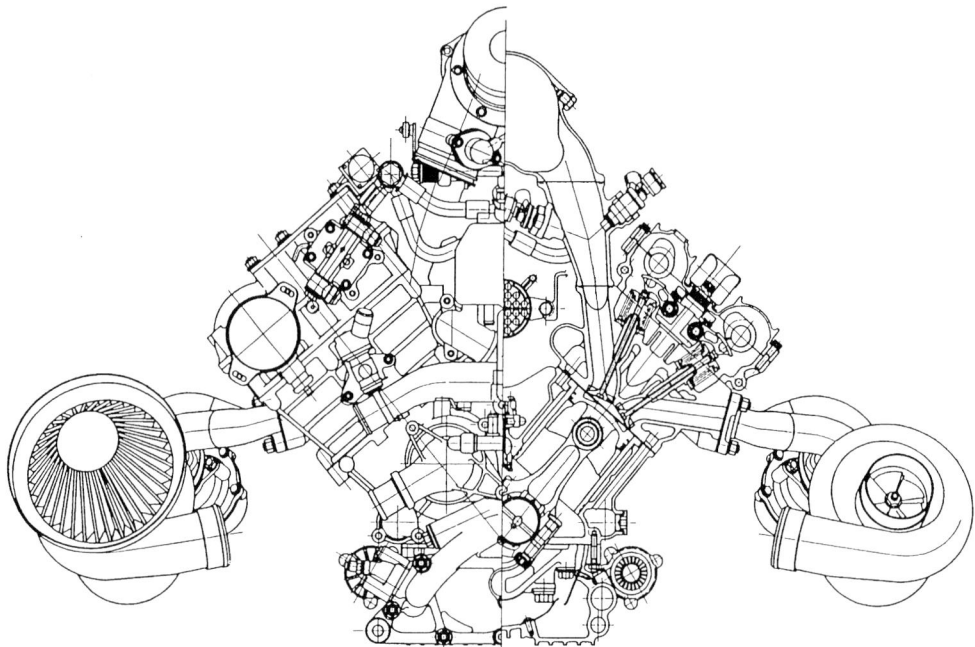

Fig. 14.10. Cross section of the RA168E 1.5 liter, 6-cylinder F1 racing engine by Honda [107]

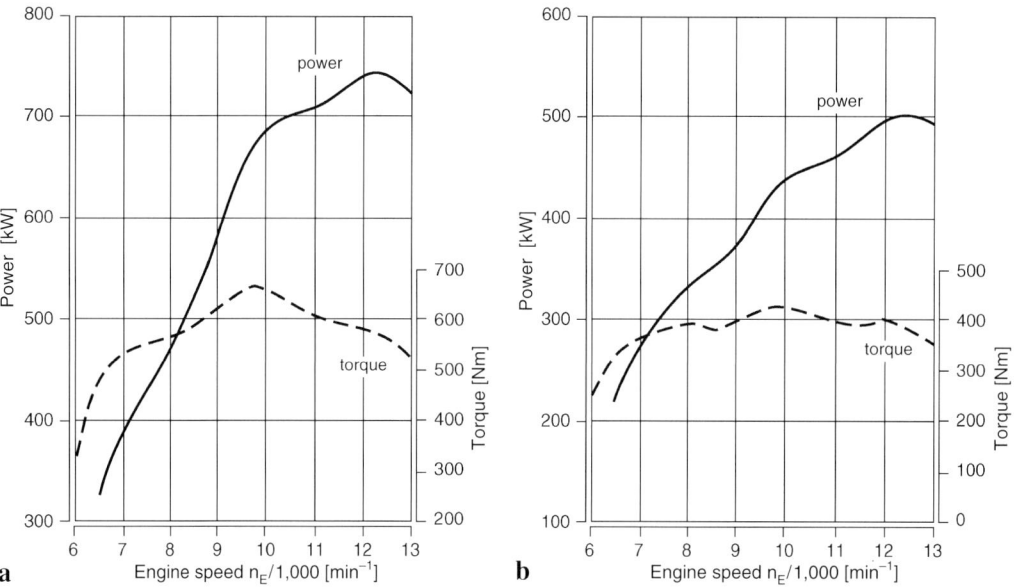

Fig. 14.11. Full-load data of the RA167E and RA168E 1.5 liter, 6-cylinder F1 racing engines by Honda [107]. **a** 4 bar maximum boost pressure; **b** 2 bar maximum boost pressure

curves of these engines are shown in Fig. 14.11, in the versions with 4 and with 2.5 bar maximum boost pressure.

The maximum engine torque of 664 Nm at 4 bar boost pressure and the rated power output at about 12,500 min^{-1} are impressive. This corresponds to a specific value of 443 Nm/l displacement.

14.1 Supercharged gasoline engines

The high quality of these engines is additionally reflected in their low fuel consumption, which is in the range of 280–300 g/kW h (in spite of the relatively high friction losses of racing engines due to their high piston velocities at rated speed).

For these racing engines it became necessary to develop ceramic turbine rotors, which could tolerate the high temperatures occurring under racing conditions. Further, rotors made of ceramic materials have lower inertia and thus improved charger load response.

In 1993, Subaru introduced its 2 liter, flat-4-cylinder engine in a version with register charging at the Tokyo Motor Show. Along with its integrated charge air cooler, the engine represents a very compact package with a power density of 92 kW/l (Fig. 14.12). The operating strategy of the

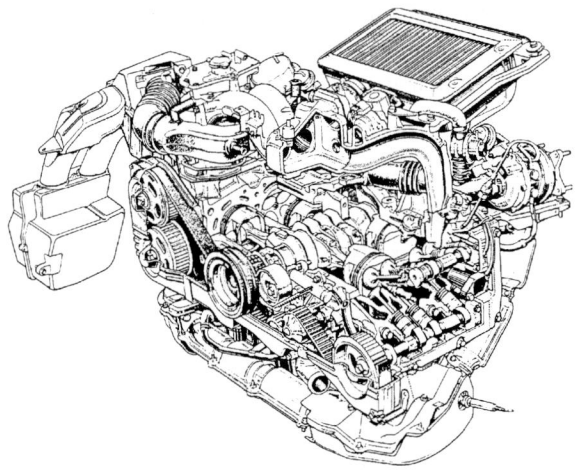

Fig. 14.12. Sectional view of the 2 liter, 4-cylinder flat-4 ("boxer") engine with register charging by Subaru [8]

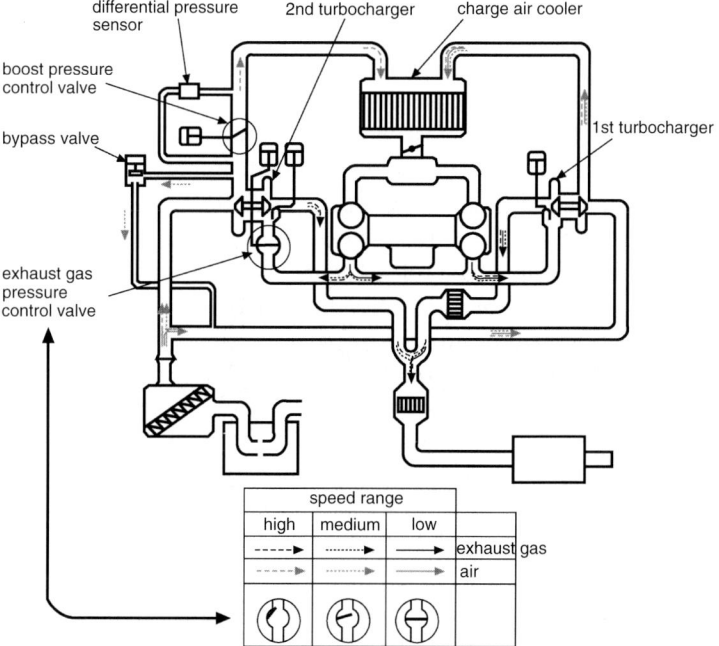

Fig. 14.13. Schematic of the register charging operation strategy of the 2 liter, flat-4-cylinder engine by Subaru [8]

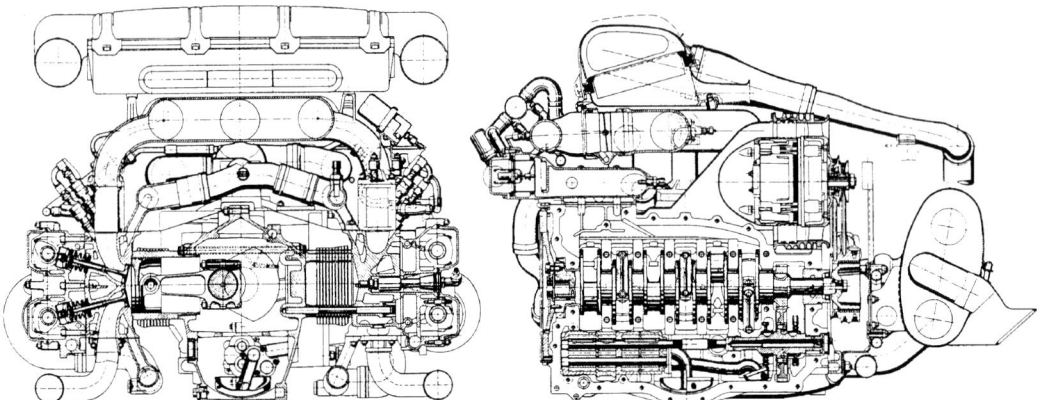

Fig. 14.14. Longitudinal and cross section of the flat-6-cylinder engine of the Porsche model 959 with Porsche register charging [15]

engine and charger unit is selected in such a way that up to about 2,500 min^{-1} the first stage alone generates the boost pressure. Then the second turbocharger is accelerated to operating speed, and above about 3,000 min^{-1} the rated boost pressure of 1.8 bar is generated with both turbochargers. In this way, the peak torque of 310 Nm is obtained at approximately 5,000 min^{-1}. Figure 14.13 shows the three operating conditions of the register charging unit in corresponding diagrams.

As a high-power version, the legendary air-cooled flat-6-cylinder of Porsche's model 959 (Fig. 14.14) was equipped with register turbocharging (see Sect. 6.3). The impressive performance data of this engine are also listed in the appendix. The specific power of this engine, 115.8 kW/l, underlines the high power potential of supercharged gasoline engines. The boost pressures generated by the compressor are in the range of 1.9 bar. By increasing the boost pressure, the power was increased to 500 kW for racing applications. The production engine, such as the racing version, was equipped with a dry-sump lubrication system.

Daimler Benz' M 119HL was also very successfully used in automobile racing. The engine was based on the aluminum 5 liter V8 engine of the E- and S-class Mercedes Benz passenger car models. By consistent advancement of the engine mechanics and the cooling system, plus the utilization of the best charger components available at that time, a power level of 700 kW and a torque of more than 1,000 Nm were obtained. By means of consistent flow division for optimized pulse exhaust gas turbocharging, selective ignition timing control, a ceramic high-performance turbine and a magnesium compressor, i.e., a rotor assembly with minimized inertia, a very harmonic torque rise could be achieved, with boost pressure response times of about 1 s (Fig. 14.15). At 255 g/kW h, the low specific fuel consumption at rated power was unrivaled. Figure 14.16 shows the fuel consumption map, Fig. 14.17 a view of the engine.

Supercharging via mechanically powered compressors was widely utilized in the early years of engine development, especially for aircraft engines. In the mid-1990s, DaimlerChrysler again applied this charging principle very successfully in its passenger car gasoline engine M111K (Fig. 14.18) for the C-class and SLK models, where the power of the basic engine was sufficient to cover wide ranges of actual driving conditions without applying the compressor. However, its pollutant emissions (catalyst light-off) and its fuel economy were improved by the downsizing effect – an aspirated engine of the same rated power would run at lower specific loads and, thus, would have a higher specific fuel consumption. Mechanical charging can also provide boost

14.1 Supercharged gasoline engines

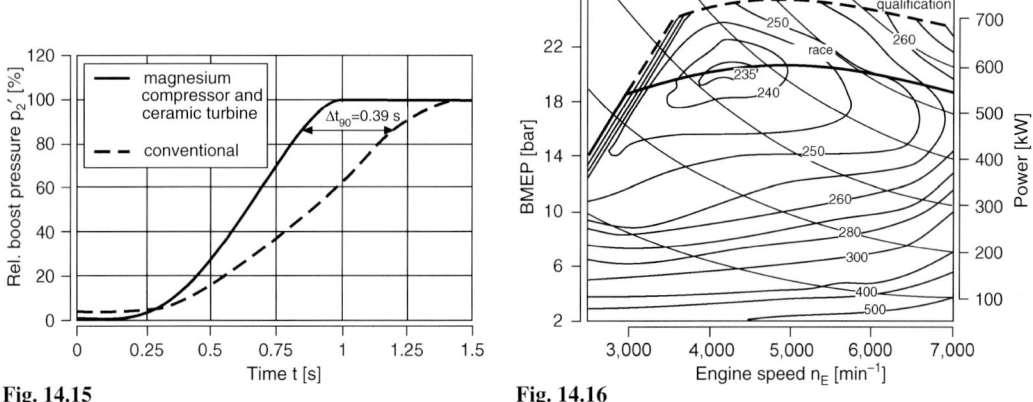

Fig. 14.15. Comparison of boost pressure response time if full-load is applied, Mercedes Benz M119HL with conventional rotor assembly and with inertia-optimized rotor assembly [62]

Fig. 14.16. Fuel consumption map of the Mercedes Benz M119HL engine [62]

Fig. 14.17. View of the Mercedes Benz M119HL engine [62]

Fig. 14.18. View of the DaimlerChrysler M111 Kompressor engine

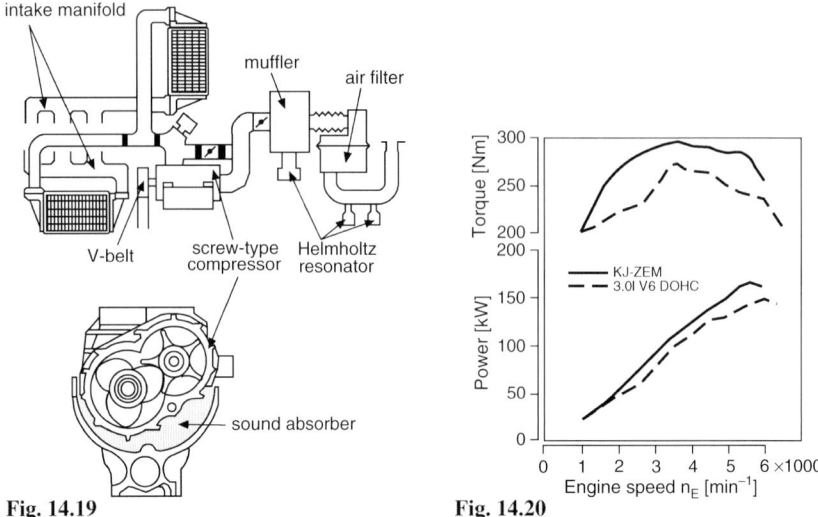

Fig. 14.19. Schematic of the mechanical charger utilized in the KJ-ZEM Miller cycle engine by Mazda [31]

Fig. 14.20. Performance characteristics of the KJ-ZEM Miller cycle engine by Mazda [31]

pressure with practically no time lag, which is demanded by the driver. This results in a driving behavior comparable to a naturally aspirated engine. The torque curve against speed is flat and does not achieve the torque back-up levels and thus the engine elasticity of exhaust gas turbocharged engines.

In the mid-1990s, Mazda too introduced a gasoline engine with mechanical supercharger, in this case a screw-type charger (Fig. 14.19). The engine was designated as a Miller-cycle engine, which would correspond to an early closing of the inlet valves and charge cooling via expansion in the cylinder down to BDC (see Sect. 6.4). The designation is not completely accurate in this case since the inlet valve was closed late, past BDC, during the compression stroke. This shortens the effective compression stroke in comparison to the expansion and thus improves the internal thermodynamic efficiency. To be correct, corresponding to its inventor this process control has to be called a Late Atkinson cycle.

However, the low gas exchange efficiency (volumetric efficiency) of the engine had to be compensated by higher charge air densities (high boost pressure and low charge air temperatures), for which the screw-type charger utilized was the ideal unit. The engine combined a suitable boost pressure control at full load with the relative extension of the expansion stroke. Compared to the basic engine, this resulted in a power increase as well as improved fuel economy in the driving cycle (part load; Fig. 14.20).

Another interesting special design is Mazda's 3-rotor Wankel engine with register charging. Figure 14.21 shows the schematic of the engine with the charger unit. In comparison to conventional turbocharging, with the register charging design chosen, it was possible to improve the torque and boost pressure behavior of this Wankel engine significantly, in the lower speed range by up to 36%.

In order to further improve the load response of a turbocharged gasoline engine, in 2005 VW presented a 1.4 liter, 4-cylinder gasoline engine (Fig. 14.22) with a combined mechanical and turbocharging system which is shown in Fig. 14.23. Thus, with the rather instantaneous boost pressure buildup possible with the mechanical supercharger, the turbo lag is completely eliminated.

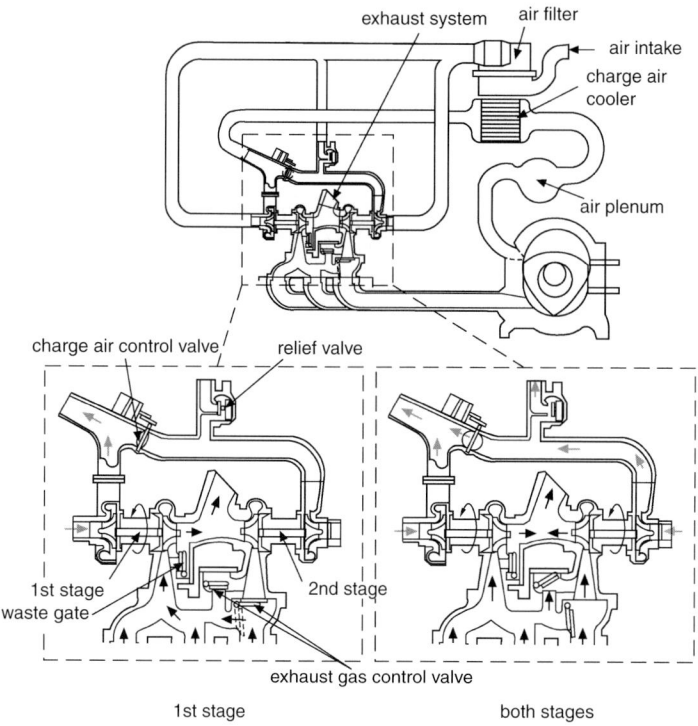

Fig. 14.21. Schematic of the register charging system of the 20B-RE 3-rotor Wankel engine by Mazda [140]

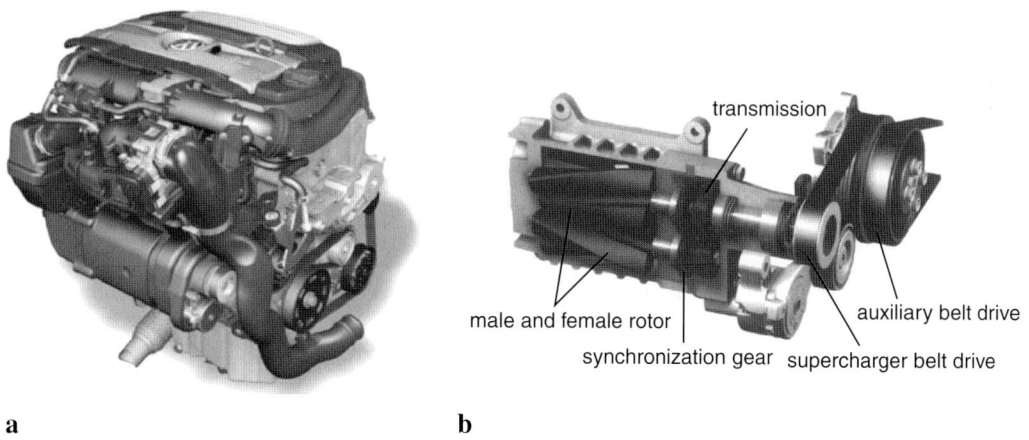

Fig. 14.22. a Rear view of the VW 1.4 liter TSI engine with combined super- and turbocharging; **b** belt-driven supercharger [84]

On the other hand, the turbocharger allows to achieve as high a power and torque output of the engine as if the boost pressure would be provided by the mechanically driven compressor (Fig. 14.24) but, of course, without the typical fuel consumption penalty at high loads and speeds of mechanically supercharged gasoline engines. The most important characteristics and performance data are summarized in Table A.1 in the appendix.

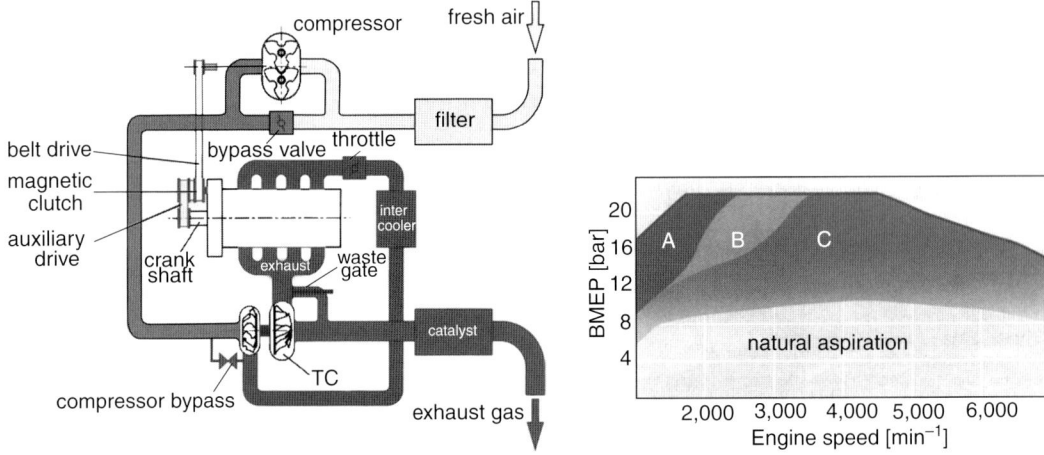

Fig. 14.23

Fig. 14.24

Fig. 14.23. Sketch of charging system of the VW 1.4 liter TSI engine [84]

Fig. 14.24. Operating ranges of combined charging system of the VW 1.4 liter TSI engine [84]. A, continuous operation of supercharger; B, intermitting operation of supercharger; C, charging with turbocharger only

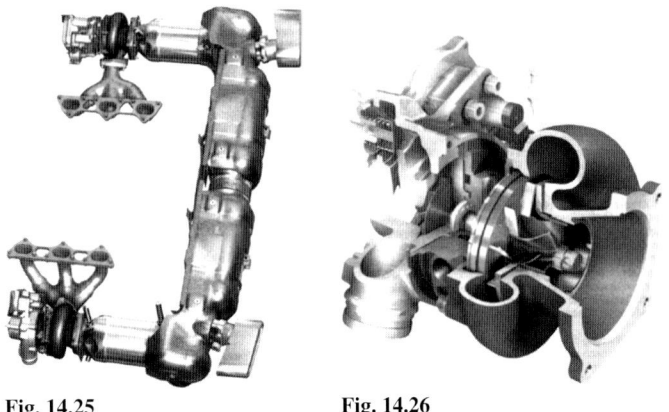

Fig. 14.25

Fig. 14.26

Fig. 14.25. Exhaust system with VTG turbochargers and exhaust manifolds of the 3.6 liter flat-6-cylinder engine by Porsche [77]

Fig. 14.26. VTG turbocharger for the 3.6 liter flat-6-cylinder engine by Porsche [77]

In 2006, the first gasoline engine with variable turbine geometry was introduced to the market by Porsche. In the exhaust system of the 3.6 liter, flat-6-cylinder engine (Fig. 14.25), two VTG turbochargers of 3K Warner (Fig. 14.26) are integrated, which allow a reliable and durable operation of the adjustable turbine inlet vanes at temperatures up to 1,000 °C. With the high boost pressure generated at low engine speeds by the turbochargers with the inlet vanes at closed position (Fig. 14.27a), torque output of the engine exceeds 600 Nm already at an engine speed of 1,900 min^{-1}. Due to the larger effective flow area of the turbines at fully open inlet vane position (Fig. 14.27b), the high torque can be maintained close to rated speed of the engine finally leading to a rated power of 353 kW. For further technical data of this engine it is referred to Table A.1 in the appendix.

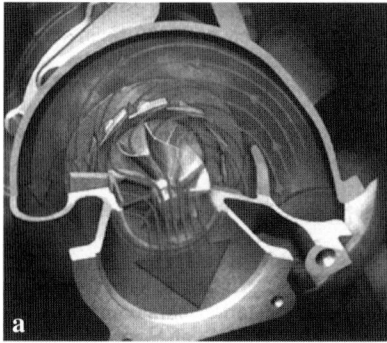 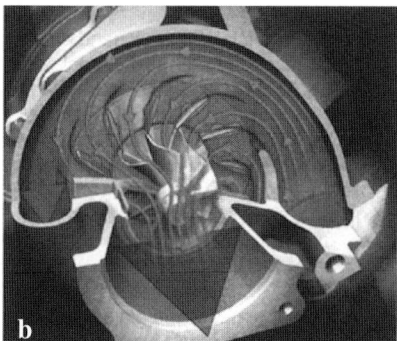

Fig. 14.27. Operating principle of the VTG turbocharger for the 3.6 liter flat-6-cylinder engine by Porsche [77]. **a** Vanes closed, small turbine. **b** Vanes open, large turbine

14.2 Passenger car diesel engines

In the late 1970s, the trend towards supercharging passenger car diesel engines became apparent. It was foreseeable that the specific power of naturally aspirated prechamber engines would not be sufficient in the medium term to cover the driving performance requirements of modern vehicles. Accordingly, Daimler Benz equipped its 3.0 liter OM617 5-cylinder engine with exhaust gas turbocharging. The performance data of this engine are summarized in the appendix. In comparison to the naturally aspirated engine (59 kW), the power was increased by 26 kW. The turbocharged engine, which was based on the naturally aspirated version, was equipped with a waste gate charger. No charge air cooling was applied. For reasons of improved fleet fuel economy, the engine was first introduced in the United States in S-class vehicles (W116). Figure 14.28 shows a charger-side view of the engine.

Subsequently, numerous turbocharged passenger car diesel engines were introduced into the market. All of these utilized the prechamber combustion (IDI) process. To further improve fuel economy, the intense development of direct-injection (DI) diesel engines started in the 1980s –

Fig. 14.28 **Fig. 14.29**

Fig. 14.28. Charger-side view of the 3 liter, 5-cylinder engine OM617A by Mercedes Benz

Fig. 14.29. Sectional view of the turbocharged direct injection 1.95 liter R4 passenger car DI diesel engine by BMW [6]

with significant involvement of AVL – and in the 1990s this technology achieved a real market breakthrough. Audi became the pioneer in this technology. In 1989 it introduced a 2.5 liter, 5-cylinder engine which was the first turbocharged passenger car mass production diesel engine with DI. By the end of the 1990s, in Europe the market share of supercharged passenger car diesel engines had increased to nearly 25%, and in some countries of the European Community their share in new vehicle registrations had reached 60% and more (Austria, Italy, France). In the following, some of these modern DI diesel engines with exhaust gas turbocharging will be discussed.

In 1998, BMW introduced its DI 4-cylinder diesel engine with turbocharging (Fig. 14.29). This 4-valve engine was the successor to the IDI 2-valve model from 1994. At a rated power of 100 kW and a displacement of 1.95 liter, this was the first passenger car DI diesel engine to exhibit a specific power density of more than 50 kW/l (51.3 kW/l). The technical data of this engine are summarized in the appendix.

The power potential of this DI diesel engine was impressively proven in endurance racing. With modifications made to the basic series engine – valve timing (late closing of the intake valves), combustion chamber (lower compression ratio), turbocharger (VTG charger for increased mass flow), reinforced crank drive – a race car with such an engine won the 24 h race on the Nuerburgring in 1998, defeating all other vehicles which were equipped with gasoline engines. Besides its estimated power of 180 kW and a maximum torque of about 400 Nm, the low fuel consumption of the engine (195 g/kW h in its best operating point and 225 g/kW h at rated power) – resulting in fewer refueling stops in comparison to cars with gasoline engines – was the major contributor to the victory in this endurance race.

While at first the DI of the fuel into the combustion chamber was performed by distributor pumps, in the mid-1980s the development of a new generation of high-pressure fuel accumulator injection systems (common rail) was initiated. Due to the inherent capability of widely influencing the fuel injection process, improvements were possible regarding pollutant emissions and the performance of DI diesel engines. The first passenger car engines with this technology were introduced into series production in 1997 by Fiat with their 4- and 5-cylinder engine models JTD. The technical data of the 5-cylinder version are summarized in the appendix.

Figure 14.30 shows the layout both of the charger unit and that of the common-rail injection system (maximum rail pressure of about 1,350 bar). With this injection system, the EU3 exhaust gas standards mandated by law could be met, and through the use of pilot-injection the combustion noise was significantly reduced – in the low-speed range by up to 8 dB(A). The comfort gained by these measures clearly increased the attractiveness of these engines for passenger car applications.

Fig. 14.30. Sectional view of the turbocharged common rail 2.4 liter, inline 5-cylinder passenger car DI diesel engine by Fiat [92]

14.2 Passenger car diesel engines

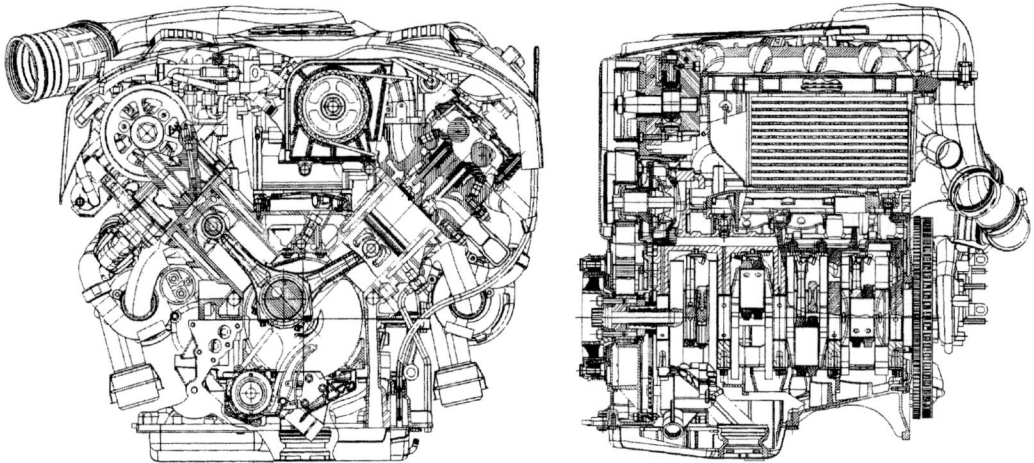

Fig. 14.31. Cross sections of the turbocharged 3.3 liter V8 TDI passenger car diesel engine by Audi [17]

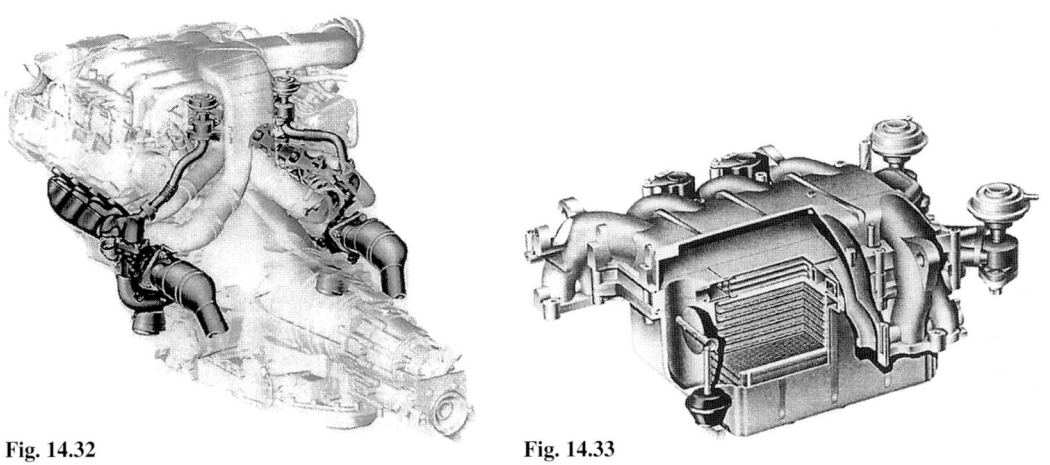

Fig. 14.32 **Fig. 14.33**

Fig. 14.32. View of the compact VTG bi-turbocharging system by Audi (3.3 liter V8 TDI) [17]

Fig. 14.33. Charge air cooling integrated into the V of the engine (Audi; 3.3 liter V8 TDI) [17]

In 1997, Audi was the first with its V6 turbocharged 4-valve engine to introduce a passenger car DI diesel V-engine into the market. In 1999, the Audi 8-cylinder V-engine with common rail and VTG bi-turbocharging followed. This engine is shown in Fig. 14.31. The extremely compact exhaust gas system including both chargers is shown in Fig. 14.32. On the air side, charge air cooling is integrated into the V of the engine (Fig. 14.33). In combination with the good fuel preparation by the common-rail injection system, the engine features high torque at very low speeds and a high power density (Fig. 14.34), but also fulfilled the stringent EU3 exhaust emission limits. Due to a careful selection of materials and production processes, a very low engine weight is achieved. Further to aluminum cylinder heads, crankcase, and cylinder block are produced in GGV thin-wall casting, allowing peak pressures of up to 160 bar. This explains the high specific fuel economy of the engine, despite the combustion controls required for compliance with EU3 emission standards.

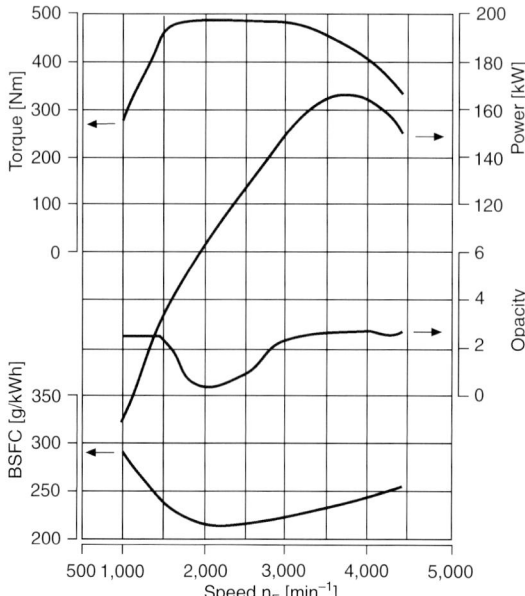

Fig. 14.34. Full-load data of the 3.3 liter V8 TDI engine by Audi [17]

In 1999, for the first time a series production car (VW Lupo TDI) was introduced with a fuel consumption of less than 3 liter/100 km (2.99 liter/100 km) in the NEDC. The powertrain consisted of a 3-cylinder, 1.2 liter DI diesel engine with VTG turbocharger, and a drivetrain especially developed for this application (automated manual transmission and engine deactivation during start-stop operation). Here, the variable-geometry turbine of the Garrett VNT12 charger enables high torque at low speeds, in such a way that – in combination with the automated manual transmission – the engine operating points in the driving cycle can be shifted into the engine map range with the best fuel economy. Figure 14.35 shows a longitudinal and a cross section of the engine. In addition to its innovative all-aluminum engine design, the connection between cylinder head and main bearing assembly is assured by integrated tie rods to allow the high ignition pressure forces. Its high-pressure injection system by cam-controlled unit injector elements marked the introduction of a

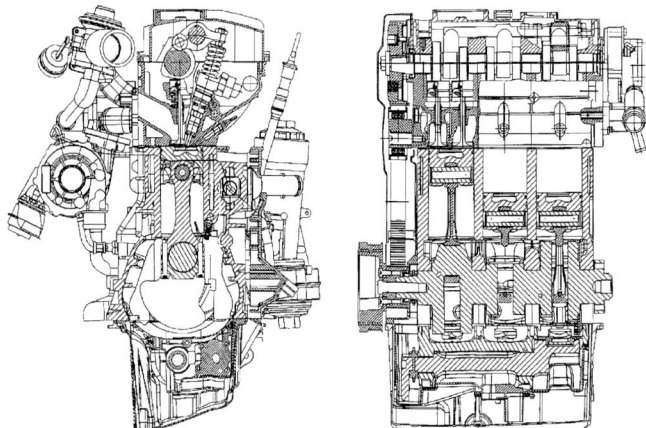

Fig. 14.35. Longitudinal and cross section of the 1.2 liter PDE diesel engine with VTG charger by VW [50]

new technology in passenger car diesel engines. Since this system can achieve injection pressures exceeding 2,000 bar, the combustion can be tuned to result not only in low fuel consumption but also in very low exhaust gas emissions (in compliance with EU3 and D4 standards).

The most important technical data of the engine are summarized in the appendix. To achieve the engine characteristics mentioned, besides a conventional air-to-air charge air cooling system the engine is equipped with a water-cooled exhaust gas recirculation system. The combination of all these measures resulted in a reduction of the CO_2 emissions in the NEDC driving cycle down to 81 g/km (Lupo 870 kg version).

One of the most modern representatives of supercharged passenger car diesel engines is the 6-cylinder inline DI diesel engine OM613 by DaimlerChrysler, the top engine of a series with 4-, 5-, and 6-cylinder engines. It is equipped with a common-rail injection system, VTG exhaust gas turbocharger, air-to-air charge air cooling, controlled and cooled exhaust gas recirculation, and a controlled flap valve in the intake port which allows the charge swirl level to be adjusted for different map-specific requirements. It represents the state of the art of passenger car diesel engine development. Figure 14.36 shows a charger-side view of the engine.

A further development stage in turbocharging technology was achieved when BMW introduced its V8 DI diesel engine (Fig. 14.37) in which the turbine inlet blade position of the VTG chargers is controlled by electric stepping motors (Fig. 14.38). With this technology, both VTG chargers (one charger for each cylinder row) can be reliably held within the stable compressor map area in the total engine operating range – especially under transient conditions and in the lower speed range – resulting in an improved utilization of the compressor map and allowing operation close to the dynamic surge limit.

In 2005, a combination of two-stage and register turbocharging of a passenger car diesel engine was also introduced by BMW in mass production (Fig. 14.39). In this case, a larger compressor of the low-pressure stage is followed downstream by a smaller compressor of the high-pressure turbocharger (Fig. 14.40).

In order to integrate a high-pressure compressor small enough for best low end torque but still not leading to any swallowing limitation in the entire engine air mass flow range, the corresponding compressor can be bypassed at higher loads. Because of the comparatively small inertia of the high-pressure stage (as the only turbocharger mainly active in such operation conditions), the

Fig. 14.36. OM 613 3.2 liter, 6-cylinder turbocharged engine by DaimlerChrysler, with VTG, EGR and common-rail injection

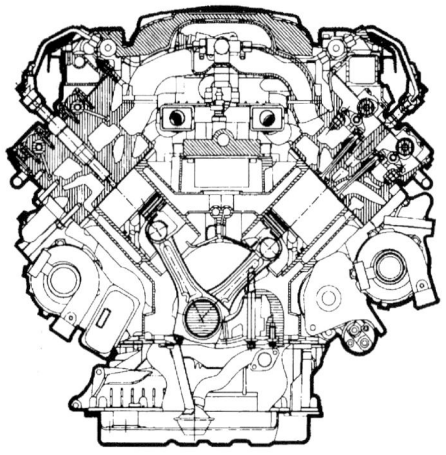
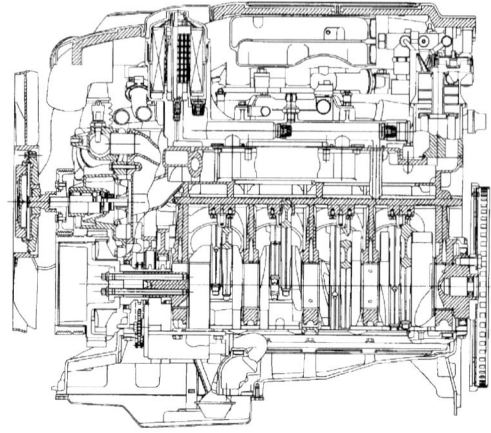

Fig. 14.37. Sectional view and cross sections of the bi-turbocharged V8 passenger car DI diesel engine by BMW [5]

Fig. 14.38. Electric stepping motor of the VTG inlet guide blade control of the V8 DI diesel engine by BMW [5]

14.2 Passenger car diesel engines

Fig. 14.39

Fig. 14.40

Fig. 14.39. BMW 3 liter, 6-cylinder DI diesel engine with mixed two-stage and register turbocharging [133]

Fig. 14.40. Turbocharging system of the BMW 3 liter, 6-cylinder DI diesel engine with mixed two-stage and register turbocharging [133]

boost pressure can be built up faster than with a conventional single-stage turbocharging system. On the other hand, at higher engine speeds and loads, the high-pressure turbine can be partly bypassed and, thus, the boost pressure controlled without opening a waste gate. Only at very high loads, i.e., near rated power, also the waste gate of the low-pressure turbine has to be opened to limit the boost pressure. The different operating modes of this advanced turbocharging system are summarized in Fig. 14.41. Further, the main engine geometry and performance data are listed in Table A.1.

Recently, two-stage turbocharging is also utilized to introduce new alternative diesel combustion processes for passenger car applications. These combustion processes require particular cylinder charge properties and, thus, higher boost pressures also at part-load operation [133]. Again, with the help of a two-stage quasi-register turbocharging system, a smaller "high-pressure" turbocharger can generate such higher boost pressures already at mid-speed part-load operation.

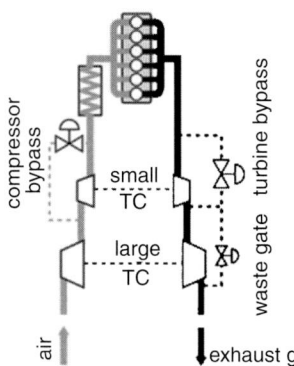
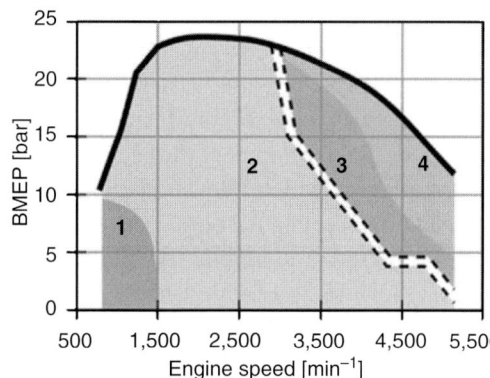

Fig. 14.41. Sketch of turbocharging system and operating strategy of the BMW 3 liter, 6-cylinder DI diesel engine with mixed two-stage and register turbocharging [133]. Operating range 1: turbine bypass closed, compressor bypass closed, waste gate closed. Operating range 2: turbine bypass controlled opened, compressor bypass closed, waste gate closed. Operating range 3: turbine bypass open, compressor bypass open, waste gate closed. Operating range 4: turbine bypass open, compressor bypass open, waste gate controlled opened

Fig. 14.42. View of the two-stroke DI diesel engine by AVL, with combined supercharging system

An engine with such a combination of a two-stage turbocharging system with an alternative diesel combustion process was presented by AVL in 2005. The advantage of this technology combination is a simultaneous reduction of engine-out emissions (-85% of NO_x and -90% of particle emissions measured in the NEDC compared with an engine with a conventional combustion system), due to the extended operation range of the engine with alternative combustion and very high power density.

Two-stroke engines also represent a logical application area for supercharging technology. They are used either if simple and low-cost engines are required (e.g., motorcycles) or if extreme power density has to be achieved at lowest weight (e.g., motorcycle racing). In the early years of the automobile, two-stroke engines were utilized due to their power density and their smoothness (double firing sequence as compared with the four-stroke engine). In 1996, AVL introduced a two-stroke DI diesel engine with combined super- and turbocharging (Fig. 14.42). The operating behavior and the charging technology of this engine was discussed in more detail in Sect. 6.6.6. The operating data achieved with this engine and the most important engine characteristics are listed in the appendix.

By combining the advantages of uniflow scavenging (with the best efficiency of all scavenging processes) with the concept of combined charging (low scavenging losses and high charge density due to turbine backpressure), it was possible to achieve maximum mean effective pressures of 11 bar. This corresponds to a four-stroke mean effective pressure of 22 bar. The engine was used as an experimental engine for the development of the charging system as well as of a novel two-stroke uniflow scavenging process which allows the cylinder spacing to be similar to that of a four-stroke engine by avoiding scavenging ports between the cylinders. A similar engine was shown by Daihatsu at the Frankfurt Automobile Exhibition in 1999 as a prototype.

As for gasoline engines (Sect. 14.2), DaimlerChrysler also published results of tests comparing mechanical charging and turbocharging using a 2.5 liter diesel engine. In addition to the charging system, the advantages and disadvantages of Roots and spiral chargers were investigated on the engine. The spiral charger exhibits the advantages of low inertia and better efficiencies at higher compression ratios (Fig. 14.43).

In the part-load range, especially relevant for passenger car engines, the conditions are reversed, resulting in lower part-load fuel consumption with the Roots blower due to the lower driving power required (Fig. 14.44). However, the overwhelming majority of today's diesel engines are designed

14.2 Passenger car diesel engines

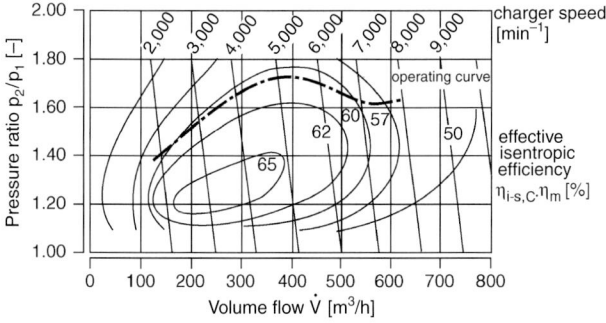

Fig. 14.43. Efficiency map of the Roots blower with engine full-load curve [113]

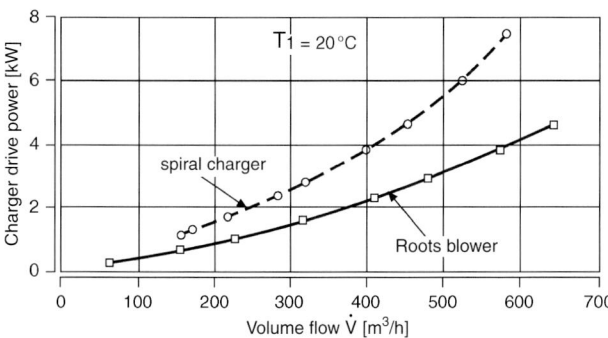

Fig. 14.44. Charger drive power of Roots blower and spiral charger [113]

in such a way that they are mainly operated with active turbocharging, so that the latter conclusion may not be generalized.

As a last example of supercharged passenger car diesel engines, we will consider the IDI diesel engine with pressure-wave charging utilized in the Mazda model 626 (Fig. 14.45). This

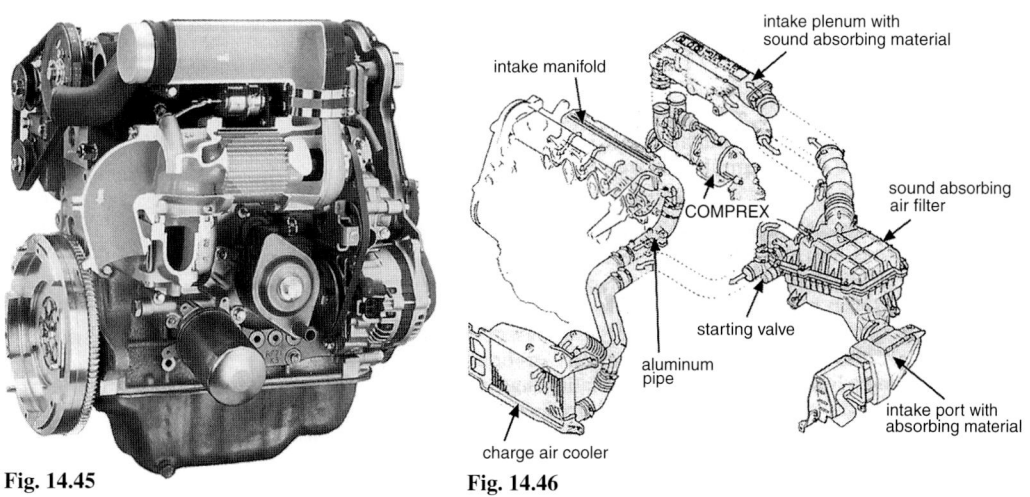

Fig. 14.45. Sectional view of the RF Comprex IDI diesel engine with pressure-wave charging by Mazda [71]

Fig. 14.46. Exhaust gas carrying components of the RF Comprex IDI diesel engine by Mazda [71]

supercharged swirl-chamber diesel engine, which was developed from a former naturally aspirated version and whose IDI technology at the time of its market introduction in 1987 was state of the art, gave the Mazda 626 model a driving performance comparable to their 2.0 liter DOHC gasoline engine. The fuel consumption was about 20% lower with the Comprex diesel engine, as compared with the gasoline version.

Figure 14.46 shows the exhaust gas carrying components of the engine. Since the Comprex charging system is especially sensitive to exhaust gas backpressure, the dimensions of the exhaust gas system have to be selected correspondingly large (manifold diameter and lowest muffler pressure loss). This can be seen in Fig. 14.46. The performance data of this engine are summarized in the appendix.

Actually, the minimum fuel consumption shown in Table A.1 underlines the significant disadvantage of prechamber engines in comparison to DI diesel engines.

14.3 Truck diesel engines

One of the first mass-produced supercharged engines was the diesel engine D1 KL by Adolph Saurer, equipped with a screw-type charger (Fig. 14.47), which was later superceded by the exhaust gas turbocharged engine D1 KT (Fig. 14.48).

In the United States, for a very long time Detroit Diesel Corporation two-stroke engines were utilized for trucks and buses, such as the 8V-92T model with exhaust gas turbocharger and upstream mechanical Roots blower. Figure 14.49 shows the view of the engine. More stringent requirements with regard to exhaust emissions and the necessity to improve the engine's fuel economy made these engines obsolete.

Some examples of modern truck engines will be presented now. The first example is the D 2876 LF by MAN, a water-cooled inline 6-cylinder engine with 4-valve cylinder head, with a maximum power of about 340 kW at 1,700–1,900 min^{-1} and a maximum torque of 2,100 Nm between 900 and 1,300 min^{-1}. Figure 14.50 shows a view of this engine.

Fig. 14.47 **Fig. 14.48**

Fig. 14.47. Mechanically charged Saurer truck engine D1 KL

Fig. 14.48. Saurer truck engine D1 KT with exhaust gas turbocharging

14.3 Truck diesel engines

Fig. 14.49. 8V-92-T two-stroke truck engine by DDC, with mechanical and exhaust gas turbocharging

Fig. 14.50. View of the MAN D-2876-LF-R-6-V4 truck engine with exhaust gas turbocharging

At the end of the 1990s, DaimlerChrysler developed their new series 900, with inline 4- and 6-cylinder engines rated between 90 and 230 kW at 2,300–2,500 min^{-1}, for applications in light-duty and medium-duty trucks. The rated power of the engine OM 904 (Fig. 14.48a) is 90–125 kW at 2,300 min^{-1}, its maximum torque is 470–660 Nm between 1,200 and 1,500 min^{-1}. The OM906 model has a rated power of 170–230 kW at 2,300 min^{-1} and a maximum torque of 810 to 1,300 Nm at 1,200 min^{-1}.

The D12C 460 by Volvo (Fig. 14.52) is a 6-cylinder inline engine with 4 valves, overhead camshaft, and fuel unit injectors. The version shown is rated at 340 kW at 1,800 min^{-1}. Its maximum

a **b**

Fig. 14.51. Views of the DaimlerChrysler OM904/6 LA inline 4- (**a**) and inline 6-cylinder (**b**) engines with exhaust gas turbocharging

Fig. 14.52. View of the Volvo FH-12-R-6-V4 truck diesel engine with turbocharger

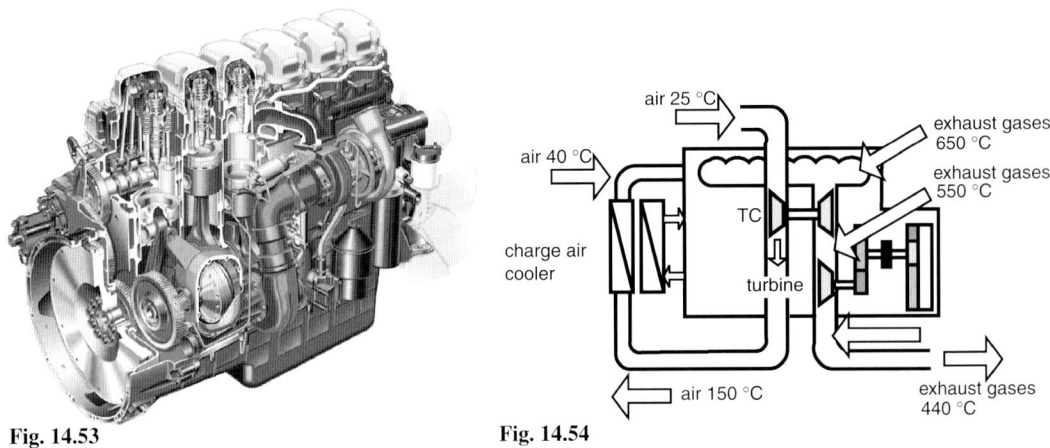

Fig. 14.53 **Fig. 14.54**

Fig. 14.53. R124-470 turbocompound truck engine by Scania [161]

Fig. 14.54. Charger unit with compound turbine of the R124-470 diesel engine by Scania [161]

torque is 2,200 Nm in the speed range between 1,000 and 1,300 $\min^{-1}$.

The R124-470 by Scania and the OM442 LAT by Daimler Benz are examples of turbocompound engines.

With the layout shown in Fig. 14.53 (the charger unit with compound turbine is shown in Fig. 14.54), the R124-470 engine by Scania is rated at 346 kW at 1,900 $\min^{-1}$. Its maximum torque is 2,200 Nm in the speed range between 1,050 and 1,350 $\min^{-1}$. With help of the compound turbine technology, the efficiency of the engine increased in its best point from 44 to 47%.

As early as 1991, Daimler Benz introduced such a compound engine on the basis of its V8 version of the model series OM440. Fuel economy improvements of around 5% were achieved in the rated power range, but this improvement was only possible in combination with an insulation of the exhaust gas manifolds [83].

All turbocompound engines with mechanical energy recovery need a step-down gear. Figure 14.55 shows a possible design, including an integrated secondary turbine with VTG.

14.5 High-performance high-speed engines

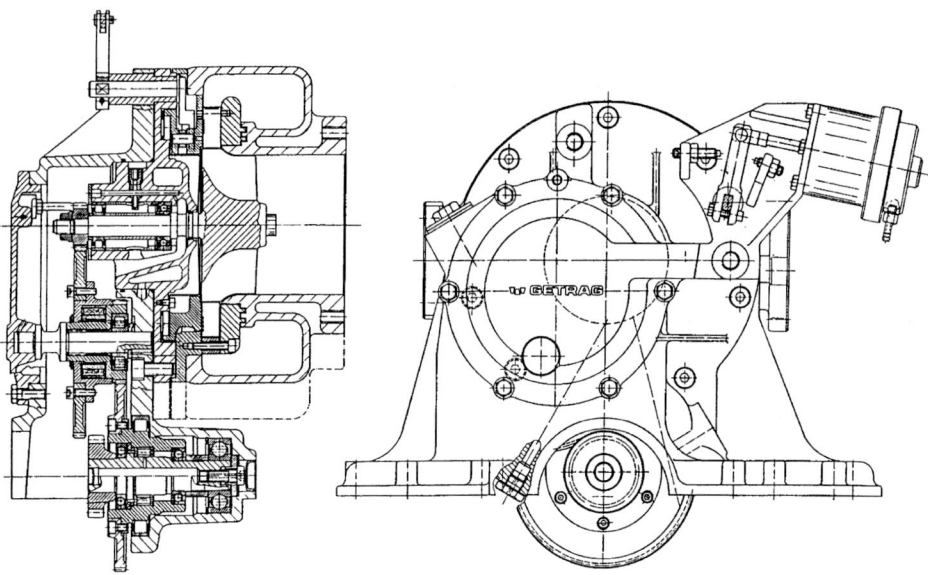

Fig. 14.55. Compound turbine of the OM442 LAT V8 turbocompound engine by Daimler Benz

Figure 14.56 shows an example of combined charging on a truck engine, the 10.9 liter, V6 engine OM441 LA by Daimler Benz. The charger unit consists of a conventional turbocharger combined with a mechanically powered Roots blower. Up to a speed of $1,250\,\mathrm{min}^{-1}$, the mechanical charger assists in supplying the engine with air. At higher speeds, the mechanical charger is bypassed and the turbocharger alone provides the charge air for the engine. With such combined charging systems, significant increases in mean effective pressures can be achieved in the lower speed range (Fig. 14.57), as well as significant improvements in transient response.

14.4 Aircraft engines

Nowadays, reciprocating piston combustion engines are only utilized in small aircraft. One of the most frequently used engines is the GSO-480 engine by Lycoming, a flat-6-cylinder engine with mechanically powered turbo compressor (Fig. 14.58). The engine generates a takeoff power of $250\,\mathrm{kW}$ at $3,400\,\mathrm{min}^{-1}$, has a continuous rating of $235\,\mathrm{kW}$ at $3,200\,\mathrm{min}^{-1}$ and weighs $225\,\mathrm{kg}$.

In 1988, Porsche also received flight certification for an exhaust gas turbocharged flat-6-cylinder engine, the PFM 3200. The engine was rated at a takeoff power of $180\,\mathrm{kW}$ at $5,300/2,343\,\mathrm{min}^{-1}$ engine/propeller speed. Figure 14.59 shows a view of this – at that time most advanced – small aircraft engine.

14.5 High-performance high-speed engines (locomotive and ship engines)

High-speed diesel engines ($n_{\max} \approx 800\text{–}2,000\,\mathrm{min}^{-1}$) for rail vehicles, fast ships, and military applications were formerly mainly two-stroke engines. The high power densities necessary for these applications demand supercharging for practically all of them. Since four-stroke engines are better suited for supercharging, the fraction of two-stroke engines has significantly declined recently. In the speed range of the engines discussed in this chapter, the two-stroke cycle shows no

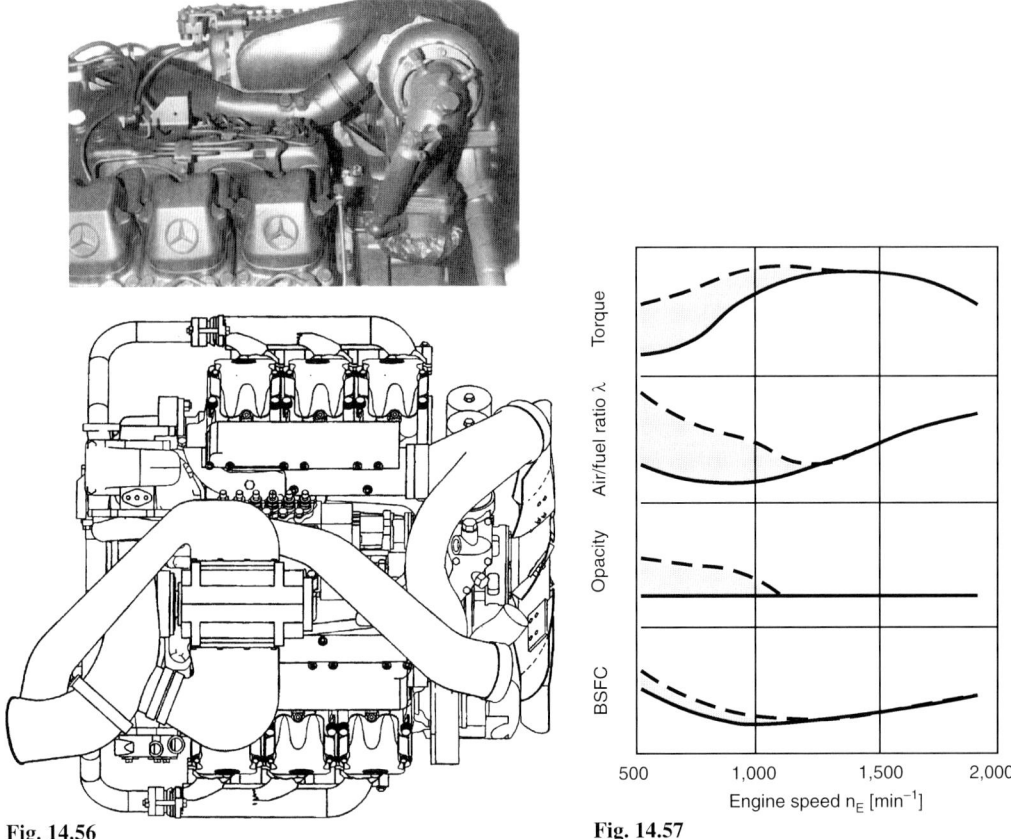

Fig. 14.56

Fig. 14.57

Fig. 14.56. Exhaust gas turbocharged OM441 LA truck engine by Daimler Benz with additional mechanical charging

Fig. 14.57. Improvement in operating behavior of the OM441 LA turbo engine by Daimler Benz by means of additional mechanical charging. Solid line, minimum BSFC or opacity, air-to-fuel ratio, and torque at minimum BSFC. Dash line, maximum torque or BSFC, opacity, and air-to-fuel ratio at maximum torque

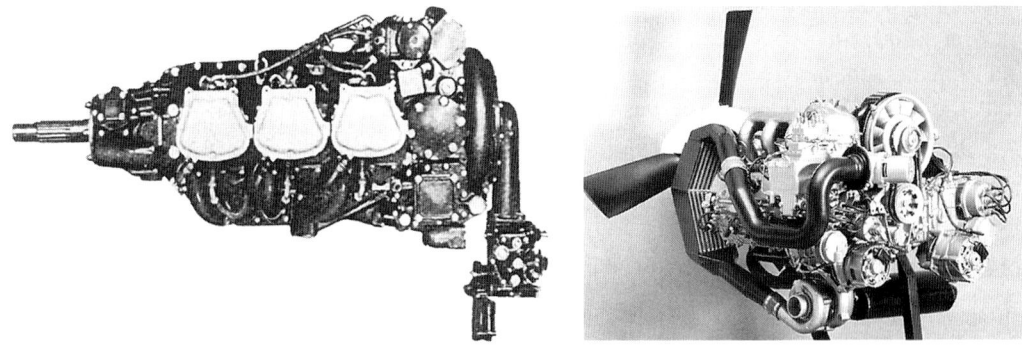

Fig. 14.58

Fig. 14.59

Fig. 14.58. GSO-480 flat-6 cylinder aircraft engine by Lycoming, with mechanically powered turbo compressor

Fig. 14.59. Porsche aircraft engine

14.5 High-performance high-speed engines

advantage either in specific power per cylinder or in its installed size or weight. Regarding thermal engine loads, the four-stroke cycle is clearly superior.

Therefore, only one two-stroke engine example, the model 16-645 E5 of the GM Electro-Motive Division, LaGrange, Illinois, will be described. For a long time, this engine dominated the American locomotive market. It is equipped with a charging system described in Sect. 7.4.4 (Fig. 7.12), powering the exhaust gas turbocharger from the engine crankshaft via a transmission and a freewheel clutch. The engine is rated at about 2,500 kW at 900 min^{-1}, its cylinder dimensions are 230 mm bore and 255 mm stroke. Figure 14.60 shows a view of the engine.

Today, the majority of engines are highly supercharged four-stroke engines, e.g., the model V 538 TB by MTU. MTU produces this model as a series of 12-, 16-, and 20-cylinder engines, with a bore of 185 mm and a stroke of 200 mm. A special feature is the installation of the exhaust gas turbochargers in the V of the engine with vertical orientation of the charger shaft. This results in especially short exhaust manifolds with small volumes and, thus, favorable conditions for pulse turbocharging. The 16-cylinder version is rated at 3,000 kW at 1,900 min^{-1}. Figure 14.61 shows a view of the engine with the vertically arranged chargers.

Fig. 14.60.   **Fig. 14.61.**

Fig. 14.60. 16-645-E5 locomotive engine by GM with mechanically assisted exhaust gas turbocharger

Fig. 14.61. 16-V-538 TB high-performance diesel engine by MTU with exhaust gas turbochargers arranged with vertical shafts

Fig. 14.62. 20V-956 TB high-performance ship diesel engine by MTU

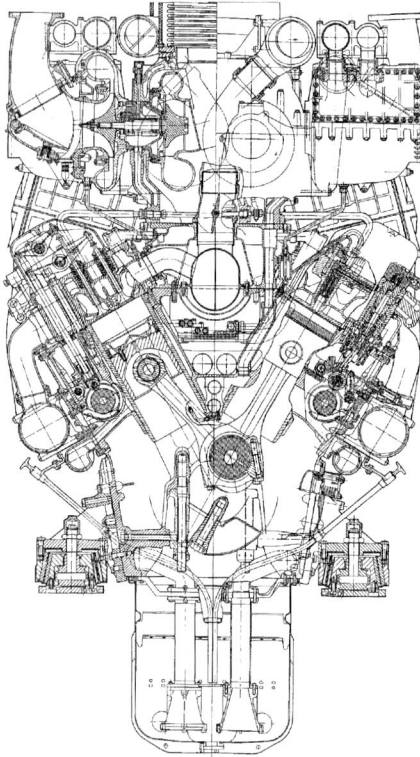

Fig. 14.63. Cross section of the MTU engine 1163 with two-stage register charging

The two engines of the V 956/1163 series by MTU have a common bore of 230 mm, but different strokes of 230 and 280 mm, i.e., individual cylinder displacements of 9.56 and 11.63 liter/cylinder, which is expressed in their model names. Figure 14.62 shows the short-stroke 20 V 956 TB type, which is rated as a ship engine at 4,900 kW at 1,500 min^{-1}, corresponding to a mean effective pressure of 20.5 bar and 245 kW power per cylinder.

Figure 14.63 shows a cross section of the longer-stroke V 1163 model. In this view, the compact layout of the high- and low-pressure exhaust gas turbochargers (two-stage register charging) can be seen. The engine generates 370 kW per cylinder at 1,300 min^{-1}, which corresponds to a very high mean effective pressure of 29.5 bar. With its maximum number of 20 cylinders, this engine series can provide up to 7,400 kW net power.

14.6 Medium-speed engines (gas and heavy-oil operation)

For stationary and marine applications, medium-speed engines ($n \approx 200$–800 min^{-1}) cover a wide speed and power spectrum. The engine layouts for this sector are correspondingly diversified. In the order of increasing power output per cylinder, the following engines are mentioned as typical examples.

The MAN B&W series 32/40 (320 mm bore, 400 mm stroke) covers inline and V engines, all of which generate the same power per cylinder, i.e., 440 kW at 720–750 min^{-1}. Figure 14.64 shows a cross section of the V engine.

The MAK M552AK engine is a good example of the design complexities necessary for the exhaust system to achieve optimum pulse turbocharging. With a bore of 450 mm and a stroke

14.6 Medium-speed engines

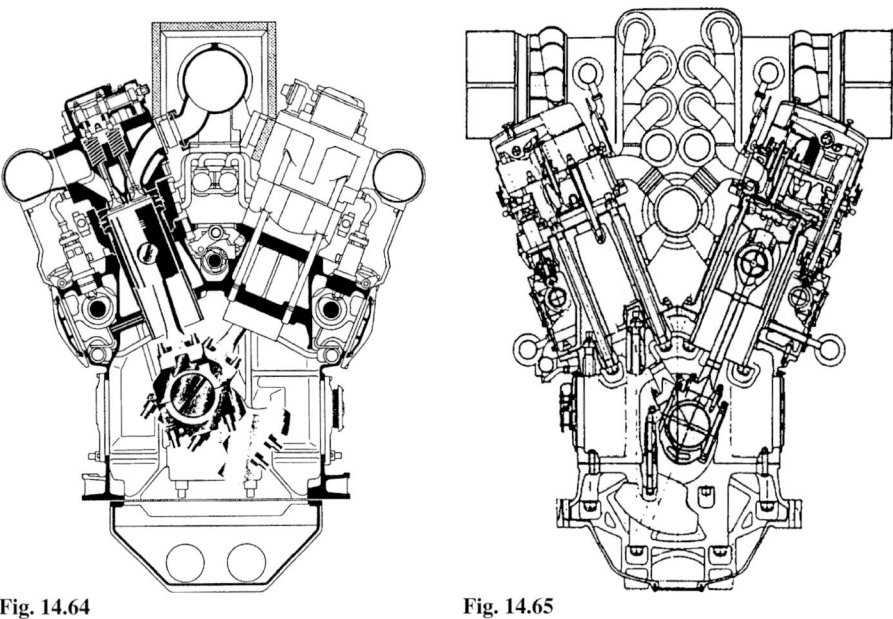

Fig. 14.64. Cross section of the MAN B&W 32/40 engine with 440 kW/cylinder

Fig. 14.65. Cross section of the MAK M552AK engine with 590 kW/cylinder

of 520 mm, this engine generates about 600 kW per cylinder at 500 min^{-1}. Intake and exhaust manifolds are arranged in the V of the engine. This results in especially short exhaust manifolds suited for pulse turbocharging. Since the charge air manifolds are also located inside the

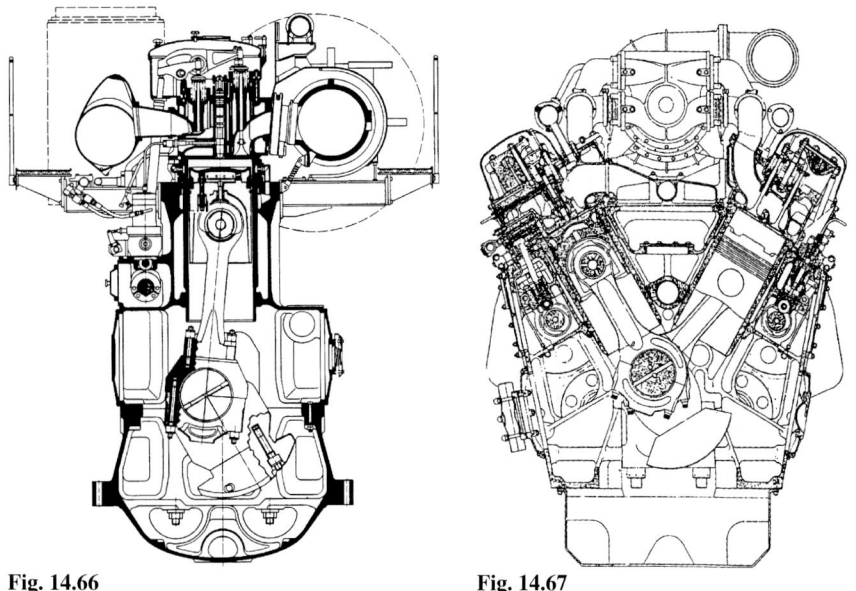

Fig. 14.66. Cross section of the MAN L 52/52 engine with 890 kW/cylinder

Fig. 14.67. Cross section of the SEMT PC4-570 V engine with 1,100 kW/cylinder

engine V, a very narrow engine width has been achieved. Figure 14.65 shows a cross section of the engine.

The MAN L 52/52 engine (Fig. 14.66) represents an inline engine with a cylinder size at the

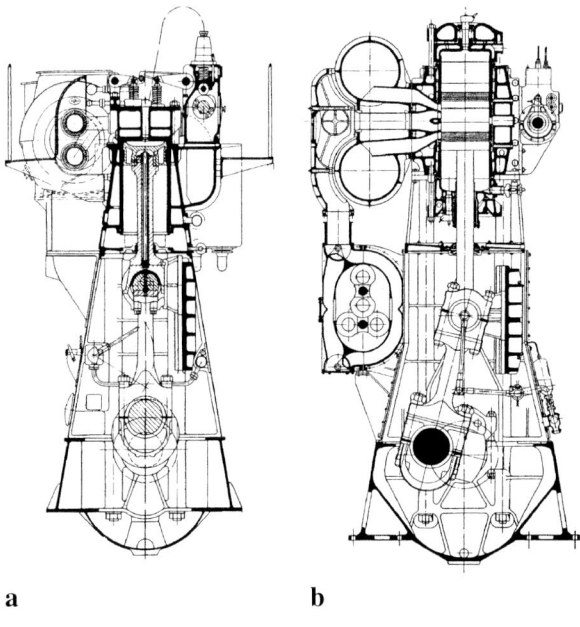

Fig. 14.68. **a** MAN four-stroke cross-head engine KV 45/66; **b** MAN two-stroke cross-head engine DZ 53/76 with mechanically powered Roots scavenging blower upstream of the exhaust gas turbocharger

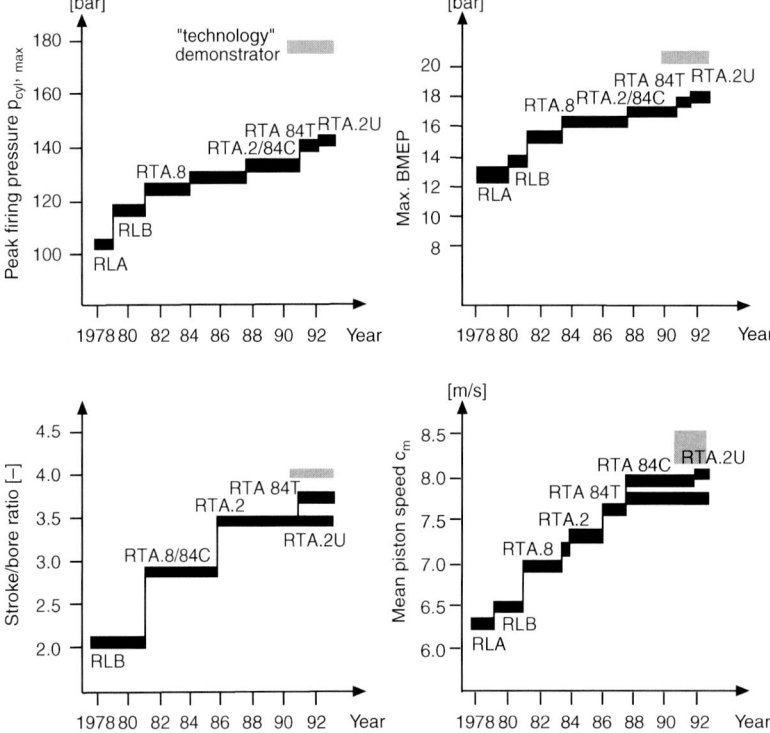

Fig. 14.69. Development trends of the main engine characteristics of slow-speed New Sulzer engines since 1978

upper end for medium-speed engines. The layout of its exhaust system emphasizes its constant-pressure turbocharging process. Its specific power is rated at 885 kW per cylinder at 500 min^{-1}.

The SEMT-Pielstick PC4-570 engine (Fig. 14.67) has a bore of 570 mm and a stroke of 620 mm. At 400 min^{-1}, it generates 1,100 kW per cylinder. This corresponds to the very high mean effective pressure of about 21 bar, which certainly can also be traced back to its effective exhaust system layout for optimum pulse turbocharging.

As an example of the fact that medium-speed engines are available in four-stroke and in two-stroke designs, the two MAN engine models KV 45/66 and DZ 53/76 can be mentioned. The KV 45/66 (Fig. 14.68a) represents a four-stroke engine with simple layout, while the DZ 53/76 (Fig. 14.68b) shows the complexity of a double-acting, loop-scavenged two-stroke engine with Roots type mechanical scavenging blower and constant-pressure turbocharging.

14.7 Slow-speed engines (stationary and ship engines)

Today, slow-speed engines ($n \approx 60\text{--}150 \text{ min}^{-1}$) are exclusively utilized for electric energy generation or as primary ship engines with direct propeller connection. For various reasons, they are all two-stroke engines, particularly due to their better power density and the fuel used. These are all operated with heavy oil, whose quality is constantly deteriorating. A two-stroke cross-head engine can tolerate such bad fuel qualities better than a four-stroke engine.

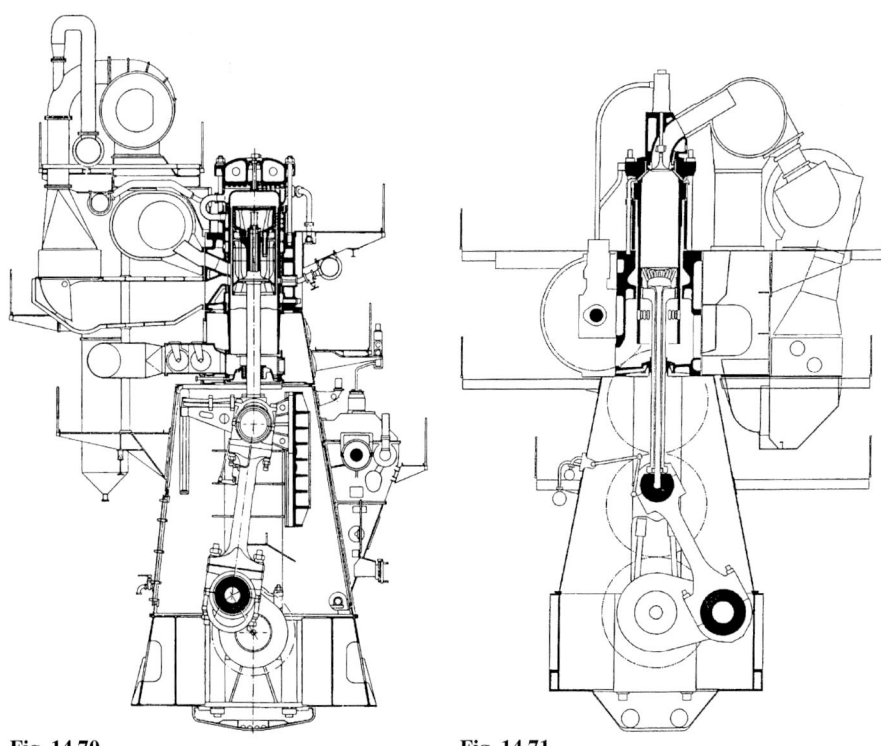

Fig. 14.70   **Fig. 14.71**

Fig. 14.70. MAN KZ 105/180 slow-speed engine with loop scavenging; about 3,000 kW/cylinder

Fig. 14.71. Cross section of the New Sulzer RT 84-T engine; about 4,000 kW/cylinder

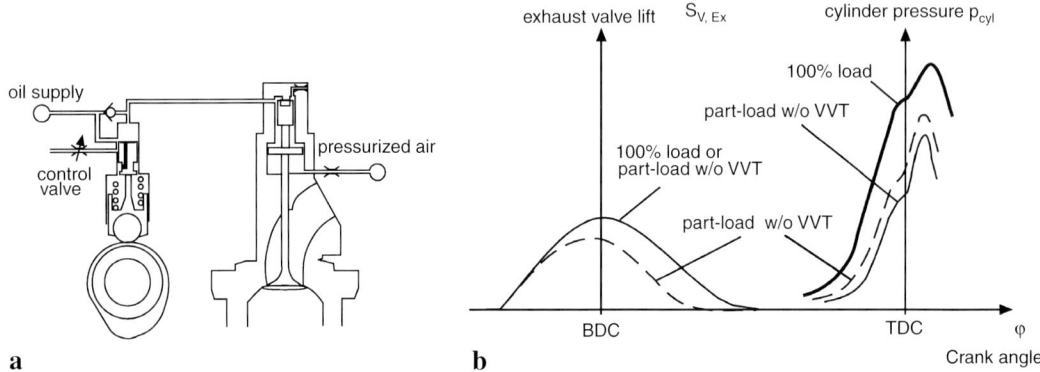

Fig. 14.72. Part-load valve timing and lift adjustment system (**a**) and corresponding cylinder high-pressure curve (**b**) with variable exhaust valve closing (VVT)

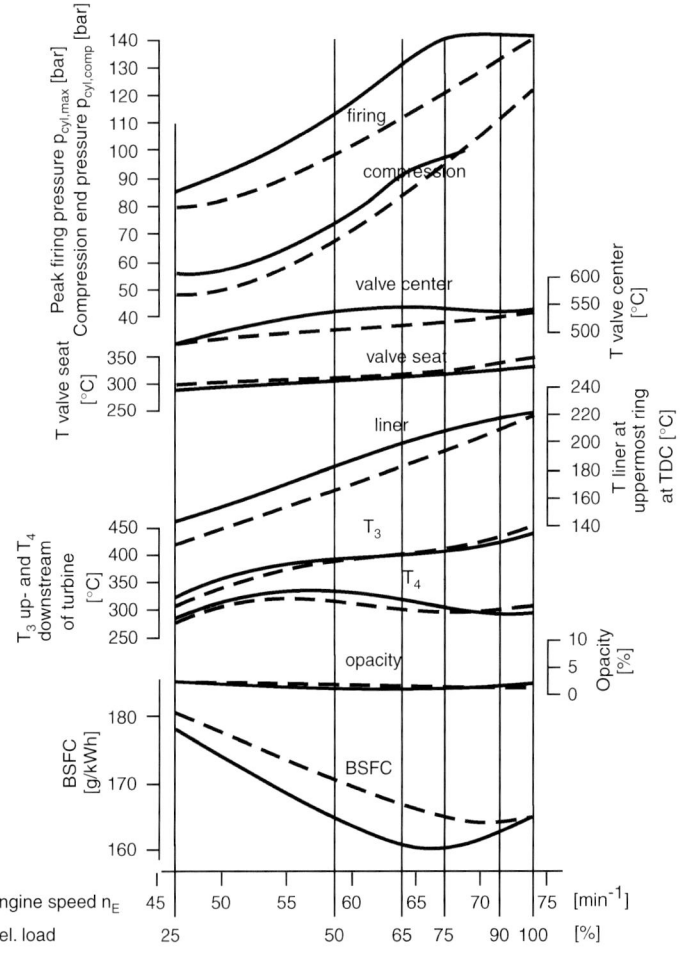

Fig. 14.73. Performance, fuel economy and exhaust gas turbocharger-specific data of the New Sulzer RT 84-T engine with (solid line) and without (dash line) variable injection timing and variable exhaust valve closing

14.7 Slow-speed engines

Further, for reasons of efficiency and thus fuel economy, very low engine speeds, down to about 60 min^{-1}, are required for direct drives of today's ship propellers which can only be achieved by two-stroke engines. The development trends which have been observed lately show continually increasing stroke lengths, with stroke-to-bore ratios greater than 4, and continually higher boost pressures with simultaneously increasing mean effective pressures.

Figure 14.69 shows the development trends from 1978 to 1992 for the most important engine parameters of the slow-speed engines by New Sulzer. The astonishing increase in peak firing pressures up to 180 bar is necessary for the achievement of best fuel economy.

In the mid-1960s, the MAN KZ 105/180 engine was the state of the art for loop-scavenged two-stroke engines. With a bore of 1,050 mm it had the largest piston diameter ever. Along with its stroke of 1,800 mm it generated 2,950 kW per cylinder. Figure 14.70 shows the cross section of this engine.

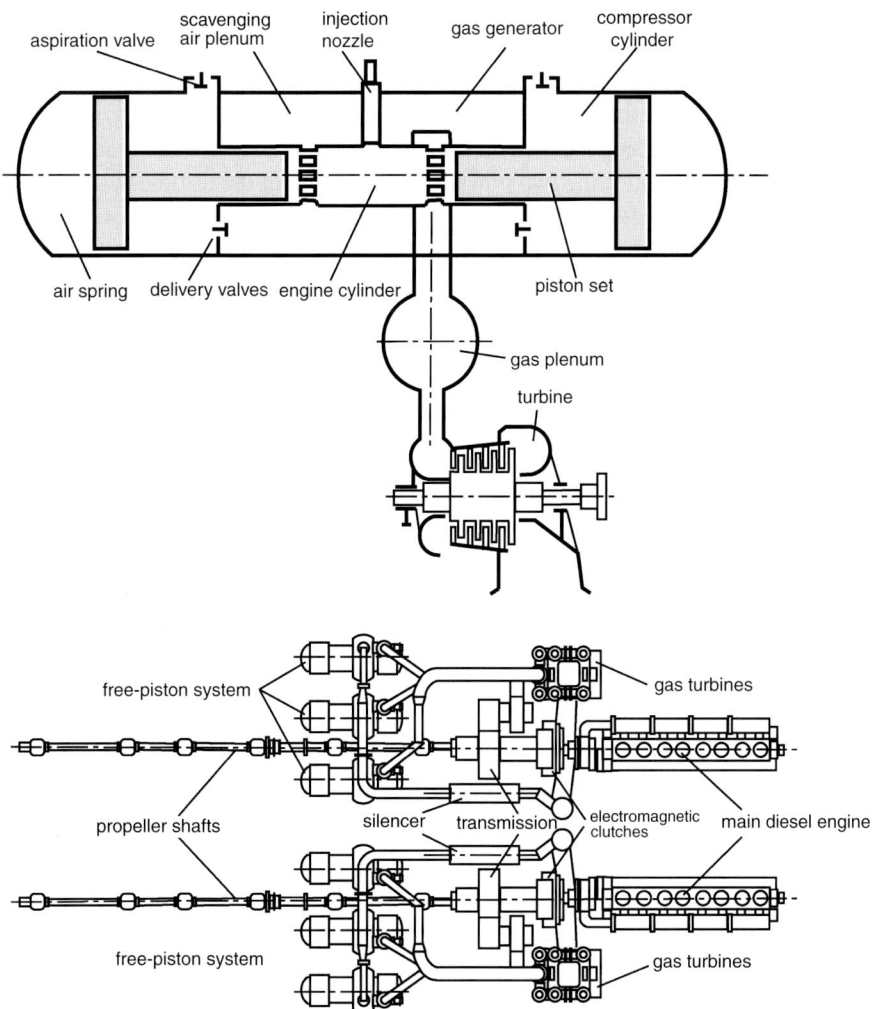

Fig. 14.74. Compound powertrain of the vessel *Fritz Heckert* by MTU; free-piston gas generators with integrated piston scavenging and charge pumps

Mitsubishi's VEC 52/105 engine was a uniflow scavenged two-stroke engine with three exhaust valves. It featured two-stage charging, with pulse turbocharging of the high-pressure turbine. Figure 6.3 shows the cross section of this engine, which generated about 1,000 kW power per cylinder at 175 min^{-1}. After that, the development trend clearly shifted to uniflow scavenged long-stroke engines with a centrally located exhaust valve (for the most part hydraulically actuated) and highly efficient constant-pressure turbocharging.

The current state of the art is shown by the RT 84 T engine by New Sulzer. It is a uniflow scavenged cross-head two-stroke engine with a central, hydraulically actuated turbolator-exhaust valve. Timing and lift of the valve can be varied (variable exhaust closing). In combination with variable injection timing, very good part-load behavior is achieved at best fuel economy. Figure 14.71 shows a cross section of this engine with its very clear design.

Figure 14.72a shows the hydraulic valve actuation system, with the arrangement for variable lift and timing adjustment. The effects of an adjustment of valve lift and timing (variable exhaust valve closing) on the cylinder pressure curve is summarized in Fig. 14.72b. Finally, the achievable engine performance data both with and without variable injection timing and variable exhaust valve closing described above are plotted in Fig. 14.73.

As a last example for a very particular layout of a powertrain with turbocharged internal combustion engines, Fig. 14.74 shows the compound system of the vessel Fritz Heckert developed by MTU, consisting of 6 free-piston gas generators with integrated piston scavenging and charge pumps which are providing the hot gas for the two power turbines. Each of the turbines were acting via a transmission on one propeller shaft. Additionally two main diesel engines could be connected to the propeller shaft via an electromagnetic clutch.

Appendix

Table A.1. Characteristics and performance data of supercharged production engines

Engine type, manufacturer, model	Layout	Displacement V_{tot} [l]	Bore × stroke [mm]	ε [–]	Valvetrain	Turbocharger	Charge air cooler/ injection system	Max. power at speed [kW at rpm]	Max. torque at speed [Nm at rpm]	Min. fuel consumption [g/kWh]
Gasoline engines										
Audi 1.8 l, 5-valve turbo	I-4	1.781	81 × 86.4	9 : 1	5V DOHC w/ VVT	3K K03	1 air-to-air	110 at 5,700	210 at 1,750	
Porsche 959	flat-6	2.85	95 × 67	8.3 : 1	4V DOHC	2 water-cooled KKK K26	2 air-to-air	331 at 6,500	500 at 5,500	
Mercedes Benz M 119 HL	V8 w/ alu crankcase, dry-sump	5.00	96.5 × 85	9 : 1	4V DOHC	2 KKK K27 K	2 air-to-air (Behr)	680 at 7,000	1,020 at 4,800	235 (map min.)
Diesel engines										
Daimler-Benz OM617	I-5	3.00	90.9 × 92.4	21 : 1	2V OHC	Garrett	inline pump charge pressure controlled	100 at 4,000	245 at 2,500	220 (full load)
BMW M47	I-4	1.95	84 × 88	19 : 1	4V DOHC	Garrett	distributor pump	100 at 4,000	280 at 1,750	202 (map min.)
Fiat JTD	I-5	2.387	82 × 90.4	18.45 : 1	2V OHC	Garrett VTG	common rail	100 at 4,200	304 at 2,000	205 (map min.)
Audi W11 bi-turbo	V8	3.328	86.4 × 78.3	18.5 : 1	4V DOHC	2 Garrett VNT 15	common rail	165 at 4,000	480 at 1,800	205 (map min.)
VW PDE	I-3 w/ balance shaft	1.196	86.7 × 76.5	19.5 : 1	2V OHC	Garrett VNT 12	cam powered unit injector	45 at 4,000	140 at 2,000	210 (map min.)
DaimlerChrysler OM613 CDI	I-6	3.224	88 × 88.34	18 : 1	4V DOHC	Garrett VNT	common rail (Bosch)	145 at 4,200	470 at 1,800–2,600	203 (map min.)
AVL-List two-stroke DI	I-3 longitud. scavenging	0.98	72 × 80	18.5 : 1	exh: 4V DOHC intake: 2 × 5 ports	Garrett VNT 15	common rail	49 at 3,500	167 at 1,500–2,500	235 (full load)
Mazda RF Comprex IDI	I-4 swirl chamber	1.998	86 × 86	21.1 : 1	2V OHC	Comprex pressure-wave	distributor pump	61 at 4,000	182 at 2,000	265 (map min.)

Appendix

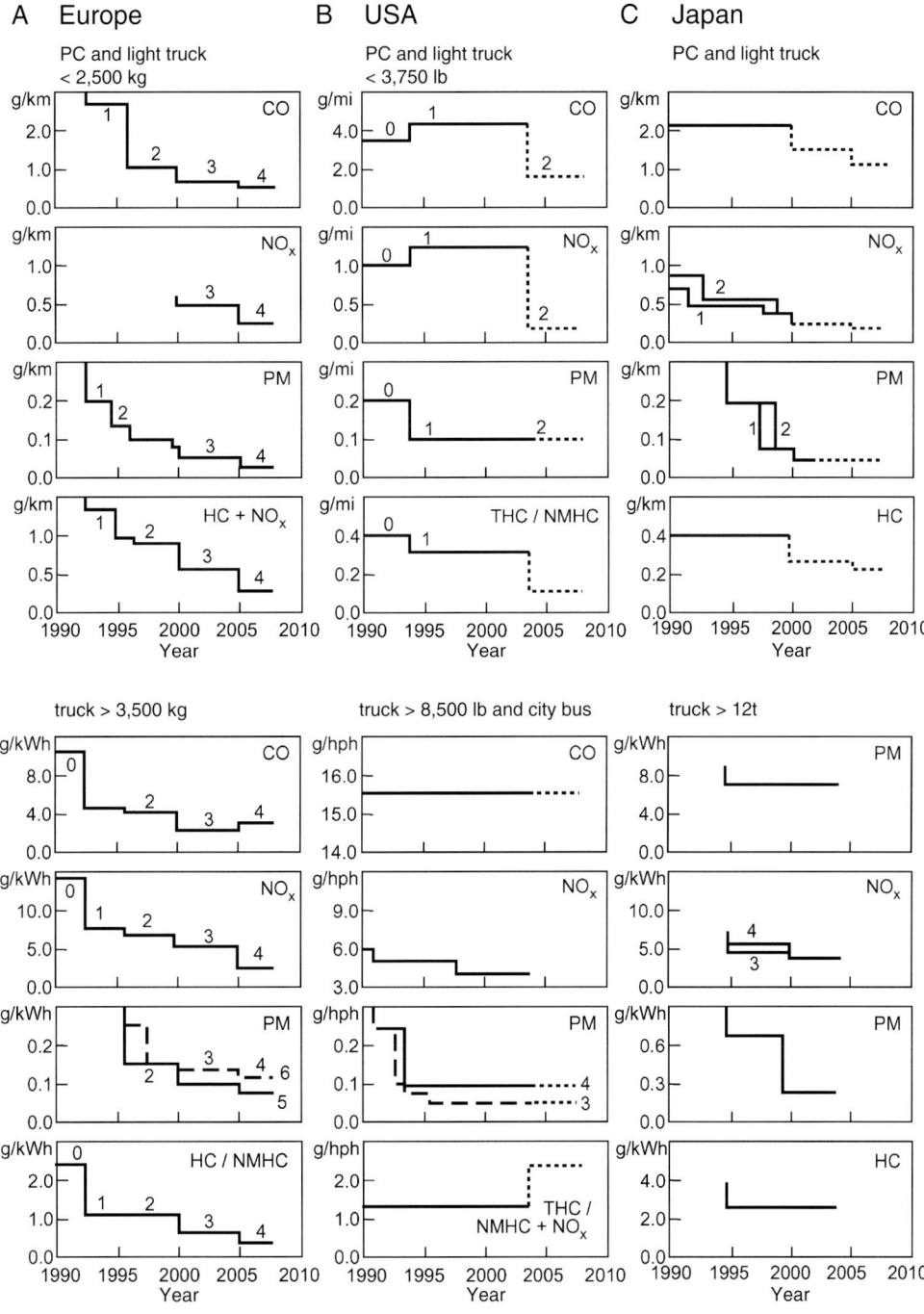

Fig. A.1. Exhaust emission standards for passenger car and truck diesel engines in Europe, the US and Japan [Bosch Kraftfahrtechnisches Taschenbuch, 23rd edn.]. **A** 0, Euro-0; 1, Euro-1; 2, Euro-2; 3, Euro-3; 4, Euro-4; 5, <85 kW; 6, >85 kW. **B** 0, Tier 0; 1, Tier 1; 2, Tier 2; 3, heavy trucks; 4, city bus. **C** 1, <1,265 kg vehicle starting weight; 2, >1,265 kg vehicle starting weight; 3, IDI; 4, DI

Table A.2. EU exhaust emission standards for gasoline engines in the ECE/EG driving cycle [Bosch Kraftfahrtechnisches Taschenbuch, 23rd edn.]

Standards	Effective date	CO (g/km)	HC (g/km)	NO_x (g/km)	HC + NO_x (g/km)
EU stage I	July 1992	2.72			0.97
EU stage II	Jan. 1996	2.2			0.5
EU stage III	Jan. 2000	2.3	0.2	0.15	
EU stage IV	Jan. 2005	1.0	0.1	0.08	

Table A.3. Exhaust emission standards for gasoline engines in the USA (49 states) and California, FTP-78 driving cycle[a] [Bosch Kraftfahrtechnisches Taschenbuch, 23rd edn.]

Agency[b]	Model year	Standards[c]	CO (g/mi)	HC (g/mi)	NO_x (g/mi)
EPA	1994	Tier 1	3.4	0.25	0.4
(49 states)	2004	Tier 2	1.7	0.125[d]	0.2
CARB		TLEV	3.4	0.125[d]	0.4
(California)[e]		LEV	3.4	0.075[d]	0.2
		ULEV	1.7	0.04[d]	0.2

[a] For updates on these standards, consult the agencies' websites
[b] EPA, Environmental Protection Agency; CARB, California Air Resources Board
[c] TLEV, transitional low-emission vehicles; LEV, low-emission vehicles; ULEV, ultralow-emission vehicles
[d] Non-methane hydrocarbons (NMHC)
[e] Model year dates depend on NMHC manufacturer fleet average (certification of individual model and total fleet)

Table A.4. Exhaust emission standards for gasoline engines in Japan, Japan driving cycle[a] [Bosch Kraftfahrtechnisches Taschenbuch, 23rd edn.]

Test procedure	CO	HC	NO_x	Evaporation (HC)
10-15 mode (g/km)	2.1–2.7 (0.67)	0.25–0.39 (0.08)	0.25–0.48 (0.08)	
11 mode (g/test)	60–85 (19.0)	7.0–9.5 (2.2)	4.4–6.0 (1.4)	
SHED (g/test)				2.0

[a] Data in parentheses are proposed standards

Table A.5. Exhaust emission standards for stationary engines according to TA-Luft (Germany; implemented 1992)

Engine type	$PM^{a,b}$ (g/m^3)	SO_2^b (g/m^3)	Form-aldehyde (g/m^3)	$NMHC^c$ (g/m^3)	CO^b (g/m^3)	$NO_x^{b,d}$ (g/m^3)	
						2-stroke	4-stroke
Diesel–gas	0.05	0.42	0.02	0.15	0.65	0.8	0.5
Natural gas		0.42	0.02	0.15	0.65	0.8	0.5

[a] PM, particulate matter
[b] 5% O_2
[c] NMHC, non-methane hydrocarbons
[d] Dry

References

1. Aeberli, K.: The Sulzer containership engines: the RTA84C and the RTA96C. New Sulzer Diesel, Winterthur, 1996
2. Aerzener Maschinenfabrik: Drehkolbengaszähler zur zuverlässigen Messung gasförmiger Medien. Manufacturer's information, Aerzener Maschinenfabrik, Aerzen
3. AG Kühnle, Kopp & Kausch: Die 2-stufige geregelte Aufladung von KKK: ein neues Aufladesystem für Nfz-Motoren. Manufacturer's information, 3K-Warner, Kirchheimbolanden
4. Anisits, F., Borgmann, K., Kratochwill, H., Steinparzer, F.: Der neue BMW Sechszylinder-Dieselmotor. MTZ 59: 698–709, 1998
5. Anisits, F., Borgmann, K., Kratochwill, H., Steinparzer, F.: Der erste Achtzylinder Dieselmotor mit Direkteinspritzung von BMW. MTZ 60: 362–372, 1999
6. Anisits, F., Borgmann, K., Kratochwill, H., Steinparzer, F.: Der neue BMW Vierzylinder-Dieselmotor. ATZ/MTZ-Sonderheft Der neue 3er: 104–116, 1998
7. Annand, J.D.: Heat transfer in the cylinders of reciprocating internal combustion engines. Proc. Inst. Mech. Eng. 177: 973–990, 1963
8. Ashby, D.: Subaru two stage twin turbo 2 litre engine. SAE Pap. 1994-10-0004
9. Bach, M., Bauder, R., Endres, H., Pölzl, H.-W.: Die konsequente Fortführung der 10-jährigen TDI Tradition bei AUDI: der neue V8 TDI. Press release, AUDI, Neckarsulm, 1999
10. Bach, M., Bauder, R., Mikulic, L., Pölzl, H.-W., Stähle, H.: Der neue V6-TDI-Motor von Audi mit Vierventiltechnik. MTZ 58: 372–382, 1997
11. Backhaus, R.: Diesel-Rennmotor mit Direkteinspritzung von BMW. MTZ 59: 578–580, 1998
12. Baets, J., Codan, E., Meier, E.: Off-design operation of large diesel engines, a challenge to the turbocharging system. CIMAC Pap. D 72, 1993
13. Baines, N.C.: A meanline prediction method for radial turbine efficiency. Inst. Mech. Eng. C554/006, 1998
14. Bammert, K., Rautenberg, M.: Radialverdichter, Messungen an beschaufelten Diffusoren, Abschlußbericht. FVV Forschungsber. 184, 1975
15. Bantle, M., Bott, H.: Der Porsche Typ 959-Gruppe B: ein besonderes Automobil – Teil 1. ATZ 88: 265–270, 1986
16. Bargende, M., Hohenberg, G., Woschni, G.: Ein Gleichungsansatz zur Berechnung der instationären Wandwärmeverluste im Hochdruckteil von Ottomotoren. In: 3. Tagung „Der Arbeitsprozess des Verbrennungsmotors". Mitteilungen des Instituts für Verbrennungskraftmaschinen und Thermodynamik, Graz, pp. 171–190, 1991
17. Bauder, R.: Der neue V8 TDI-Motor mit Common Rail von AUDI. In: Bargende, M., Essers, U. (eds.): Dieselmotorentechnik 2000. Expert Verlag, Renningen-Malmsheim, pp. 160–182, 2000 (Kontakt & Studium, vol. 580)
18. Beineke, E., Woschni, G.: Rechnerische Untersuchung des Betriebsverhaltens ein- und zweistufig aufgeladener mittelschnellaufender Viertaktdieselmotoren. MTZ 39: 93–98, 1978
19. Biaggini, G., Buzio, V., Ellenson, R., Knecht, W.: Der neue Dieselmotor Cursor 8 von Iveco. MTZ 60: 640–649, 1999
20. Blair, G.P.: The correlation of theory and experiment for scavenging flow in 2-stroke cycle engines. SAE Pap. 881265, 1988
21. Borila, Y.G.: Some aspects of performance optimization of the sequentially turbocharged highly-rated truck diesel engine with turbochargers of unequal size and a pulse converter. Inst. Mech. Eng. C105/86, 1986
22. Bozung, H.G.: Die MAN-NA und -NA-VP-Turboladerbaureihe zur ein- und zweistufigen Abgasturboaufladung. MTZ 41: 125–136, 1980
23. Bozung, H.G., Nachtigal, J.: An electrical auxiliary drive system for turbochargers. Motorship Sept. 1979: 71–76, 1979
24. Bromnick, P.A., Pearson, R.J., Winterbone, D.E.: Intercooler model for unsteady flows in engine manifolds. Proc. Inst. Mech. Eng. D 212: 119–132, 1998

25 Büchi, A.: Exhaust turbocharging of internal combustion engines. Franklin Institute, Philadelphia, Pa., 1953 (Journal of the Franklin Institute Monograph, Nr. 1)

26 Büchi, A.: Über die Entwicklungsetappen der Büchi-Abgasturbo-Aufladung. MTZ 13: 25–28, 1952

27 Bulaty, T.: Spezielle Probleme der schrittweisen Ladungswechselrechnung bei Verbrennungsmotoren mit Abgasturboladern. MTZ 35: 177–185, 1974

28 Bulaty, T., Skopil, M., Codan, E.: A flexible program system for the simulation of tubocharged diesel engines. ABB Rev. 1994/95: 28–35, 1994

29 Buratti, R., Carlo, A., Lanfranco, E., Pisoni, A.: DI Diesel engine with variable turbine turbocharger (VTG): a model based boost pressure control strategy. Meccanica 32: 409–421, 1997

30 Chesse, P., Hetet, J.-F., Tauzia, X., Frayret, J.-P.: Influence of the alteration of the compressor surge line on the operation limit of a turbocharged marine diesel engine. In: Proceedings of the 17th annual fall technical conference of the ASME Internal Combustion Engine Division, September 24–27, 1995, Milwaukee, Wisconsin, vol. 2. American Society of Mechanical Engineers, New York, pp. 51–57, 1995 (ICE-Vol. 25-2)

31 Choshi, M., Asanomi, K., Abe, H., Okamoto, S., Shoji, M.: Development of V6 Miller cycle engine. JSAE Rev. 15: 195–200, 1994

32 Christensen, H.H.: Two problems and how to „bypass" them (part load by-pass-systems for Diesel engines). ABB Turbo Mag. I/1998: 13–15, 1998

33 Christensen, H.H.: Variable turbine geometry turbocharger in gas engines. ABB Turbo Mag. I/1997: 11–12, 1997

34 Codan, E.: Optimierung des Aufladesystems und Betriebsverhaltens von Großdieselmotoren durch Computersimulation. In: 5. Aufladetechnische Konferenz, 11.–12. 10. 1993, Augsburg, pp. 19–36, 1993

35 AVL: CRUISE, user manual. Manufacturer's information, AVL List, Graz, 1999

36 Cser, G.: Resonanz-Saugsysteme für moderne Pkw-Motoren. In: 4. Aufladetechnische Konferenz. VDI, Düsseldorf, pp. 277–296, 1991 (VDI-Berichte, vol. 910)

37 Brown Boveri: Das Comprex-Funktionsprinzip. BBC, Baden, Druckschrift Nr. CH-Z123220D

38 Dibelius, G.: Teilbeaufschlagung von Turbolader-Turbinen. BBC Mitt. 52: 180–189, 1965

39 Dommes, W., Naumann, F.: Der aufgeladene Fünfzylindermotor des Audi 200. ATZ 82: 49–58, 1980

40 Durst, B., Thams, J., Görg, K.: Frühzeitige Beurteilung des Einflusses komplexer Bauteile auf den Ladungswechsel mittels gekoppelter 1-D-3D-Strömungsberechnung. MTZ 61: 218–223, 2000

41 Eberle, M.: Beitrag zur Berechnung des thermodynamischen Zusammenwirkens von Verbrennungsmotor und Abgasturbolader. Dissertation, Eidgenössische Technische Hochschule Zürich, Zürich, Switzerland, 1968

42 Eckardt, D.: Untersuchung der Laufradströmung in hochbelasteten Radialverdichterstufen. FVV Forschungsber. 154, 1974

43 Eckert, B., Schnell, E.: Axial- und Radialkompressoren, 2nd edn. Springer, Berlin Göttingen Heidelberg, 1961

44 Eiser, A., Erdmann, H.-D., Grabow, J., Mikulic, L.: Der neue AUDI V6-biturbo-Ottomotor. Press release, AUDI, Neckarsulm, 1998

45 Eitel, J.: Ladeluftkühlung mit Niedertemperatur-Kühlmittelkreisläufen für Kfz.-Verbrennungsmotoren. MTZ 53: 122–126, 1992

46 Engels, B.: Abgasturboauflagung. FVV Forschungsber. 287, 1981

47 Engels, B.: Untersuchung zur Verbesserung des Drehmomentverhaltens abgasturboaufgeladener Fahrzeugdieselmotoren. Dissertation, TH Aachen, Aachen, Federal Republic of Germany, 1981

48 Ermisch, N., Dorenkamp, R., Neyer, D., Hilbig, J.: Das Antriebsaggregat des 3-L-Lupo. In: Bargende, M., Essers, U. (eds.): Dieselmotorentechnik 2000. Expert Verlag, Renningen-Malmsheim, pp. 1–7, 2000 (Kontakt & Studium, vol. 580)

49 Ermisch, N., Neyer, D., Hilbig, J., Scheliga, W.: Der neue 1,2-l-Zylinder-Dieselmotor von Volkswagen. In: Technologien um das 3-Liter-Auto. VDI, Düsseldorf, pp. 461–485, 1999 (VDI-Berichte, vol. 1505)

50 Esch, H.-J., Zickwolf, P.: Comparison of different exhaust gas turbocharging procedures on Porsche engines. Inst. Mech. Eng. C112/86, 1986

51 Gersdorff, K. von, Grassmann, K., Schubert, H.: Flugmotoren und Strahltriebwerke, 3rd edn. Bernhard & Graefe, Bonn, 1995

52 Giannattasio, P., Dadone, A.: Applications of a high resolution shock-capturing scheme to the unsteady flow computation in engine ducts. Inst. Mech. Eng. C430/055, 1991

53 Glamann, P.W.: Das Aufladesystem mittels Verteilergetriebe in seiner Entwicklung zum schaltungsfreien Treibwerk hoher Leistungskonzentration. MTZ 26: 151–159, 1965

54 Görg, K.A.: Berechnung instat. Strömungsvorgänge in Rohrleitungen an Verbrennungsmotoren unter besonderer Berücksichtigung der Mehrfachverzweigung. Dissertation, Ruhr-Universität Bochum, Bochum, Federal Republic of Germany

55 Gunz, K.: Dynamisches Verhalten von Dieselmotoren in einem CODOG Antriebssystem am

Beispiel einer Fregatte. In: 7. Aufladetechnische Konferenz, Oktober 1997, Dresden, pp. 73–80, 1997
56. Haider, G.: Die mechanische Aufladung, 2nd edn. Published by the author, Wien, 2000
57. Harr, T., Mack, E., Schulze, R.: Der neue Sechszylinder-Dieselmotor OM 906 LA von Daimler Benz. MTZ 59: 526–539, 1998
58. Harten, A., Engquist, B., Osher, S., Chakravarthy, S.R.: Uniformly high order accurate essentially non-oscillatory schemes, III. J. Comput. Phys. 71: 231–303, 1987
59. Hawley, J.G., et al.: Comparison of a variable geometry turbocharging (VTG) over conventional wastegated machines to achieve lower emissions. Inst. Mech. Eng. C524/070/97, 1997
60. Hiereth, H.: Untersuchung über den Einsatz aufgeladener Ottomotoren zum Antrieb von Personenkraftwagen. Dissertation, Technische Universität München, München, Federal Republic of Germany, 1978
61. Hiereth, H., Eisele, E., Polz, H.: Investigations into the use of the Comprex® supercharger-system on a high speed Diesel-car-engine. SAE Pap. 75.994, 1975
62. Hiereth, H., Müller, W., Withalm, G.: The Mercedes-Benz Group C engines for the World-Sports-Car-Prototype Racing Championships 1989 and 1990. SAE Pap. 92.0674, 1992
63. Hiereth, H., Withalm, G.: Some special features of the turbocharged gasoline engine. SAE Pap. 79.0207, 1979
64. Hiereth, H., Withalm, G.: Das Instationärverhalten des aufgeladenen Ottomotors. MTZ 43: 71–75, 1982
65. Hiereth, H.: Eignungsabschätzung neuerer Aufladesysteme für Fahrzeugmotoren. MTZ 46: 397–402, 1985
66. Hiereth, H.: New results of passenger car Diesel engine pressure-wave-charges with and without a particulate trap. SAE Pap. 88.0005, 1988
67. Hiereth, H.: Registeraufladung bei Pkw-Motoren. Lecture at Technischen Universität Hannover, Hannover, 1982
68. Hiereth, H.: Testing methods for the transient behaviour of charged vehicle engines. SAE Pap. 86.0451, 1986
69. Hiereth, H.: Car tests with a free-running pressure-wave-charger: a study for an advanced supercharging system. SAE Pap. 89.0453, 1989
70. Hiroyasu, H., Kadota, T., Arai, M.: Development and use of a spray combustion modeling to predict Diesel engine efficiency and pollutant emissions; part 1, combustion modeling; part 2, computational procedure and parametric study. Bull. JSME 26: 569–575, 1983
71. Hitomi, M., Yuzuriha, Y., Tanaka, K.: The characteristics of pressure wave supercharged small Diesel engine, SAE Pap. 890454, 1989
72. Hoefer, C.: Renault Gordini EF 1: erster Formel-1-Motor mit Abgasturboaufladung. MTZ 45: 415–418, 1984
73. Hohenberg, G.: Experimentelle Erfassung der Wandwärme von Kolbenmotoren. Habilitationsschrift, Technische Universität Graz, Graz, Austria, 1983
74. Jenny, E., et al.: BBC-Sonderheft, Nr. 3, Aufladung, August 1987
75. Jenny, E.: Der BBC-Turbolader, Geschichte eines Schweizer Erfolges. Birkhäuser, Basel, 1993
76. Jungbluth, G., Noske, G.: Ein quasidimensionales Modell zur Beschreibung des ottomotorischen Verbrennungsablaufes, Teil 1 und Teil 2. MTZ 52: 262–267, 318–328, 1991
77. Kerkau, M., Knirsch, S., Neußer, H.-J.: Der neue Sechszylinder-Biturbo-Motor mit variabler Turbinengeometrie für den Porsche 911 Turbo. In: Lenz, H.P. (ed.): 27. Internationales Wiener Motorensymposium. VDI, Düsseldorf, pp. 111–136, 2006 (Fortschritt-Berichte, series 12, vol. 622)
78. Kern, J.: Neue Wärmeübertrager: kompakt und vollständig rezyklierbar. ATZ 101: 670–674, 1999
79. Kessel, J.A., Schaffnit, M., Schmidt, M.: Modelling and real-time-simulation of a turbocharger with variable turbine geometry (VTG). SAE Pap. 98.0770, 1998
80. Knoll, R., Prenninger, P., Feichtinger, G.: Zweitakt-Prof. List Dieselmotor, der Komfortmotor für zukünftige kleine PKW. In: Lenz, H.P. (ed.): 17. Internationales Wiener Motorensymposium. VDI, Düsseldorf, pp. 96–110, 1996 (Fortschritt-Berichte, series 12, vol. 267)
81. Kolb, W.: Einstufige Zentripetalturbinen. MTZ 25: 103–105, 1964
82. Kollmann, K., Heil, B., Bruchner, K., Klein, R.: Das 3-Zylinder-Motorenkonzept für den Smart: eine kompetente Lösung für eine außergewöhnliche Aufgabenstellung. In: Lenz, H.P. (ed.): 19. Internationales Wiener Motorensymposium. VDI, Düsseldorf, pp. 103–145, 1998 (Fortschritt-Berichte, series 12, vol. 348)
83. Körner, W.D., Bergmann, H. et al.: Neue Wege beim Turbocompoundantrieb. ATZ 93: 242–248, 1991
84. a) Krebs, R., Szengel, R., Middendorfer, H., Fleiß, M., Laumann, A., Voeltz, S.: Neuer Ottomotor mit Direkteinspritzung und Doppelaufladung von Volkswagen, Teil 1, Konstruktive Gestaltung. MTZ 66: 844–856, 2005

b) Krebs, R., Szengel, R., Middendorfer, H., Sperling, H., Siebert, W., Theobald, J., Michels, K.: Neuer

Ottomotor mit Direkteinspritzung und Doppelaufladung von Volkswagen, Teil 2, Thermodynamik. MTZ 66: 979–986, 2005
85. Behr: Ladeluftkühlung. Behr, Stuttgart
86. Lange, K.H., et al.: Ein aufgeladener BMW-Sechszylinder-Ottomotor. MTZ 40: 575–578, 1979
87. Laustela, E., Müller, R.: Turbocharging modern heavy duty diesel engines. In: Heavy duty engines: a look at the future. American Society of Mechanical Engineers, New York, pp. 99–106, 1994 (ICE-Vol. 22)
88. Lee, M.S.K., Watanabe, S., Nagakura, H., Shiratuchi, M., Sugihara, H., Ienaga, M.: Improvement of a turbocharged and intercooled Diesel engine powered vehicle's startability by means of a three wheel turbocharger. SAE Pap. 945018, 1994
89. Lindner, E.: Ein Entwurfsverfahren für diagonale Turboverdichterlaufräder. VDI-Ber. 947, 1992
90. Lustgarten, G., Hashimoto K., Brown, D.T.: Sulzer RTA84T, the modern VLCC-engine. New Sulzer Diesel, Winterthur, 1993
91. Lustgarten, G., Miculicic, N.: Zweitakt – langsam laufende Dieselmotoren – eine Potentialabklärung. In: 4. Aufladetechnische Konferenz. VDI, Düsseldorf, pp. 225–253, 1991 (VDI-Berichte, vol. 910)
92. Maiorana, G., Sebastiano, G., Ugaglia, C.: Die Common-Rail-Motoren von Fiat. MTZ 59: 582–588, 1998
93. Malobabic, M.: Das Betriebsverhalten leitschaufel- und bypassgeregelter PKW-Abgasturbolader. Dissertation, Universität Hannover, Hannover, Federal Republic of Germany, 1989
94. Mayer, A., El-Nasar, I., Komauer, C.: Kennfeldverhalten und Auslegungsmethode beim Druckwellenlader Comprex, Teil 1 und 2. ATZ 87: 149–154, 291–298, 1985
95. Mayer, M.: Abgasturbolader: sinnvolle Nutzung der Abgasenergie. Moderne Industrie, Landsberg/Lech, 1994 (Bibliothek der Technik, vol. 103)
96. Meier, E.: Die Anwendung von Pulse-Convertern bei Viertakt-Dieselmotoren mit Abgasturboaufladung. BBC-Mitt. 55: 420–428, 1968
97. Meier, E.: Zweistufige Aufladung. BBC-Mitt. 52: 171–179, 1965
98. Meier, E., Czerwinski J., Streuli, A.: Vergleich verschiedener Aufladesysteme mit Hilfe von Kenngrößen. MTZ 51: 54–62, 1990
99. Meier, E., Czerwinski, J.: Turbocharging systems with control intervention for medium speed four stroke Diesel engines. ASME 89-ICE-12, 1989
100. Melchior, J., Thierry, A.T.: Hyperbar system of high supercharging. SAE Pap. 74.0723, 1974
101. Meurer, S.: Der neue MAN-Größtmotor mit 4000 PS Zylinderleistung. MTZ 28: 415–419, 1967
102. Mezger, H.: Turbocharging engines for racing and passenger cars. SAE Pap. 78.0718, 1978
103. Miller, R., Weberkerr, H.U.: The Miller supercharging system for diesel and gas engines. Proc. CIMAC 1957
104. Morel, T., et al.: A new approach to integrating performance and component design analysis through simulation. SAE Pap. 88.0131, 1988
105. N.N.: New Fuji four stroke engine with two-stage turbocharging. Motorship July 1971: 155, 1971
106. Obländer, K., Fortnagel, M., Feucht, H-J., Conrad, U.: The turbocharged five-cylinder Diesel engine for the Mercedes Benz 300 SD. SAE Pap. 78.0633, 1978
107. Otobe, Y., Goto, O., Miyano, H., Kawamoto, M., Aoki, A., Ogawa, T.: Honda Formula One turbocharged V-6 1.5L engine. SAE Pap. 89.0877, 1989
108. Pfleiderer, C., Petermann, H.: Strömungsmaschinen, 5th edn. Springer, Berlin Heidelberg New York Tokyo, 1986
109. Pflüger, F., AG Kühnle, Kopp & Kausch: Die 2-stufige geregelte Aufladung: ein neues Aufladesystem für NFZ-Motoren. In: 2. Stuttgarter Symposium, 18.–20. Februar 1997, pp. 268–286, 1997
110. Pischinger, R., Kraßnik, G., Taucar, G., Sams T.: Thermodynamik der Verbrennungskraftmaschine. Springer, Wien New York, 1989 (Die Verbrennungskraftmaschine, N.S., vol. 5)
111. Pucher, H.: Vergleich der programmierten Ladungswechselrechnung für Viertaktdieselmotoren nach der Charakteristiken-Theorie und der Füll- und Entleermethode. Dissertation, Technische Universität Braunschweig, Braunschweig, Federal Republic of Germany, 1975
112. Pucher, H., Eggert, T., Schenk, B.: Experimentelle Entwicklungswerkzeuge für Turbolader von Fahrzeugmotoren. In: 6. Aufladetechnische Konferenz, 1.–2. Oktober 1997, Dresden, pp. 227–240, 1997
113. Pütz, W.: Untersuchungen zur mechanischen Aufladung am PKW-Dieselmotor. In: 5. Aufladetechnische Konferenz, 11.–12. Oktober 1993, Augsburg, pp. 271–291, 1993
114. Rautenberg, M., et al.: The charging of Diesel engines for passenger cars using turbochargers with adjustable turbine guide vans. ASME 82-GT-41, 1982
115. Reimold, H.W.: Bauarten und Berechnung von Ladeluftkühlern für Otto- und Dieselmotoren. MTZ 47: 151–157, 1986
116. Rohne, K.H., Hinden, H., et al.: VTR/C..4P: a turbocharger with high pressure ratio for highly supercharged 4-stroke Diesel engines. CIMAC Pap. D 62, 1991
117. Rudolph, H.-J., Königstedt, J., Brunken, R., Teufel, H., Binder, T.: Der Vierzylinder-Turbomotor für den AUDI A3. MTZ 58: 416–420, 1997

118 Rupp, M.: Turbocharging marine diesel engines. ABB Turbo Mag. II/96: 20–26, 1996

119 Rupp, M., Nissen, M.: Power turbines for an energy bonus from diesel engines. ABB Rev. 1994/95: 22–27, 1994

120 Scherenberg, H.: Abgasturbo-Aufladung für Personenwagen-Dieselmotoren. ATZ 79: 479–486, 1977

121 Schmidt, E.: Einführung in die technische Thermodynamik und in die Grundlagen der chemischen Thermodynamik, 7th edn. Springer, Berlin Göttingen Heidelberg, 1958

122 Schmitz, T., Holloh, K.D., Jürgens, R.: Potentiale einer mechanischen Zusatzaufladung für Nutzfahrzeugmotoren. MTZ 55: 308–313, 1994

123 Schorn, N.: Potentialabschätzung von Abgasturboaufladeverfahren. In: 5. Aufladetechnische Konferenz, 11.–12. 10. 1993, Augsburg, supplement to proceedings, 1993

124 Schrott, K.H.: Die neue Generation der MAN-B&W-Turbolader. MTZ 56: 596–601, 1995

125 Schwarz, C., Woschni, G., Zeilinger, K.: Die Gesamtprozessanalyse für Fahrzeuge mit aufgeladenen Dieselmotoren: ein schnelles Simulationswerkzeug durch die Darstellung der Arbeitsprozessrechnung als N-dimensionales Kennfeld. In: 5. Aufladetechnische Konferenz, 11.–12. 10. 1999, Augsburg, pp. 315–343, 1993

126 Schweitzer, P.H.: Scavenging of 2-stroke cycle Diesel engines. Macmillan, New York, 1949

127 Seifert, H., und Mitarbeiter: Die Berechnung instationärer Strömungsvorgänge in den Rohrleitungssystemen von Mehrzylindermotoren. MTZ 74: 421–428, 1972

128 Seifert, H.: Instationäre Strömungsvorgänge in Rohrleitungen an Verbrennungskraftmaschinen: die Berechnung nach der Charakteristikenmethode. Springer, Berlin Göttingen Heidelberg, 1962

129 Sieber, R.: Bewertung von Kühlsystemen mit wasser- und luftgekühlten Ladeluftkühlern. MTZ 34: 512, 1984

130 Smith, A.: Integrierte Programmwerkzeuge zur Simulation bei der rechnergestützten Motorenentwicklung. MTZ 58: 702–705, 1997

131 Sod, G.A.: A survey of several finite difference methods for systems of nonlinear hyperbolic conservation laws. J. Comput. Phys. 27: 1–31, 1978

132 Spinnler, G: „Ecodyno®": a new supercharger for passenger car engines. ABB Techn. Beschreibung, 1991

133 Steinparzer, F., Stütz, W., Kratochwill, H., Mattes, W.: Der neue BMW Sechszylinder-Dieselmotor mit Stufenaufladung. MTZ 66: 334–344, 2005

134 Stephan, K., Mayinger, F.: Thermodynamik, vol. 1, 14th edn. Springer, Berlin Heidelberg New York Tokyo, 1992

135 Stihl, M.: Flexible manufacturing of ABB turbochargers. ABB Rev. 1994/95: 36–44, 1994

136 Swain, E.: A simple method for predicting centrifugal compressor performance characteristics. Inst. Mech. Eng. C405/040, 1990

137 Swain, E.: Turbocharging the submarine Diesel engine. Mechatronics 4: 349–367, 1994

138 Swain, E., Elliott, C.: Controlling a variable-geometry turbine nozzle on a turbocharger fitted to a diesel engine in a submarine environment. Inst. Mech. Eng. C484/026, 1994

139 Swain, E., Meese, H.: Extension of a centrifugal compressor performance prediction technique. Inst. Mech. Eng. C554/008, 1998

140 Tashima, S., Taqdokora, T., Niwa, Y.: Development of sequential twin turbo system for rotary engine. SAE Pap. 910624, 1991

141 Tauzia, X., Hetet, J.F., Chesse, P., Grosshans, G., Mouillard, L.: Computer aided study of the transient performances of a highly rated sequentially turbocharged marine diesel engine. Proc. Inst. Mech. Eng. A 212: 185–196, 1998

142 Traupel, W.: Die Theorie der Strömung durch Radialmaschinen. Braun, Karlsruhe, 1962

143 Traupel, W.: Thermische Turbomaschinen, vols. 1 and 2, 3rd edn. Springer, Berlin Heidelberg New York, 1988 and 1982

144 Truscott, A., Porter, B.C.: Simulation of a variable geometry turbocharged diesel engine for control algorithm development. Inst. Mech. Eng. C524/127/97, 1997

145 N.N.: Variable Ansaugsysteme. Krafthand 13/14 (1994): 906–910, 1994

146 Von der Nüll, W.T.: Zunehmende Einführung des Abgasturboladers auch für Ottomotoren. MTZ 24: 321–325, 1963

147 Watson, M., Janota, M.S.: Turbocharging the internal combustion engine. Macmillan, London, 1982

148 Will, G.: Modellvorstellungen zur Strömung in radialen Laufrädern. Dissertation, Technische Universität Dresden, Dresden, German Democratic Republic, 1970

149 Winkler, G.: Ein geschlossenes Diagramm zur Bestimmung der Betriebspunkte von Abgasturboladern an Viertakt-Motoren. MTZ 41: 451–457, 1989

150 Woschni, G.: A Universally applicable equation for the instantaneous heat transfer coefficient in the internal combustion engine. SAE Pap. 67.0931, 1976

151 Woschni, G.: Einfluss von Rußablagerungen auf den Wärmeübergang zwischen Arbeitsgas und Wand im

Dieselmotor. In: 3. Tagung „Der Arbeitsprozess des Verbrennungsmotors". Mitteilungen des Instituts für Verbrennungskraftmaschinen und Thermodynamik, Graz, pp. 149–169, 1991

152. Woschni, G.: Verbrennungsmotoren. Lecture notes, Technische Universität München, München, 1980

153. Woschni, G., Anisitis, F.: Eine Methode zur Vorausberechnung der Änderung des Brennverlaufs mittelschnellaufender Dieselmotoren bei geänderten Betriebsbedingungen. MTZ 34: 106–115, 1973

154. Woschni, G., Bergbauer, F.: Verbesserung von Kraftstoffverbrauch und Betriebsverhalten von Verbrennungsmotoren durch Turbocompounding. MTZ 51: 108–116, 1990

155. Wunsch, A.: Aufladung von Fahrzeug-Dieselmotoren mit Abgasturbolader und mit der Druckwellenmaschine Comprex®. MTZ 31: 17–23, 1970

156. Zehnder, G., Mayer, A.: Comprex pressure-wave supercharging for automotive diesels. SAE Pap. 84.0132, 1984

157. Zehnder, G., Mayer, A.: Supercharging with Comprex to improve the transient behaviour of passenger car Diesel engines. SAE Pap. 86.0450, 1986

158. Zeitzen, F.: Hightech, Vorstellung SCANIA R124-470 mit Turbocompound und Hpi-Einspritzung. Lastauto Omnibus 2001(4): 20–22, 2001

159. Zellbeck, H., Friedrich, J., Berger, C.: Die elektrisch unterstützte Abgasturboaufladung als neues Aufladekonzept. MTZ 60: 386–388, 1999

160. Zinner, K.: Aufladung von Verbrennungsmotoren, 3rd edn. Springer, Berlin Heidelberg New York Tokyo, 1985

161. Zinner, K., Eberle, M.K.: Die Leistungsumrechnung und Prüfung aufgeladener Diesel- und Gasmotoren bei geänderten atmosphärischen Bedingungen. MTZ 34: 67–71, 1973

Subject index

ABB 57, 117, 118, 121, 122, 217
Acceleration behavior 133
Acceleration support 140
Acquisition of measurement data 184
Action turbine 70
Agricultural applications 146
Aircraft engines 245
Air delivery ratio 27
Air filter 48
Air-to-air charge air cooling 212
Air volumetric efficiency 28
Amount of residual gas 29
Assembly 202
Atkinson 114, 230
Audi 2, 222, 223, 224, 234, 235, 236, 256
Auto Union 222
AVL 13, 48, 49, 132, 158, 187, 234, 240, 256
AVL-Cruise 49
AVL-Fame 48
Axial bearing 200
Axial compressor 61
Axial turbine 65

Backward bent impeller blades 65
BBC see Comprex
 Bearing 199, 200
Bearing housing 196
 machining 202
Behr 213, 256
Bernoulli 61
Blowoff 54, 151
BMW 222, 223, 233, 234, 237, 238, 239, 256
Bosch 256
Buechi, Alfred 4, 79, 81
Bypass 53
Bypass valve 167
Bypass valve at air intake side 165

Carbon monoxide 192
CFD (computational fluid dynamics) 48, 92
Charge air cooler 46, 208
 designs 209
Charge air cooler made totally of aluminum 212

Charge air cooling 8
Charge air cooling system 208
Charge density 6
Charge mass flow 6
Charger speed control 54
Chevrolet 2
Choke limit 64
CO emissions 33
Combined charging and special charging
 processes 121
Combustion chamber shape 160
Command response 162
Compressor 61
Compressor blade pitch 91
Compressor control possibilities 90
Compressor housing 195
Compressor impeller 198, 201, 205
Compressor impeller–turbine rotor diameter
 ratio 84
Compressor outlet diffuser 90
Compressor power control 162
Compressor selection 89
Comprex pressure wave charging system (BBC) 4, 9,
 125, 146, 150, 241, 256
Constant-pressure turbocharging 76
Control interventions 162, 173
Controlled two-stage turbocharging 106
Control strategies for VTG chargers 173
Cooling 194, 196, 205
Cooling surfaces 210
Cser 9
Curtiss Wright 2, 116
Cycle efficiency factor 27
Cycle simulation 37, 92
Cylinder charge 5
Cylinder work 5

Daihatsu 240
Daimler, Gottlieb 2
Daimler Benz 113, 225, 228, 233, 244, 245, 256
DaimlerChrysler (DC) 2, 58, 237, 240, 243, 256
Detroit Diesel Corporation (DDC) 242
Diesel, Rudolf 2, 3

Diesel engine 3, 179
   with twin-flow turbine 99
   with variable turbine geometry 98
Differential compound charging 121
Direct charge air cooling 213
Disengagement 54
Displacement compressor 9, 14, 51, 194
Disturbance response 162

Eaton 56
Effective efficiency 26
Efficiency chain 27
Efficiency of density recovery 209
Electric energy recovery 119
Electronic waste gate and VTG control systems 179
Emission control parameters 170
Emission data 31, 191, 257, 258
Energy balance of charging system 74
Engine air mass flow 188
Engine blowby 189
Engine braking performance 177
Engine efficiencies 26
Engine power output 5, 6
Engine speed 186
Engine torque 185
European driving cycle (NEDC) 50, 236, 240
Exhaust brake 173
Exhaust gas aftertreatment 34
Exhaust gas analyzer 192
Exhaust gas catalyst 48
Exhaust gas emissions 31, 191, 257, 258
Exhaust gas energy recovery 25
Exhaust gas plenum 205
Exhaust gas opacity 192
Exhaust gas recirculation (EGR) 48
Exhaust gas turbocharger 9, 195, 216
Exhaust gas turbocharger control 162
Exhaust gas turbocharger layout for automotive application 151
Exhaust gas turbocharger with variable turbine geometry 173
Exhaust gas turbocharging 2, 60
Exhaust system design 75
External mixture formation 38

Ferrari 10,11
Fiat 234, 256
Fixed-geometry exhaust gas turbocharger 163
Flat-tube intercooler 211
Floating bushing 199
Flow compressor 15
Flow division in twin-flow housing 87
Flow processes 48
Flow-stabilizing measures 90

Foettinger 54
Four-stroke engine 18
Fuel combustion rate 26
Fuel mass flow 189
Future developments in supercharging 215
FVV 49

Garrett 125, 220, 236, 256
Gas exchange cycle 24, 27
Gas exchange phase 39
Gas exchange work 25
Gasoline engine 2, 179
Gasoline engine with fixed-geometry turbocharger and waste gate 97
General Motors Company (GMC) 141, 247
Generator operation 138

HC emissions 34, 192, 257, 258
Heat transfer coefficient 210
Heat transfer value 210
Helmholtz resonance charging 12
Helmholtz resonator 9, 11, 12
High-altitude behavior 135
High-performance high-speed engines 245
High-pressure process 24
Hispano-Suiza 105
Honda 225, 226
Housing 196
Hydrocarbons 192
Hyperbar charging process 128

Ignition timing 160
IHI 58
Indicated efficiency 26
Indicated engine power output 7
Indirect charge air cooling 214
Intake manifold resonance charging 9
Intake muffler 48
Intercooler efficiency 208
Internal mixture formation 38
Iveco 180

KKK, now 3K-Warner 57, 64, 82, 91, 106, 164, 167, 217, 218, 222, 223, 256
Knocking combustion 159

Large chargers 202
Laval turbine 70
Load response 133
Locomotive applications 146
Locomotive engines 245
Low-pressure processes 24
Lubrication 195, 199

Lycoming 245
Lysholm 9, 15, 58, 59, 216

Main turbocharger equation 75
MAK 248
MAN 4, 65, 89, 105, 141, 202, 242, 250, 251, 253
MAN B&W 248
Manifold flow 42
Manifold separation 79
Maritime applications 146
Maschinenfabrik Winterthur 4
Mass flow function 18, 40, 67
Matching the turbine 84
Maybach, Wilhelm 2
Mazda 58, 230, 241, 256
MCC 224
Measurement of turbocharger speed 187
Measuring point layout 185
Mechanical auxiliary supercharging 122
Mechanical efficiency 27
Mechanical recovery 117
Mechanical scavenging auxiliary supercharger 221
Mechanical stress 35
Mechanical supercharging 51, 215
Mechanic charger 46
Mechanics of superchargers 194
Medium-speed engines 248
Mercedes Benz 56, 123, 228, 256
Miller process 114, 230
Mitsubishi 106, 254
Mixture-related volumetric efficiency 28
MTU 109, 111, 112, 247, 248, 254
Muffler 48
Multipulse layouts 79

Napier 117
NEDC (New European Driving Cycle) 50, 236, 240
New Sulzer 253, 254
Nitrogen oxide 192
NO_x emissions 33
Numeric process simulation 36
Numeric simulation of engines with exhaust gas turbocharging 97
Numeric simulation of engine operating behavior 158
Numeric 3-D CFD (computational fluid dynamics) simulation 48, 92

Ogura 56
Opel 2, 11, 224
Oxidation or NO_x-storage catalyst 35

Particulate filter 34
Particulate matter emissions 34, 196, 259, 260
Part-load boost pressure 172

Part-load waste gate 170
Passenger car diesel engines 233
Perfect mixing scavenging 41
Perfect scavenging 41
Performance characteristics of supercharged engines 133
Perkins 122, 123
Pierburg 57, 58
Plenum elements 44
Porsche 109, 228, 232, 245, 246, 256
Pressure and temperature data 189
Pressure ratio 23
Pressure–volume flow map
  of piston engine 17
  of supercharger 13
Preswirl 90
  control 90
Process efficiency 6, 27
Production 200
Propeller operation mode 139
Pulse converter 80
Pulse turbocharging 77

Quantity control 161

Radial compressor 62
Radial turbine 66
Radiator 48
Real engine scavenging 41
Register charging 108, 143, 218
Renault 225
Requirement specifications 144
Resistance thermometer 190
RGM Messtechnik GmbH 185
Roots blower 2, 9, 13, 15, 56, 216, 240, 241, 242, 245
Rotor assembly 198
Rotor dynamics 200
Rotors 194
Round tube intercooler 211

Saab 2, 222
Saurer 12, 242
Scania 118, 244
Scavenging efficiency 29
Scavenging ratio 29
Sealing 194
Sealing system 197
SEMT Pielstick 251
Ship engines 245
Ship engine with register charging 143
Shortcut scavenging 41
Siemens 142
Single bushing bearing 199
Single-stage register charging 108

Slow-speed engines 137, 251
Smart 224
Soot and particulate matter emissions 192
Spiral charger 57
Stationary engine 137
Steady-state behavior 173
Steady-state layout 151
Subaru 227
Supercharged gasoline engines 222
Supercharging systems 217
Supported exhaust gas turbocharging 124
Supporting the exhaust gas turbocharger 219
Surge limit 64
Svenska Rotor Maskiner 58
Swallowing capacity function of turbine 68

Test bench 185
Thermal stress 34
Thermocouples 190
Throttle upstream of compressor 168
Torque behavior 134
Torque increase 145
Transient control strategies 166
Transient layout 154
Transient operation of ship engine 143
Transient response of exhaust gas turbocharged engine 146
Trim 84
Turbine blade speed ratio 85
Turbine blade vibrations 200
Turbine design 82
Turbine entry cross-sectional area-to-distance radius ratio (*A/R* ratio) 84
Turbine flow losses 87
Turbine housing 196
Turbine map 68
Turbine performance 23
    control 162

Turbines 65, 82, 198, 206
Turbine swallowing capacity function 71, 73
Turbocharger 44
Turbocharger speed 187
Turbocharger test benches 71
Turbocharger total efficiency 44
Turbocompound operation 219
Turbocompound process 116
Turbo compressor 15
Turbocooling and Miller process 113
Turbodyne/Honeywell 220
Twin-flow turbine housing 87
Two-stage controlled supercharging 219
Two-stage register charging processes 110
Two-stage turbocharging 106
Two-stroke engines 17, 22, 141

Variable charger speed control 54
Variable intake systems 9
Variable turbine geometry (VTG charger) 68, 152, 162
Vibe parameter 39
Volkswagen (VW) 57, 123, 180, 221, 230, 231, 232, 236, 256
Volumetric efficiency 19, 29
Volvo 221, 243, 244

Wall friction losses 42
Wall heat loss 38
Wankel, Felix 9, 15, 56, 57, 230
Waste gate 163
Waste gate boost pressure control 151, 170
Water-cooled charge air cooler 211
Water injection 34
Werkspoor 4
Woschni, Gerhard 39, 40

ZF 20, 55
Zinner, Karl 138

SpringerEngineering

Kevin L. Hoag

Vehicular Engine Design

2006. X, 223 pages. 170 illustrations.
Hardcover **EUR 78,–**
Recommended retail price. Net-price subject to local VAT.
ISBN 978-3-211-21130-4
Powertrain

This book provides an introduction to the design and mechanical development of reciprocating piston engines for vehicular applications. Beginning from the determination of required displacement, coverage moves into engine configuration and architecture. Critical layout dimensions and design trade-offs are then presented. Coverage continues with material and casting process selection for the cylinder block and heads. Each major engine component and sub-system is then taken up in turn.

From the contents:
The internal combustion engine: an introduction • Engine maps, customers, and markets • Engine validation and durability • Engine development process • Determining displacement • Engine configuration and balance • Cylinder block and head materials and manufacturing • Block layout and design decisions • Cylinder head layout design • Block and head development • Engine bearing design • Engine lubrication • Engine cooling • Gaskets and seals • Pistons and rings • Crankshafts and connecting rods • Camshafts and the valve train • Subject index

P.O. Box 89, Sachsenplatz 4 – 6, 1201 Vienna, Austria, Fax +43.1.330 24 26, books@springer.at, **springer.at**
Haberstraße 7, 69126 Heidelberg, Germany, Fax +49.6221.345-4229, SDC-bookorder@springer.com, springer.com
P.O. Box 2485, Secaucus, NJ 07096-2485, USA, Fax +1.201.348-4505, orders@springer-ny.com, springer.com
Prices are subject to chance without notice. All errors and omissions excepted.

SpringerTechnik

Rudolf Pischinger, Manfred Klell, Theodor Sams

Thermodynamik der Verbrennungskraftmaschine

Zweite, überarbeitete Auflage.
2002. XVII, 475 Seiten. 283 Abbildungen.
Gebunden **EUR 189,95**, sFr 309,50*
ISBN 978-3-211-83679-6
Der Fahrzeugantrieb

Aufbauend auf die Zusammenstellung relevanter Grundlagen der Thermodynamik und die Darstellung idealisierter Motorprozesse werden aktuelle null-, quasi-, ein- und mehrdimensionale Methoden zur Analyse und Simulation des realen Motorprozesses besprochen, wobei Fragen des Wärmeübergangs, der Verbrennung, der Schadstoffbildung und des Ladungswechsels inklusive Aufladung erörtert werden. Der enge Bezug zur Praxis ist u.a. durch die Analyse des Arbeitsprozesses einer Reihe charakteristischer moderner Verbrennungsmotoren gegeben. Das Buch eignet sich als Lehrbuch für Studenten und angehende Ingenieure ebenso wie als Nachschlagewerk für Fachleute in der Praxis.

Aus dem Inhalt:
Allgemeine Grundlagen
Verbrennung
Idealisierte Motorprozesse
Analyse und Simulation des Systems Brennraum
Analyse des Arbeitsprozesses ausgeführter Motoren
Anwendung der thermodynamischen Simulation
Anhänge

P.O. Box 89, Sachsenplatz 4–6, 1201 Wien, Österreich, Fax +43.1.330 24 26, books@springer.at, **springer.at**
Haberstraße 7, 69126 Heidelberg, Deutschland, Fax +49.6221.345-4229, SDC-bookorder@springer.com, springer.com
P.O. Box 2485, Secaucus, NJ 07096-2485, USA, Fax +1.201.348-4505, orders@springer-ny.com, springeronline.com
Preisänderungen und Irrtümer vorbehalten. * Unverbindliche Preisempfehlung

Springer and the Environment

WE AT SPRINGER FIRMLY BELIEVE THAT AN INTERnational science publisher has a special obligation to the environment, and our corporate policies consistently reflect this conviction.

WE ALSO EXPECT OUR BUSINESS PARTNERS – PRINTERS, paper mills, packaging manufacturers, etc. – to commit themselves to using environmentally friendly materials and production processes.

THE PAPER IN THIS BOOK IS MADE FROM NO-CHLORINE pulp and is acid free, in conformance with international standards for paper permanency.

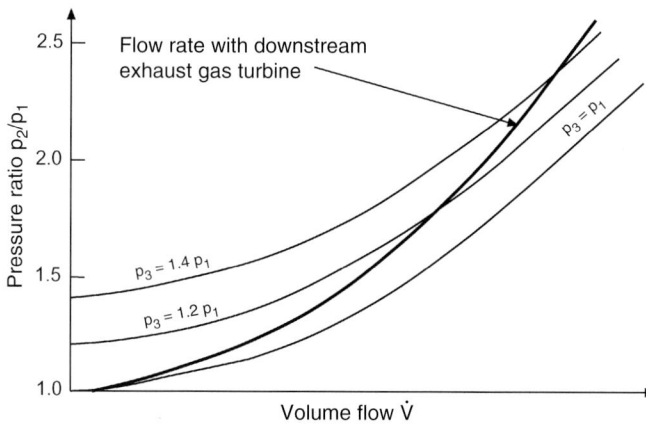

Fig. 2.13. Volume flows through the two-stroke engine, depending on the boost pressure ratio p_2/p_1 and the backpressure p_3

calculated from the aspirated air or charge, as well as the air or charge scavenged during valve overlap.

Approximately, the following equation applies:

$$\dot{V}_1 = V_{cyl} \frac{n_E}{2} \frac{\rho_2}{\rho_1} \lambda_{vol} + \psi_{23} \frac{\rho_2}{\rho_1} \sqrt{2RT_2} \mu_{red} \frac{\int A_{red}\, d\varphi}{720}. \qquad (2.19)$$

In addition to the equation for the two-stroke engine, here λ_{vol} designates the volumetric efficiency.

For supercharged four-stroke engines with larger valve overlap, the volumetric efficiency can be calculated with good approximation by the following, empirical, equation:

$$\lambda_{vol} \sim \frac{\varepsilon}{\varepsilon - 1} \frac{T_2}{313 + \frac{5}{6}t_2}, \qquad (2.20)$$

where ε is the compression ratio, T_2 is the temperature upstream of the inlet valve in kelvin, and t_2 in degrees Celsius. The function takes into account the fact that with valve overlap there is no reverse expansion of the residual gases, and it considers the heating of the charge air during the intake process. The first term of Eq. (2.19) is proportional to the engine speed, the second is dependent on the pressure ratio and the valve overlap, which is addressed via A_{red}. A map of a four-stroke engine with typical operation (swallowing) lines is shown in Fig. 2.14 with the engine speed as parameter, for engines with and without relevant valve overlap.

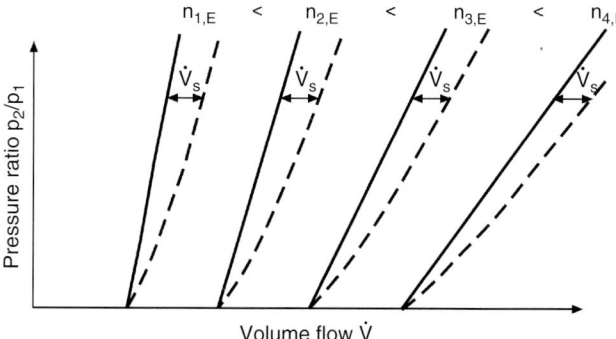

Fig. 2.14. Operation (swallowing) characteristics of a four-stroke engine, as a function of engine speed, with (dash lines) and without (solid lines) valve overlap. The horizontal gap between the two lines at a specified speed corresponds to the scavenge part $\dot{V}_s$ of the total volume flow.

2.6.2 Interaction of two- and four-stroke engines with various superchargers

Since now the maps of both chargers and engines have been defined in a compatible way, it becomes easy to show the interaction of various charger systems with two- and four-stroke engines and then to evaluate the characteristics of each particular combination.

Four-stroke engine with mechanically powered displacement compressor

As can be seen in Fig. 2.15, at constant speed ratio between charger and engine, points of intersection between charger and engine speed curves result in clearly defined pressure relations. On the one hand, these increase slightly with increasing engine speed, on the other hand they depend on the valve timing of the engine (small or large valve overlap with changed scavenging quantity through the cylinder). Overall, the described combination results in an acceptable boost pressure in the entire load and speed range of the engine and, with an approximately constant torque curve in the engine speed range, also satisfies the requirements for automotive applications.

In order to cover the total load range of the engine, the boost pressure must be continuously adjustable between ambient and maximum possible pressure. Regarding the control mechanisms it should only be mentioned here that the displacement compressor, due to the fact that its characteristic curves are very similar to those of the engine, offers good control conditions, since only relatively small differential quantities between charger delivery and engine air demand have to be blown off at partial load or have to be governed. The corresponding control aspects are covered in depth in Sect. 4.3.

Four-stroke engine with mechanically powered turbo compressor

Here the combined pressure–volume flow map (Fig. 2.16) also provides information about the engine characteristics that can be expected. At an assumed constant ratio of charger to engine speed, it can be recognized that only very limited load demands can be met with such a combination of engine and supercharger.

With increasing engine speed, boost pressure increases parabolically, which is suitable for applications where the engine is used in combination with an aero or hydro propeller drive (e.g., a ship or aircraft propeller) or in steady-state operation close to its rated speed.

Applications with engine operation in a wide map range, e.g., automotive applications, are only reasonable with the use of a variable speed ratio for the charger drive, as it is shown in Fig. 2.17 with a continuously variable ZF-Variomat transmission.

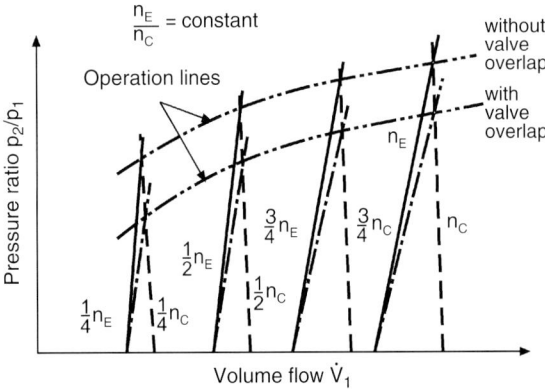

Fig. 2.15. Combined pressure–volume flow map of a four-stroke engine with mechanically powered displacement compressor

2.6 Interaction between supercharger and internal combustion engine 21

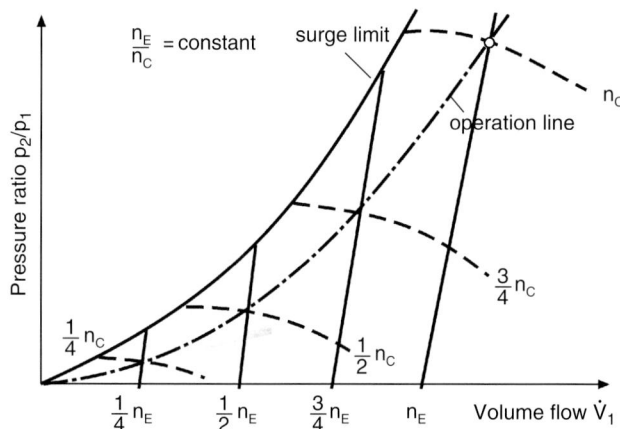

Fig. 2.16. Combined pressure–volume flow map of a four-stroke engine with mechanically powered turbo compressor with constant speed ratio

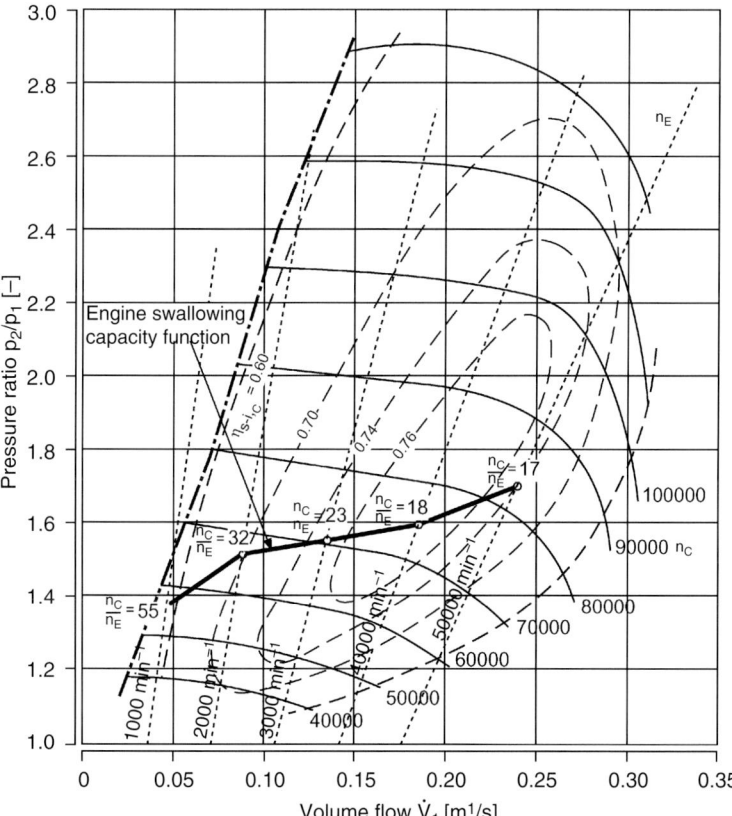

Fig. 2.17. Pressure–volume flow map of a four-stroke engine with turbo compressor and variable speed ratio of the charger drive via ZF-Variomat

With turbo compressors, control measures may become necessary due to their instable map area. However, they are in any case necessary to adapt the boost pressure for part-load operation. They are far more complex than for displacement compressors, since boost pressure changes can only be achieved via changing the charger speed, e.g., by a change of the charger transmission

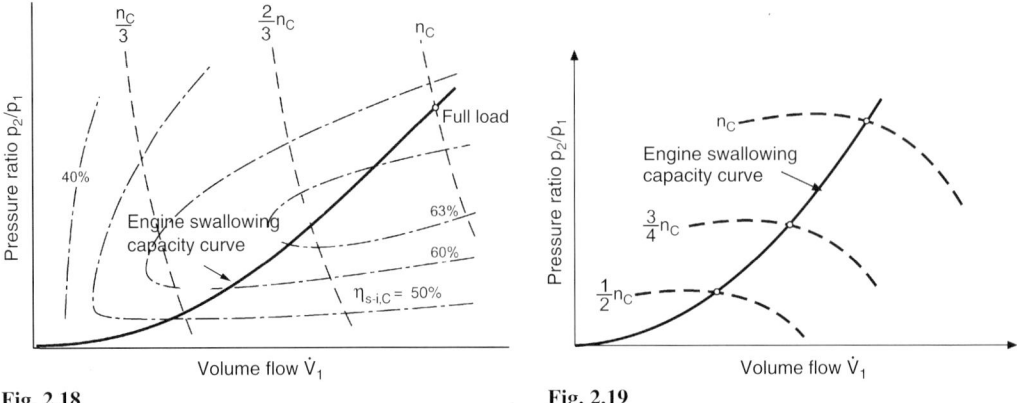

Fig. 2.18. Pressure–volume flow map of a two-stroke engine with mechanically powered displacement compressor

Fig. 2.19. Pressure–volume flow map of a two-stroke engine with mechanically powered turbo compressor

ratio. (In Sect. 4.3, the corresponding control measures and mechanisms are described for charger types in production today.)

Two-stroke engine with mechanically powered displacement compressor

In the past, the combination of a two-stroke engine with a mechanically powered displacement compressor (Fig. 2.18) was frequently realized by using the lower side of the piston of large crosshead engines as a scavenging or supercharge pump. Today, this design is applied only in very rare cases, since its complexity is significantly higher than in the case of other supercharging concepts.

Two-stroke engine with mechanically powered turbo compressor

As Fig. 2.19 shows, the combination of a two-stroke engine with a mechanically driven turbo compressor meets the requirements of various applications, e.g., either in a propeller drive or for stationary gen sets. The torque characteristics and the required torque demand from a ship's propeller as a flow engine correspond by principle very well. It has to be considered, however, that any acceleration creates an additional need for torque, which can hardly be covered with the possible operations curves of this engine-charger combination.